BUILDING ECONOMICS

appraisal and control of building design cost and efficiency

By the same author:

Building Quantities Explained
Civil Engineering Quantities
Municipal Engineering Practice
Civil Engineering Specification
Planned Expansion of Country Towns

BUILDING ECONOMICS

appraisal and control of building design cost and efficiency

by

IVOR H. SEELEY

BSc, MA, PhD, FRICS, CEng,
FIMunE, FIQS, AIOB

Head of Department of Surveying
Trent Polytechnic, Nottingham

MACMILLAN

First published 1972

Published by
THE MACMILLAN PRESS LTD
London and Basingstoke
Associated companies in New York, Toronto,
Melbourne, Dublin, Johannesburg and Madras

SBN 333 11560 0

Text set in 10/12pt. Monotype Baskerville, printed by letterpress,
and bound in Great Britain at The Pitman Press, Bath

PREFACE

THIS BOOK CRITICALLY examines and applies to practical situations the various methods of controlling the cost of buildings at the design stage, which have as their main aim the securing of maximum value for money and a tender which is within the agreed cost limit. In order to implement effective cost control it is necessary to understand fully the various factors which bear upon building prices and to take both present and future costs into account. The subject area thus forms an important facet of quantity surveying, being an activity which is concerned primarily with economic efficiency and value for money.

Tony Brett-Jones has described (Chapter 8, reference 27) how quantity surveyors are now able to exercise an important and valuable professional skill in giving independent cost advice, which can be used for budgeting, cost planning and cost control. The quantity surveyor's role, as defined in the RICS report on the future role of the quantity surveyor issued in 1971, is to ensure that the resources of the construction industry are utilised to the best advantage of society by providing *inter alia* the financial management for projects and a cost consultancy service to client and designer during the whole construction process. The distinctive competence of the quantity surveyor is a skill in measurement and valuation in the field of construction, in order that such work can be described and the cost and price be forecast, analysed, planned, controlled and accounted for.

As long ago as 1957, William James described how the continually rising cost of building, growing at a faster rate than that of manufactured goods, caused clients to feel that the building industry, and its allied professions, was inefficient. In consequence, a building client was inclined to be incredulous of realistic cost forecasts and to require from his professional advisers both an increasing measure of cost explanation and an ever greater attention to cost efficiency in design. Since that time building clients have become more knowledgeable with more sophisticated requirements, and they now demand a much wider range of expert cost advice than quantity surveyors were accustomed to giving in the past. In particular they require an overall appraisal of the alternatives available to them, to enable decisions to be soundly based economically. Hence there has developed a pressing need to refine the tools of cost prediction and control, and for the quantity surveyor to possess a much wider knowledge of factors influencing costs and other related development aspects; this book seeks to meet these needs. It also aims to cover the current examination requirements of the Royal Institution of Chartered Surveyors and the Institute of Quantity Surveyors in Building Economics and Cost Planning, and to give extensive coverage of Project Cost Control and Project Development in the 1972 RICS final examination syllabuses. It will be of assistance to those studying Economic and Quantitative Analysis in the new associate membership Parts I and II examinations of the Institute of Building. It should also prove to be of value to those proceeding to degrees and diplomas in quantity surveying, building economics and building.

In addition, it is felt that the book will be of considerable use to practising quantity surveyors as a handy means of reference, whilst architects, building surveyors, property managers and contractors may find much of interest within its pages.

Metric units have been used throughout and where rationalised metric sizes have not yet been decided, equivalent metric dimensions have been incorporated. Readers wishing to familiarise themselves with the relative values of metric and imperial measures may find the metric conversion table in appendix 6 helpful. On the drawings, all dimensions in metres are shown with a decimal marker, while all other sets of figures represent millimetres. This procedure eliminates need for the use of the 'm' and 'mm' symbols for dimensions on drawings.

Economists might argue with some justification that the book is more concerned with *building economy* than *building economics*. Nevertheless, the term building economics has been widely used by the quantity surveying profession to describe the investigation of factors influencing building cost, with particular reference to the interaction of building design variables. Furthermore, this study extends much further to embrace such matters as the economics of building development, costs in use, land use and value determinants, and environmental economics, and in addition investigates the methods available for controlling building prices.

I. H. SEELEY

Nottingham
Spring 1972

ACKNOWLEDGEMENTS

THE AUTHOR acknowledges with gratitude the willing co-operation and assistance received from the many organisations and individuals over a period of many years, so many that it is not possible to mention them all individually.

The author has been influenced by the valuable work undertaken by the pioneers in building design cost control techniques and has profited immensely from their investigations. In this connection the names of William James, CBE, FRICS, PPIArb; James Nisbet, FRICS; P. W. Grafton, FRICS; and Cyril Sweett, FRICS immediately come to mind. It would also be appropriate to make reference to the valuable groundwork in costs in use carried out by Dr. P. A. Stone. In addition, the author has benefited immeasurably from the work of the Wilderness Group, the former RICS Cost Research Panel, the BCIS management committee, the former Ministry of Education, and the Department of the Environment, and has drawn freely from their findings.

The RICS Building Cost Information Service, though the good offices of Douglas Robertson and Patrick Amos, kindly agreed to the use of cost analyses and other relevant cost information, and the following design teams have also made valuable information available, without which it would have been extremely difficult to have incorporated really meaningful case studies:

T. H. Thorpe and Partners, Architects and Quantity Surveyors, Derby.

White, Cooper and Turner, Architects, 314 High Holborn, London W.C.1.

S. Hardy, Dip.Arch, ARIBA, Regional Architect, British Rail, Eastern Region and R. P. H. Brind, FRICS, Chief Quantity Surveyor; in collaboration with Turner and Holman, F/ARICS, F/FIQS, FIArb, Chartered Quantity Surveyors, Museum Street, York.

H. Rackham, CEng, FIMunE, MIHE, City Engineer, Surveyor and Planning Officer, City of New Sarum.

Brian Bunch, ARIBA, MRTPI, Chief Architect and R. Taylor, FRICS, Chief Quantity Surveyor to Redditch Development Corporation.

Auburn, Ainsley and Partners, Chartered Quantity Surveyors, Rotherham, Yorkshire.

The author is also indebted to the Controller of Her Majesty's Stationery Office for permission to reproduce figures 13, 14 and 15 from *Flats and Houses*, 1958, and to Ronald Sears for producing the final drawings of such outstanding quality. Certain data in chapter 14 have been extracted from other works by the author, namely *Municipal Engineering Practice* and *Planned Expansion of Country Towns*. Finally, grateful thanks are due to the publishers for abundant help and consideration throughout the production of the book, and the thoughts and ideas expressed by members of the staff of the Department of Surveying at Trent Polytechnic were most welcome.

CONTENTS

FIGURES

TABLES

1 THE CONCEPT OF COST CONTROL

COST CONTROL AIMS at ensuring that resources are used to the best advantage. In these days of ever-increasing costs the majority of promoters of building projects are insisting on jobs being designed and executed to give maximum value for money. Hence, quantity surveyors are employed to an increasing extent during the design stage to advise architects on the probable cost implications of their design decisions. As buildings become more complex and building clients more exacting in their requirements, so it becomes necessary to improve and refine the cost control tools. Rising prices, restrictions on the use of capital and high interest rates have caused building clients to demand that their professional advisers should accept cost as an element in design, and that they should provide a balanced cost in all parts of the building, as well as an accurately forecast overall cost. The report of the Special (Future of the Profession) Committee of the Royal Institution of Chartered Surveyors[1] describes how many building clients now adopt cost limits for projects and are instrumental in spreading an awareness of efficiency and value-for-money in building; and this presages greater use of the quantity surveyor in establishing cost targets, in the appraisal of alternative solutions, and in cost control as a continuous process.

An RICS study group[2] also highlights that part of the quantity surveyor's function concerned with ensuring that his client receives value for money in building work. Thus advice may be given on the strategic planning of a project which will affect the decision whether or not to build, where to build, how quickly to build and the effect of time on costs or prices and on profitability. During the design stage, advice is needed on the relationship of capital costs to maintenance costs and on the cost implications of design variables and differing constructional techniques. The cost control process should be continued through the construction period to ensure that the cost of the building is kept within the agreed cost limits. Haddock[3] has described how building cost can be considered as a medium relating purpose and design and it certainly forms an important aspect of design. A client is very much concerned with quality, cost and time: he wants the building to be soundly constructed at a reasonable cost and within a specified period of time. The term 'cost' which is used extensively throughout this book signifies cost to the client as distinct from cost to the contractor.

HISTORICAL DEVELOPMENT OF COST CONTROL PROCESSES

Grafton[29] has described how cost planning, as at present operated, is the logical extension of a process which has continued since the days of the eighteenth-century measurers, who were employed to measure and value the cost of work after it was both designed and executed. Thus the measurers were involved after the building was erected, to measure and value, and to argue with the client and architect on behalf of groups of tradesmen, who at that time had not been brought together under a main contractor. The main contractor system, which became fully operative in the early nineteenth century, implied price competition before construction which previously was a rare happening. The measurers soon realised that a new function was required and that they possessed the necessary skills to undertake it. It was in response to this situation that they developed the skills of premeasuring, of taking off quantities from the drawings before construction started and assembling them in a bill of quantities to provide a rational basis for competition. At that time it constituted an extremely valuable contribution to the building process. Hence, with the development of the quantity surveying profession, the work was measured and priced before execution, but after design.

The next development was the introduction of approximate estimating techniques which attempted to give a forecast of the probable tender figure, although the basis of the computations often left much to be desired. It was subsequently realised that by the use of cost planning techniques and the methods of cost analysis on which they depend, probable cost can be reasonably accurately determined early in the design process, and sometimes even before the design is commenced, and can be identified with the client's own required limit of cost in the sure knowledge that the development of the design can be controlled to accord with it. Grafton[29] has described how 'cost planning is a process of precosting which attempts to represent the total picture of anticipated cost in a way which provides a clear statement of the issues and isolates the courses of action and their relative cost, so as to provide a guide to decision making'.

Cannell[30] has traced the increasing complexity of building projects and the many specialist skills now associated with their design. The host of alternative solutions occasioned by the many specialisations now involved have created vastly different problems from those of prewar days. Clients require, and are entitled to expect, positive assurance that their money is being spent wisely and well. In the wider sphere, if this country is to continue to prosper in a highly competitive world, we must as a nation increase our earning power and to this end our national resources must be used to their optimum effectiveness. The quantity surveyor has a vital role to play as financial adviser to the building industry in producing economically viable solutions to many kinds of building and development problems; preparing budget estimates; cost planning; the appraisal of alternative cost solutions, contractual arrangements and tendering procedures; preparing contract and subcontract documents; advising on acceptable financial relationships between contractor and client; price negotiation for both traditional and system building; cost checking; forecasting anticipated final costs; settlement of contractors' claims; cost control; cost analysis; costs-in-use and cost benefit studies.

Kenyon[31], expressing the architect's viewpoint, considered that the quantity surveyor was now required to be an encyclopedia of information on every aspect of building costs. With new methods of construction and new materials continuously being introduced, he must be fully informed on them and able to advise on comparative costs. He must work with the architect and other specialists and plan the costs simultaneously with the planning of the building. To be a building economist, Higgin[32] is of the opinion that the quantity surveyor will need to develop understandings and techniques of a kind that will deal, not just with the items that go into the accountancy of a particular building, but with the forces, economic and others, which have determined the nature and relationships of building costs, and which determine the trends they show. Economics, in effect, is a study of all the forces which determine the present functioning and probable future trends of a whole industrial or financial system. The building economist will need to know the effects of private and public investment policies and aesthetic and planning factors, all of which play some part in determining the whole system of economic forces which lie behind the building process.

NEED FOR COST CONTROL

It is vital to operate an effective cost control procedure during the design stage of a project to keep the total cost of the job within the building client's budget. Where the lowest tender is substantially above the initial estimate, the design may have to be modified considerably or, even worse, the project may have to be abandoned. Pressures from five main sources have combined to stress the importance of effectively controlling building costs.

(1) There is greater urgency for the completion of projects, and few building clients have sufficient time for the redesign of schemes consequent upon the receipt of excessively high tenders.

(2) Building clients' needs are becoming more complicated, more consultants are being engaged and the estimation of probable costs becomes more difficult. Furthermore, Dunstone[40] has described how individual developments tend to be block-sized rather than plot-sized in scope.

(3) Employing organisations, both public and private, are becoming larger and are themselves adopting more sophisticated techniques for the forecasting and control of expenditure, and they in their turn expect a high level of efficiency and expertise from their professional advisers for building projects.

(4) The introduction of new constructional techniques, materials and components creates greater problems in assessing the capital and maintenance costs of buildings.

(5) Rising prices, restrictions on the use of capital and high interest rates all make effective cost control that more important.

The method of approach and form of cost control are often dictated by the type of development and nature of the promoter. Some examples will serve to illustrate this point.

(1) *Single house.* A prospective owner usually has a reasonably clear picture of the amount and form of accommodation that he requires and of the price that he is prepared to pay, but he may be open to persuasion on such matters as fittings, finishings and form of central heating. The designer's task of equating accommodation and cost is frequently made more difficult by the client having a preconceived idea of cost, often based on information extracted from house magazines which can easily be misleading. Single houses built to individual clients' requirements are bound to be more expensive than those forming part of an estate and, for this reason, it is incumbent upon an architect, who may or may not be supported by a quantity surveyor, to exercise the greatest care and skill in the design of the project with constant checks on costs.

(2) *Housing estate developments.* In speculative housing work the aim is to provide the type of accommodation that the majority of purchasers want at a reasonable price. This entails a considerable amount of market research to establish current demand. The estate developer's task is made difficult by the majority of prospective purchasers seeking value for money – wanting the maximum amount of accommodation and fittings at the lowest possible price and yet, at the same time, expecting a reasonable standard of quality. The developer is often faced with considerable competition from other builders and from existing properties, and should be continually examining the possibilities offered by the introduction of new materials, components and constructional techniques, and the comparative economics of alternative layouts. This latter aspect is considered in more detail in chapter 5.

(3) *Local authority development.* Local authorities and other public bodies are frequently subject to strict control of expenditure on new projects by the sanctioning government departments. Local authorities require loan sanction from central government for the majority of new building projects and consent will be withheld if the estimated cost exceeds the recognised cost limit. Typical examples are 'cost per place' limits for schools and housing yardsticks (possibly metresticks of the future) for local authority houses and flats. This form of cost control will be considered more fully in chapter 8.

(4) *Commercial and industrial development.* The circumstances vary tremendously with this class of development. Sometimes the owner is anxious to keep the initial constructional costs to an absolute minimum when available funds are restricted, and he may not be unduly concerned with maintenance costs on which some measure of tax relief will be forthcoming. In other cases clients are very much concerned with 'total costs' and wish to see a sensible relationship between initial costs and maintenance costs.

(5) *Development companies.* Many large scale developments, such as blocks of flats and offices, are financed by development companies whose primary objective is to erect the building for investment or sale.

There are two basic procedures here. Firstly, the property may be erected for a specific occupier and the development company agrees accommodation requirements and an acceptable rent figure with the occupier. The development company often borrows money from a large financial organisation, such as a bank or insurance company, and needs a reasonable margin of profit. Secondly, the property may be erected as part of a speculative venture, and the developer and his advisers need to exercise great care in selecting and acquiring likely sites and should aim at putting them to the most profitable permitted use. 'Permitted use' means uses of land permitted under planning regulations. The rent or profit received will be determined largely by the extent and type of accommodation, its location and the state of the property market. Expert advice is needed on the most desirable pattern of accommodation and quality of finish, and on current rent levels. The general practice or valuation surveyor is well fitted to advise on these aspects.

Extensive blocks of offices in London may let well, whereas groups of shops on local authority housing schemes outside London may not. It is imperative that a realistic assessment of the probable demand for the particular type of building is made before any provision takes place. In all cases the quantity surveyor should be able to supply accurate forecasts of costs based on cost records and to ensure that the money allocated to the job is spent wisely.

MAIN AIMS OF COST CONTROL

The implementation of effective cost control procedures enables the architect to be kept fully informed of the cost implications of all his design decisions. It necessitates close collaboration between the architect and quantity surveyor throughout the design stage. In public offices and integrated practices it is probably easier to secure the desired closeness of working, than with separate architectural and quantity surveying practices possibly located some distance apart. A concerted effort should enable any difficulties of physical separation to be overcome. There is an overwhelming need for close working relationships throughout the design stage and of mutual confidence and understanding between the various professional advisers. From these desiderata may flow the establishment of more integrated or multidisciplinary practices and consortia with the prime aim of offering a better and more comprehensive service to the client. Waters[38] has described how the multidisciplinary concept can be applied to a large architectural practice.

The main aims of cost control are probably threefold.

(1) To give the building client good value for money – a building which is soundly constructed, of satisfactory appearance and well suited to perform the functions for which it is required, combined with economical construction and layout.

(2) To achieve a balanced and logical distribution of the available funds between the various parts of the building. Thus the sums allocated to cladding, insulation, finishings, services and other elements of the building will be properly related to the class of building and to each other.

(3) To keep total expenditure within the amount agreed by the client, frequently based on an approximate estimate of cost prepared by the quantity surveyor in the early stages of the design process. There is a need for strict cost discipline throughout all stages of design and execution to ensure that the initial estimate, tender figure and final account sum are all closely related. This entails a satisfactory frame of cost reference (estimate and cost plan), ample cost checks and the means of applying remedial action where necessary (cost reconciliation).

THE IMPORTANCE OF BUILDING

Approximately one-half of the value of fixed capital produced in this country each year emanates from the construction industry, and this amounts to about £4500m of which about twenty per cent relates to repair and maintenance work. The industry employs about one and a half million people which is equivalent to about seven per cent of the working population. Furthermore, if employees in allied industries, such as manufacturers and suppliers of building materials and components, and professional advisers and administrative personnel on building projects were to be included, the total number of persons involved would be in the order of two and a half million. These statistics give an indication of the immensity of the work undertaken by the construction industry, which embraces both building and civil engineering work, and its relative importance in the national economy.

Employers in the construction industry embrace a wide range of interests from local authorities, government departments and nationalised industries to industrialists, development companies and private individuals. Public authorities account for about one-half of the work undertaken by the construction industry and this can have serious effects for the industry in times of financial crisis. Brett-Jones[4] has described how the Government has in this way acquired the power, even without the licensing of private building, to influence thousands of projects in the preconstruction formulation stage at any time. He has postulated, with considerable justification, that to put the brake on large projects after inception leads to inefficient use of resources because these are the buildings which require the longest time for planning. Hence any curtailing or restriction of building work should have regard to the preplanning periods required for different types and sizes of project, with a view to mitigating, or at least minimising, the harmful effects of such restrictions.

Building work embraces a wide range of activities in both the types and sizes of job undertaken. There are surprisingly large variations in the size of projects within any particular use class. Residential buildings, for instance, can range from small, comparatively simple single-storey elderly persons' dwellings to complex schemes of high rise development to secure residential densities in the order of 400 persons per hectare. Similarly, industrial buildings can range from small workshops of about 200 sq m floor area to large factories occupying many hundreds of thousands of square metres. Buildings also vary in the form of the benefits that they generate. A residential building has a direct value to the occupier in the satisfaction that it gives him, whereas the value of an industrial building is more related to the products which can be manufactured within it.

BUILDING OUTPUT AND COSTS

Significance of Building Costs

Many clients are concerned about the rising cost of building but, as Hillmore[5] has shown, many builders have experienced financial difficulties in the late sixties and early seventies. The Ministry of Housing and Local Government[6] found that the cost of the average local authority house in England and Wales, inclusive of land, rose by more than £1000 between 1964 and 1969 (£3140 to £4190). In London the rise was even greater – from

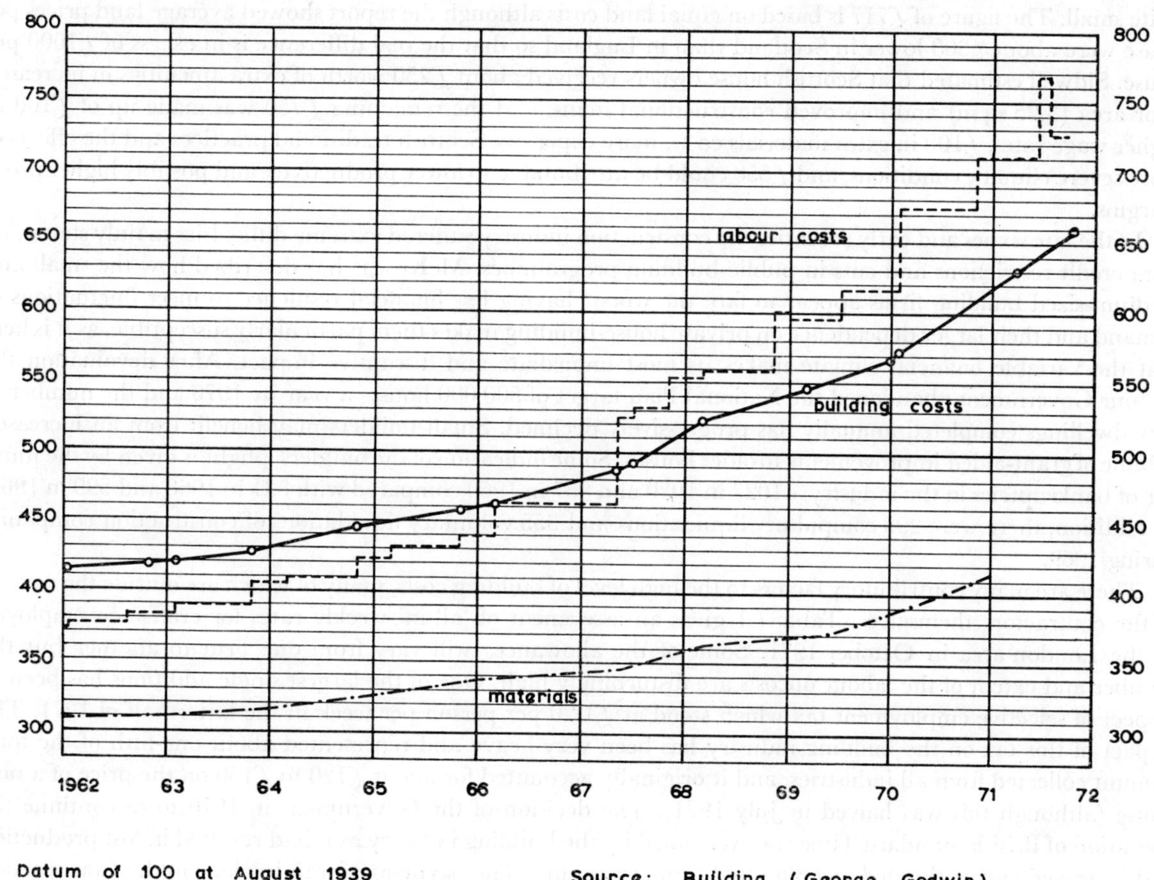

Figure 1.1 Fluctuations in building, labour and material costs – 1962–1971

£4590 to £6280, despite the operation of the Ministry cost yardstick. Some of the Ministry statistics demonstrate the increasingly heavy pressure of local authority housing on public funds. The exchequer subsidy rose from £67m to £110m in the five-year period ending March 1969, while in the same period contributions from local rates more than doubled from £21m to £44m.

Figure 1.1 shows estimated average building costs together with average labour and materials costs over the period 1962 to 1971, based on a datum of 100 in August 1939. There was a steadying of costs from 1956 to 1960 with materials prices falling during the latter part of the period. Sharp increases in building costs occurred in the late sixties and early seventies. It is significant that labour costs have tended to rise at a faster rate than materials costs while building costs tended to rise more sharply than both. This causes activities which are highly labour intensive, such as painting and most repair and maintenance work, to be proportionately more expensive. The materials figures are based on information obtained by *Building* from the Central Statistical Office and the labour rates include an allowance for emoluments such as guaranteed week, sick pay, holidays with pay, national insurance, selective employment tax and training board levy (but not incentive payments, plus rates, overtime and fares), and are based on the conventional average of one labourer to one craftsman. There are also regional price differences, and Reiners[7] found that in 1958 prices of three to five-storey flats in inner London were twenty to thirty per cent higher than those in the provinces, although these regional differences in price were not uniform in the various trades.

The Sidwell Report[36] on private house building costs in Scotland found that the average cost of a house in Scotland in 1970 was £717 more than a house in England, although the differences in size and construction were quite small. The figure of £717 is based on equal land costs although the report showed average land prices per house were about £300 lower in Scotland than in England so that the real difference is in excess of £1000 per house. Sidwell estimated that Scottish house owners received about £250 worth of extra amenities in increased floor area (3.25 sq m) and improved constructional items, and the remaining £750 was made up of £100 in higher wage rates, £100 in extra materials costs, more expensive Scottish traditional practices and the effects of more severe climatic conditions, and £550 could be attributable to lower productivity and possibly higher profit margins.

In the late sixties and early seventies the construction industry suffered extreme difficulties mainly stemming from credit restrictions and cuts in public building programmes. McKown[8] has described how the small and medium-sized building firms appear to fare the worst; having less financial resilience to meet fluctuations of demand and their larger dependence on private house-building makes them particularly susceptible, as it is here that the variable financial climate makes its most immediate and disruptive impact. After devaluation the Labour Government abandoned the National Plan target of 500 000 houses a year by 1970 and the number of new dwellings completed annually has progressively declined. Small builders may benefit from an increased volume of grant-aided improvements to older houses. Some indication of the builders' plight is given by the number of bankruptcies in the industry – 1027 in 1970 and 957 in 1969 compared with 831 in 1968 and 599 in 1964. In addition, there were 295 compulsory liquidations and 355 voluntary liquidations of construction companies during 1969.

There are many contributory factors to the high level of building costs, many of which are outside the control of the contractors themselves. Table 1.1 gives an assessment of 'all-in' weekly rates for craftsmen employed in the London area in October 1971. Some of the allowances will vary from one firm to another but the number and extent of the labour oncosts are disturbingly high. One of the largest single additions has been in respect of selective employment tax which stood at £1.20 per person per week in the latter half of 1971. The impact of this tax on the building industry has been very heavy and represented about one-fifth of the total amount collected from all industries, and it originally accounted for about £120 to £150 on the price of a new house (although this was halved in July 1971). The decision of the Government in 1970 to discontinue the operation of British Standard Time was welcomed by the building industry as it had resulted in lost production and increased costs estimated at £30m to £40m per annum. The payment of training levy to the Construction Industry Training Board has caused considerable resentment particularly amongst the smaller building firms who have for the most part received little in the way of grants. The situation deteriorated sharply in 1970 when the Board suffered a deficit of £12m, due to underestimates of expenditure and was forced to seek additional borrowing facilities, to make economies in its administration and to introduce proportional grants. It was estimated that £700m was outstanding to the construction industry in 1971 as a result of the operation of periods for honouring of certificates, retention provisions and periods of final measurements. In view of the high interest rates paid by many contractors, the present time would seem opportune for a review of the financial arrangements of building contracts.

TABLE 1.1

LABOUR COSTS OCTOBER 1971

Average estimated weekly cost of craftsman		
Direct costs		
Forty hours (London and Liverpool rates) (labourer's rate is £17.20)		£20.20
One hour overtime per day at time-and-a-quarter six-and-a-quarter hours @ 50½p		3.15½
Four hours (Saturday) at time-and-a-half six hours @ 50½p		3.03
Allowances (average) fifty-two-and-a-quarter hours @ 1p		0.52
Tool money (carpenter)		0.20
Four public holidays $\frac{£16.16}{49}$		0.33
		£27.43½
Indirect costs		
Selective employment tax	£1.200	
Industrial injuries	0.060	
National health service	0.083	
National insurance contribution	0.744	
Redundancy fund	0.063	
		£2.15
Graduated pension		0.84
Annual holidays with pay		1.25
Proportion of employer's national health insurance contributions, SET and graduated pension to cover four public holidays		0.05
Three weeks national insurance and SET for annual holidays		0.13
Sick pay scheme		0.10
CITB levy		0.38½
Redundancy, provision for disbursement		0.28
		£5.18½

A committee of the Institute of Quantity Surveyors[39] recognised the existence of a widespread liquidity problem within the building industry and considered that the main reasons for this were as follows.

(1) The level of retention monies held against work in progress under the conditions of contract.

(2) The further amount of 'hidden retention' represented by the usual timelag between the execution of work and the receipt of the corresponding payment by the contractor or subcontractor concerned.

(3) Delays in certificates and payment of final accounts.

(4) Delays caused by claims.

Building Output

The growth in the rate of house and flat building since the second world war is shown in figure 1.2. In the late forties and early fifties private house building was restricted by a building licensing system and the bulk of residential development was undertaken by local authorities. In 1959 the numbers of private houses built exceeded those provided by the public sector. The total number of completions has fluctuated considerably over the years and reached its peak in 1968 at 413 715. In the late sixties the total value of work undertaken by the construction industry annually approached £4000m, roughly subdivided as shown below.

New housing	£m 1300
Other new work	1800
Repairs and maintenance	800
Total	£m 3900

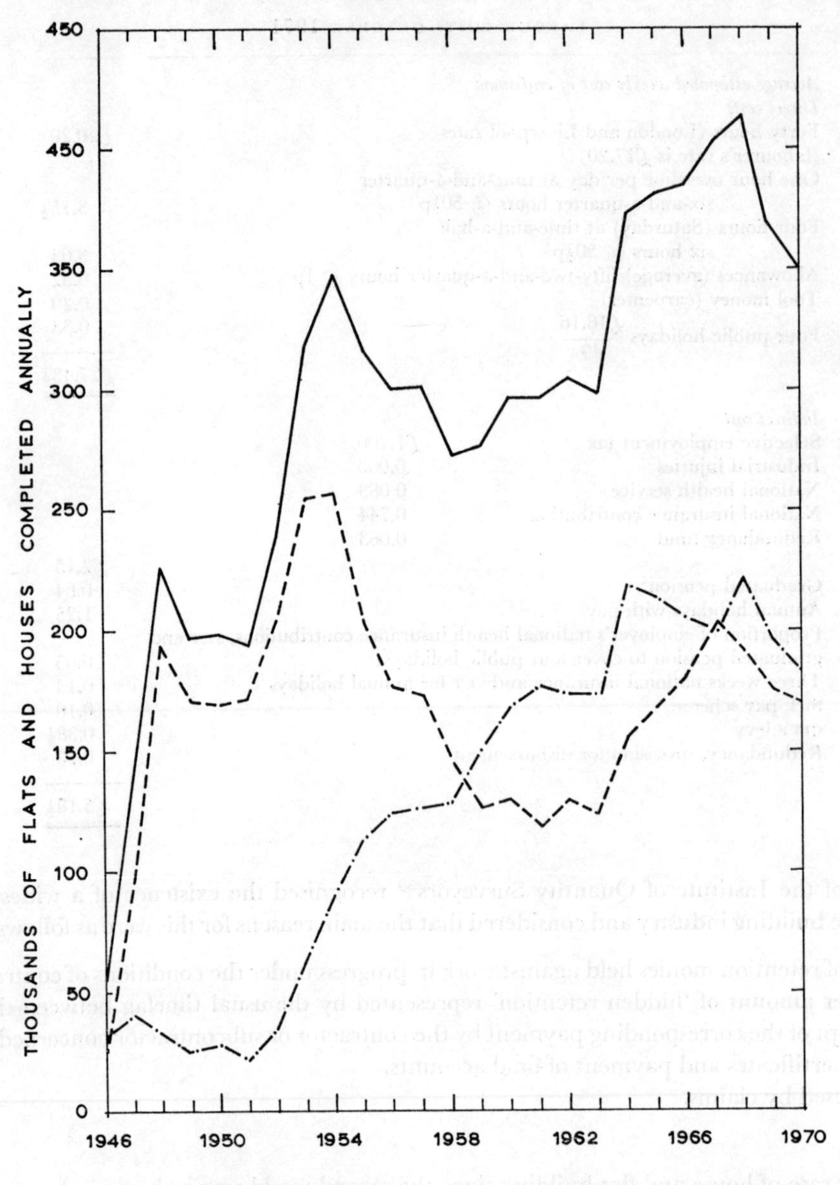

Figure 1.2 House and flat completions in Britain – 1946–1970

Building Productivity

It has been estimated that the building industry will be faced with an increase in demand of between fifty and sixty per cent in the next decade, with quite a modest increase in the labour force. This creates the need for substantial increases in output and the better use of all resources. The Ministry of Housing and Local Government produced some disturbing reports after the Second World War which showed that in 1947, forty-five per cent more man-hours were required to perform a given quantity of house-building work as compared with 1938–9, and that this represented a decline in productivity of thirty-one per cent. By 1949 there had been a considerable improvement, although the man-hours were still twenty-six per cent above the 1938–9 level, with productivity twenty per cent below prewar level, and in 1951 the position was still much the same.[9] The Ministry made recommendations to local authorities with a view to reducing house-building costs. These recommendations included reducing room heights by six inches (150 mm) making possible a saving of £15–£20 per house; reducing the floor areas of three-bedroom houses by 100 sq ft (9.3 sq m), saving about £90 per house and omitting the second WC. Towards the end of 1950 a number of local authorities were considering ways of building houses to let at lower rentals, as it was becoming apparent in some areas that many persons on the housing lists were unable to afford the rents flowing from the current house plans.

Bishop[10] held the view that schemes for improved productivity were as relevant to the design team as to the contractor. Designers of individual buildings, by considering the way each stage of construction may be handled, can ensure designs that are relatively easy to build. Designers who concentrate on a particular building type, or always employ similar construction techniques, have an opportunity to encourage, and sometimes promote, the introduction of new techniques and new components. Finally there is the considerable potential of those concerned with industrialised building for innovation and for the integration of design and production. Bishop has also illustrated ways in which detailed design can simplify work, or at least eliminate work which can be tackled by a gang without interruption from other gangs.

(1) *Simplification.* For example, arranging brickwork in long runs between quoins.

(2) *Ensuring continuity.* For example, using joist hangers so that bricklayers' work is not interrupted while carpenters set joists; or arranging precast concrete stair flights to be self-supporting from one floor to another rather than by building in half-landings.

(3) *Separating the work of gangs.* For example, detailing claddings so that joints and flashings can be completed independently of the erection gang.

(4) *Reducing the number of separate operations.* For example, detailing reinforcement and construction joints so that concrete floors and walls may be cast in one operation.

(5) *Making mechanisation feasible.* For example, specifying flooring of uniform thickness so that the screeds may be finished at one level, thus making the use of power floats more practicable.

It is vital that the bills of quantities are so framed to be immediately apparent to an estimator that the building has been designed for the use of cheap and easy construction processes.

Since the 1952 report of the Ministry of Housing and Local Government[9] there has been a steady reduction in labour requirements for house building, with the average labour commitment per house dropping from about 2650 to 1800 man-hours, equivalent to an improvement of about two-and-a-half per cent per annum. Between 1963 and 1968 output per head in the building industry increased by between three and four per cent per annum. This improved efficiency is the cumulative effect of a variety of factors: increased mechanisation; improved management techniques; increased prefabrication and industrialisation; greater utilisation of incentive bonus schemes, and improved training methods. Surveys of the Building Research Station have shown that on the least efficient sites the man-hours per house were three times as great as on the most efficient. It was also found that there is no strong relationship between the plan area of a dwelling and the man-hour requirements. Productivity was found to be high in firms paying a target bonus; where subcontractors were employed on the work rather than the main contractor's own labour; in small firms in which the builder himself worked with the tools; and in firms working regularly on house building.[11]

Forbes[12] has described the problems involved in the measurement of productivity on building jobs. If man-hours of each trade for the whole job are considered sufficient, then the information can usually be obtained from weekly time-sheets. It is however likely that it will be necessary to apportion the men's time between specific parts of the work and/or to different activities, when special arrangements will have to be made to collect the information by direct observation.

The output of work performed will normally be measured by a functional unit varying with the level of detail at which the study is being made, for example, number of houses built or number of bricks laid. The great variety of building types and techniques makes it difficult to define the work content in a way to give one physical measurement of the work done. At the coarser levels of detail the dwelling or bed-space can serve as a suitable unit in house building. At the other end of the scale the output must be considered in terms of the work done, for example, the output of two gangs of bricklayers erecting similar buildings could be compared on the man-hour requirements per thousand bricks laid. On the other hand the work content of, for instance, carpenters' second fixings is less easy to compare as so much depends upon design detail and specification. At an even finer level of detail this unit of output would be broken down to a list of tasks, such as 'hang door' or 'fix cupboard'. Furthermore, when comparing the effect of different forms of construction on productivity it is always important to determine how far the observed differences arise both from the particular design and specification or from organisational aspects, including the ability and motivation of different operatives.[12]

Many local authorities have endeavoured to reduce building costs and increase efficiency by the use of industrialised building and consortia arrangements. Industrialised building aims at producing building systems and methods that permit rapid assembly on site with the minimum use of labour, and this subject will be considered further in chapter 4. The establishment of consortia enables bulk orders to be placed for components with the advantage of discounts for large quantities, serial tendering, development of designs and feedback of information from sites. Similarly the National Building Agency was set up to help the smaller housing authorities to pool and collate their requirements into phased programmes and to obtain more efficient and more economical working arrangements.

In the past builders and designers have experienced difficulties in keeping abreast of new materials and components and there has often quite naturally been a reluctance to use products which were neither well-established nor officially approved. This impediment to progress was largely removed by the Minister of Public Building and Works setting up the Agrément Board in 1966 with the use of the Building Research Station as its technical agent. The Board's principal objective is to assist innovation in the building industry in respect of building materials, products, components and processes, by providing an assessment on the basis of examination, testing and other investigation. Products, processes, etc., which, in the opinion of the Board, are likely to give a satisfactory performance will qualify for a statement or certificate identifying the product and its method of use or of installation, together with a summary of appropriate design data. In general, these certificates will cover the period before a British Standard Specification can be devised and issued and will, in appropriate cases, indicate compliance with the relevant part of the Building Regulations.

The Role of the Quantity Surveyor

It is highly desirable that the quantity surveyor should be involved in the planning and design stages of a project. He can illuminate and amplify quality decisions by bringing to bear the full spectrum of his professional knowledge of what is available in the market, to reinforce the architect's range of experience, synthesise the professional skills of the various members of the design team, help to incorporate features which the contractor can provide efficiently and competitively in favour of those which create greater difficulties; and, by the careful balancing of alternatives in terms of quality and economics, he can provide a sound basis for decision-making by the architect and client. The quantity surveyor's prime aim should always be to assist building clients in obtaining optimum value for money. Nisbet[13] has described how quantity surveying activities are expanding rapidly in the fields of feasibility studies, cost advice during the early stages of projects and cost control during the design and post-contract stages.

We have seen how the quantity surveyor can give advice on the strategic planning of a project which may affect the decision whether or not to build, where to build, how quickly to build and the effect of time on costs or prices and on profitability. Cost control throughout the design and construction stages may be extended to the giving of advice on taxation matters and cash flow, while the optimum use of production resources is important when projects using industrialised building systems are negotiated. Building clients are certainly becoming more knowledgeable and are demanding a much wider range of expert cost advice than quantity surveyors have been accustomed to giving in the past. An RICS study group[14] has expressed the view that much needs to be done to develop contract documents appropriate to performance specifications and to produce bills that can be used during the course of projects for management purposes and to monitor costs.

In the exercise of cost planning techniques the quantity surveyor is concerned with many issues of building economics, some involving returns as well as costs, and some of these are listed below.

(1) Substitution between capital and running costs to secure the minimum total cost.

(2) Investigation of the different ways of producing the same building at lower cost.

(3) Finding ways of slightly altering a building so that for the marginally greater use of resources, the returns are more than proportionately increased.

(4) Investigating methods of using the same resources to produce a different building which could give greater returns.

TENDERING ARRANGEMENTS

Conventional tendering procedures have been criticised on the grounds that they fail to take full advantage of modern techniques and may inhibit the wider industrialisation of building processes. Against this background, a symposium on tendering procedure[15] attempted to clarify the factors which determine the most suitable tendering method to be adopted and the circumstances under which the client's interests are best served by negotiation. All tendering procedures aim at selecting a suitable contractor and obtaining from him at an appropriate time an acceptable offer, or tender, upon which a contract can be let.

Governmental Committee Recommendations

The Simon Committee[16] in 1944 drew attention to the fact that 'low prices resulting from indiscriminate tendering result in bad building' and that 'resources are wasted when many firms tender for the same job'. The Banwell Report[17] suggested that invitations to tender should be limited to a realistic number of firms, all of whom were capable of executing the work to a recognised standard of competence. The Banwell Committee appeared to favour the general use of standing approved lists of contractors and that *ad hoc* lists should be used mainly when the work was of a specialist or one-off nature. The Committee further recommended that the period allowed for tendering should be adequate for the type of project and welcomed 'firm-price' contracts. The Ministry of Housing and Local Government issued revised model standing orders to local authorities in 1964 to facilitate the wider use of selective tendering procedures, and in 1965 the Ministry gave guidance to local authorities on the operation of selective tendering.[18]

In 1965 a working party was established by the Economic Development Committee for Building to examine the Banwell Report and its implementation and its report was submitted in 1967.[19] This report considered that insufficient attention was paid to the importance of time and its proper use and that clients seldom define their requirements in sufficient detail at the start of negotiations. It favoured the main contractor joining the design team at an early stage. The report further urged the wider adoption of the practices detailed in *Code of Procedure for Selective Tendering*,[20] *Selective Tendering for Local Authorities*[18] and *Early Selection of Contractors and Serial Tendering*,[21] although they recognised that in the public sector this would require a more flexible approach to satisfy standards of accountability. Although the working party saw the merit in 'firm-price' contracts, they stressed the difficulties involved in producing firm tenders in a market where materials prices tend to fluctuate and contractors are often invited to tender on incomplete documentation.

Types of Contractual Arrangement

There are a variety of client/contractor relationships and the choice will be influenced considerably by the particular circumstances. They range from *cost reimbursement* or *cost plus* contracts at one end of the scale to truly lump sums, such as non-variable package deals, at the other. Cannell[22] has pointed out that the essential difference between these two extremes devolves upon which party to the contract is to carry the risk of making a loss (or profit) and the incentives which are built into the contract to encourage the contractor to provide an efficient and economic service to the client. A brief examination and comparison of the main categories of contractual arrangement follows.

Cost plus contracts. These contracts are sometimes referred to as *cost reimbursement* contracts or *prime cost* contracts. In practice they can take three quite different forms.

(1) Cost plus percentage contracts are those in which the contractor is paid the actual cost of the work plus an agreed percentage of the actual or allowable cost to cover overheads, profit, etc. They are useful in an emergency, when there is insufficient time available to prepare detailed schemes prior to commencement of the work, but it will be apparent that an unscrupulous contractor could increase his profit by delaying the completion of the works. No incentive exists for the contractor to complete the works as quickly as possible or try to reduce costs.

(2) Cost plus fixed fee contracts are those in which the sum paid to the contractor will be the actual cost incurred in the execution of the work plus a fixed lump sum, which has been previously agreed upon and does not fluctuate with the final cost of the job. No real incentive exists for the contractor to secure efficient working, although it is to his advantage to earn the fixed fee as quickly as possible and so release his resources for other work. This type of contract is superior to the cost plus percentage type of contract.

(3) Cost plus fluctuating fee contracts are those in which the contractor is paid the actual cost of the work plus a fee, with the amount of the fee being determined by reference to the allowable cost by some form of sliding scale. Thus the lower the actual cost of the works, the greater will be the value of the fee that the contractor receives. An incentive then exists for the contractor to carry out the work as quickly and as cheaply as possible, and it does constitute the best form of cost plus contract from the employer's viewpoint.[23]

Target cost contracts. These contracts have been introduced in recent years to encourage contractors to execute the work as cheaply and efficiently as possible. A basic fee is generally quoted as a percentage of an agreed target estimate often obtained from a priced bill of quantities. The target estimate may be adjusted for variations in quantity and design and fluctuations in the cost of labour and materials, etc. The actual fee paid to the contractor is obtained by increasing or reducing the basic fee by an agreed percentage of the saving or excess between the actual cost and the adjusted target estimate. In some cases a bonus or penalty based on the time of completion may also be applied. Cannell[22] has described how target cost contracts can be useful when dealing with unusual or particularly difficult situations, but that the real difficulty lies in the agreement of a realistic target.

Measure and value contracts. These contracts include those based on schedules of rates, approximate quantities and bills of quantities. The great merit of these contracts lies in the predetermined nature of the mechanism for financial control provided by the precontract agreed rates. The risk of making a profit or loss is with the contractor. Another adaption of the orthodox bill of quantities contract is serial tendering whereby a series of contracts are let to a single contractor. Serial contracting is based on a standing offer, made by a contractor in competition, to enter into a number of lump sum contracts in accordance with the terms and conditions set out in a master or notional bill of quantities. Morris[24] has described how this procedure is particularly suitable for a known programme of building work over a period of time in a specific locality, and where there is a high degree of standardisation of construction. The advantages claimed for the system include improved relationships with the contractor, more effective cost control and faster and more economical planning and execution of jobs.[41] The Institute of Building[37] holds the view that 'the present organisation of design/construction relationships is perpetuated by the competitive tendering system and inhibits communication between the different functions involved in a building project'.

Contracts based on drawings and specification. These are often described as 'lump sum' contracts although they may be subject to adjustment in certain instances. They form a useful type of contract where the work is limited in extent and reasonably certain in its scope. They have on occasions been used where the works are uncertain in character and extent, and by entering into a lump sum contract the employer hoped to place the onus on the contractor for deciding the full extent of the works and the responsibility for covering any additional costs which could not be foreseen before the works were commenced. The client would then pay a fixed sum for the works, regardless of their actual cost, and this constitutes an undesirable practice from the contractor's point of view.[23]

Package deal contracts. These constitute a specialised form of contractual relationship in which responsibility for design as well as construction is entrusted to the contractor. Cannell[22] has described how the less developed the design, the less detailed the specification and hence the less precise must be the calculation of the price. Contingencies must be included to provide for the unknown. Madden[25] has shown how package deal contracts have been employed for local authority housing, often incorporating heavy industrialised systems, a new hospital unit, a major defence installation, and factories. In housing schemes about six contractors may be invited by a local authority to submit a complete development scheme for a large site. The contractors may use their own design teams or private architects to prepare schemes within the specified requirements of densities and costs, and the successful contractor will subsequently be required to collaborate with the authority's architect. This form of approach is particularly favoured where special factors operate, such as the use of building systems.

Parrinder[26] has shown that contractor-sponsored systems make it essential for the contractor to be brought into the design team. It has also become necessary to evolve procedures which will allow competition between contractors offering different systems. This has involved two separate stages: first, a competition to find the contractor who can best satisfy the functional, aesthetic and economic needs; and second, a period of negotiation with the selected contractor using data, especially prices, derived from the first stage. An attempt is made in *The Chartered Quantity Surveyor and Package Contracts*[29] to summarise the services that quantity surveyors can offer in connection with package contracts at the precontract, tender and contract stages.

Brett-Jones[28] has described in a most lucid and comprehensive manner an ideal procedure for maintaining adequate financial control of contractor-design projects. He favoured not more than three or four contractors, all of whom must have adequate resources and experience to undertake the project. The brief submitted to contractors should not be so detailed as to preclude contractors from using their initiative and know-how. He suggested the compiling of a cost benefit analysis against the client's requirements to put them in true perspective; an example being the inclusion of balconies – are they for instance being introduced as an amenity feature or to assist in maintenance? The brief is accompanied by all relevant documents including surveys, consents, statutory requirements, 1:100 plans, sections and elevations, and larger typical detail drawings. Contractors would be required to submit lump figures to cover site management costs, head office overheads, design costs and profit, and superstructure work. Substructures, external works and services will be evaluated in a provisional bill of quantities, and provisional sums will be listed by the client. In the final assessment regard has to be paid to aesthetic as well as financial aspects. To control costs on the job, lump sum figures for building work will have to be broken down and checked.

COMPARISON OF COST PLANNING AND APPROXIMATE ESTIMATING

In the past there has been some confusion on occasions over the two processes and there are still a few quantity surveyors who believe the terms to be identical. Admittedly many quantity surveyors were first introduced to cost planning through approximate estimating. Cost planning aims at ascertaining costs before many of the decisions are made relating to the design of a building. It provides a statement of the main issues, identifies the various courses of action, determines the cost implications of each course and provides a comprehensive economic picture of the whole. The architect and quantity surveyor should be continually questioning whether a specific item of cost is really necessary, whether it is giving value for money or whether there is not a better way of performing the particular function.

Cost planning does therefore differ significantly from approximate estimating. Approximate estimating aims at providing a preview of the probable tender figure with the method employed often being influenced by the amount of information available. Approximate estimating methods will be examined and compared in chapter 6.

Cost planning, on the other hand, does not merely estimate the tender sum but probes much deeper into the cost implications of each part of the building, whereby each design decision is analysed and costed, and the final decision maintains a sensible relationship between cost, quality, utility and appearance. In addition cost planning permits control of expenditure throughout the design and construction stages.

Hence approximate estimating often plays a largely passive role after the major design decisions have been made, whilst cost planning bears upon the decisions themselves and plays an active part in the formulation of the design. Indeed, the cost plan should be regarded as one of the elements of design.

COST CONTROL TERMINOLOGY

A number of terms are used widely in cost control work and it is deemed advisable to define and explain these terms prior to their use.

Cost plan. A statement of the proposed expenditure on each section or element of a new building related to a definite standard of quality. Each item of cost is generally regarded as a 'cost target' and is usually expressed in terms of cost per square metre of floor area of the building as well as total cost of the element.

Cost check. The process of checking the estimated cost of each section or element of the building as the detailed designs are developed, against the cost target set against it in the cost plan.

Cost analysis. The systematic breakdown of cost data, often on the basis of elements, to assist in estimating the cost and in the cost planning of future projects. Cost analyses are often supplemented with specification notes, data concerning site and market conditions and various quantity factors, such as wall to floor ratios. The form and method of use of cost analyses are dealt with in chapter 8. Cost analysis aims at examining the cost of buildings already planned or built and for which priced bills of quantities and tenders are available. It has been suggested that it is in the nature of a post-mortem,[33] but in practice it is more valuable than this as it can assist materially in the design and cost evaluation of new projects.

Approximate estimating. Computing the probable cost of new building works at some stage before the bill of quantities is produced. It is an essential and integral part of the cost planning process.

Element. A component or part of a building that fulfils a specific function(s) irrespective of its design, specification or construction, such as walls, floors and roofs. Many cost plans and cost analyses are prepared on an elemental basis.

Cost research. All methods of investigating building costs and their interrelationship, including maintenance and running costs, in order to build up a positive body of information which will form basic guidelines in planning and controlling the cost of future projects.

Costs-in-use. Investigating the total costs of building projects – initial capital costs and maintenance and running costs throughout the predicted lives of the buildings. It provides the only way of obtaining the overall cost picture but does present a number of difficulties in practice, and these are examined in chapter 11.

Cost study. Breaking down the total cost of buildings with the following objectives:

(1) to reveal the distribution of costs between the various parts of the building;
(2) to relate the cost of any single part or element to its importance as a necessary part of the whole building;
(3) to compare the costs of the same part or element in different buildings;
(4) to consider whether costs could have been apportioned to secure a better building;
(5) to obtain and use cost data in planning future buildings; and
(6) to ensure a proper balance of quantity and quality within the appropriate cost limit.[33]

Cost control. All methods of controlling the cost of building projects within the limits of a predetermined sum, throughout the design and construction stages.

Cost planning. This is often interpreted as controlling the cost of a project within a predetermined sum during the design stage, and normally envisages the preparation of a cost plan and the carrying out of cost checks. *Building Bulletin No. 4*[33] describes how cost planning uses the information gained by cost analysis to maintain a surer control over costs of future projects.

A more comprehensive and apt definition was devised at the RICS postgraduate cost planning course at Brixton,[34] namely: 'A systematic application of cost criteria to the design process, so as to maintain in the first place a sensible and economic relation between cost, quality, utility and appearance, and, in the second place, such overall control of proposed expenditure as circumstances might dictate.' This definition is so important that it warrants further consideration.

Emphasis is directed towards the proper consideration of design criteria other than cost to produce a properly balanced design, and it is advisable that the quantity surveyor does not lose sight of this. For instance, the cost of a project could be reduced merely by using cheaper materials, finishings and fittings, regardless of the fact that maintenance and running costs would probably be increased considerably in consequence. Furthermore the lower quality materials and components may be quite out of keeping with the class of building in which they are being incorporated. An economically priced job is required, but not necessarily the cheapest as a certain standard of quality has to be maintained. A building must also be designed so that it can satisfactorily perform its required functions. It might, for example, be possible to cheapen the cost of a factory roof by incorporating more columns and thus reducing the roof spans. This approach would be quite fruitless if the factory needed large unrestricted floor areas for its successful operation. Costs can often be reduced by disregarding the aesthetic quality of the building to be erected. Plain façades devoid of any form of embellishment could reduce costs but provide most uninspiring elevations. The author believes that Conservation Year 1970 taught us that every building client has a duty to assist in improving the environment and, through it, the well-being and general satisfaction of the community.

Building economics. Ward[35] has postulated that building economics is concerned with two main objectives:

(1) to ensure the efficient use of available resources; and
(2) to seek to increase the rate of growth of construction work in the most efficient manner.

RICS COST RESEARCH PANEL/BCIS MANAGEMENT COMMITTEE

With the increasing emphasis on cost planning, the Council of the Royal Institution of Chartered Surveyors established in 1956 the Cost Research Panel, administered by the Quantity Surveyors Committee; in fact cost research and advice is now administered by the Building Cost Information Service Management Committee and the Quantity Surveyors Research and Information Committee. The terms of reference of the Panel were as follows.

(1) To keep under review the sources of cost information available to the industry, and to consider the desirability of creating a focal point for the collection of information.
(2) To initiate research into matters affecting the cost of building, including the effect of standardisation and repetition in design.
(3) To draw conclusions from the results of research, to make recommendations and to publish such results.
(4) To stimulate generally an interest in the cost aspects of building.

The Cost Research Panel set about pursuing its various activities with enthusiasm. The Panel prepared a report for the Minister of Housing and Local Government on the cost of flats and houses in England and Wales and numerous papers on cost aspects were prepared or sponsored and published in *The Chartered Surveyor*. Reference will be made to many of these papers in later chapters. The Panel also liaised with the Brixton School of Building (now a constituent college in the South Bank Polytechnic) in the holding of postgraduate cost planning courses and with the Royal Institute of British Architects. An extensive programme of cost research work was arranged for implementation by study groups of quantity surveyors in the branches, with each branch being allocated a particular project. Subsequently, meetings and conferences on cost planning subjects were arranged throughout the country.

In 1964 the Building Cost Information Service was established with the object of exchanging building cost information between chartered quantity surveyors. Individual members provide data from their own resources and in return receive the information made available by all. The distribution from Institution headquarters takes the form of regular mailings of information sheets for members to file in looseleaf binders; this information consists of brief and amplified costs analyses of building projects and general building cost data, such as building cost indices. Membership of the service has in the past been restricted to chartered quantity surveyors who are principals of firms or heads of departments in the public service, but its scope is now being widened. Each subscriber, excluding educational establishments, must undertake to submit cost analyses or other cost information for dissemination to all members, and an annual subscription is payable by each firm or branch office to cover operating costs.

COST IMPLICATIONS OF METRICATION AND DIMENSIONAL CO-ORDINATION

Contractors are affected by the change to SI units in a variety of different ways, due to changes in the following activities and equipment.

(1) Measurement – distances, areas, volumes, capacities and pressures all given in metric terms or SI units.
(2) Measuring equipment such as rules, tapes, surveying staffs and weighing equipment in metric units.
(3) Office machines to be replaced or modified.
(4) Recalibration of plant.
(5) Costing, bonusing, estimating and work study data all to be rewritten in metric terms.
(6) Building, safety and other regulations in unfamiliar terms.
(7) Rewriting computer programmes.
(8) Longterm planning in metric.

The contractor is also faced with many problems as a result of the change and some of the most significant are now listed.

(1) Human problems resulting from an inbuilt resistance to change, particularly with older workers.
(2) Problems of retraining personnel to think metric with the varying requirements of different categories of operatives and staff and the need to introduce training programmes at the right time.
(3) Time taken to obtain the 'feel' of new units and become proficient in their use.
(4) Increased risk of errors and need for more checking.
(5) General administration in the changeover period.
(6) Changeover problems such as mixed supplies of materials and components in metric and imperial terms, which will also present difficulties for suppliers.
(7) Difficulty of co-ordination with subcontractors and statutory undertakers.

It seems evident that contractors will incur increased costs to cope with the problems previously outlined, which could be in the order of five to eight per cent, but will vary with the type of organisation and the nature of the particular job. It is to be hoped that in the longterm these increased costs will be more than offset by savings resulting from dimensional co-ordination, which will in its turn lead to rationalisation and large scale production with accompanying economies of scale. Secondary benefits will also accrue from the reduced number of simpler units and the use of the same terminology as continental and other countries. A metric conversion table in included in appendix 6.

REFERENCES

1. ROYAL INSTITUTION OF CHARTERED SURVEYORS. Surveyors and their future: Report of the special Committee (Future of the Profession), 1970

2. ROYAL INSTITUTION OF CHARTERED SURVEYORS STUDY GROUP. The future role of the quantity surveyor. *The Chartered Surveyor*, **102.1** (1969)
3. M. HADDOCK. Quantity surveying: the evolving profession. *Building* (14 March 1969)
4. A. T. BRETT-JONES. Project management. *The Chartered Surveyor* (January 1966)
5. P. HILLMORE. The rising cost of building. *The Guardian* (18 June 1970)
6. MINISTRY OF HOUSING AND LOCAL GOVERNMENT. Housing statistics. HMSO (1970)
7. W. J. REINERS. Research, cost planning and cost control. *The National Builder*, **38.9** (1959), 300–303.
8. R. MCKOWN. Outlook grey – but not black. *The Guardian* (3 July 1968)
9. MINISTRY OF HOUSING AND LOCAL GOVERNMENT. The cost of house-building: Third Report of Committee of Inquiry. HMSO (1952)
10. D. BISHOP. Architects and productivity. *Current Papers: Design Series 57*. Building Research Station (1966)
11. W. J. REINERS. The study of operations and economics at the Building Research Station. *The Chartered Surveyor* (April 1962)
12. W. S. FORBES. Some aspects of the measurement of productivity in the building industry. *Current Papers: Construction Series 28*. Building Research Station (1966)
13. J. NISBET. It's elemental my dear bill said the q.s. to the builder. *Building Design* (1 August 1970)
14. RICS STUDY GROUP. The future role of the quantity surveyor. *The Chartered Surveyor* (July 1969)
15. COLLEGE OF ESTATE MANAGEMENT. Tendering procedure: a symposium (1966)
16. SIMON COMMITTEE. The placing and management of building contracts. HMSO (1944)
17. MINISTRY OF PUBLIC BUILDING AND WORKS (BANWELL REPORT). The placing and management of contracts for building and civil engineering work. HMSO (1964)
18. MINISTRY OF HOUSING AND LOCAL GOVERNMENT. Selective tendering for local authorities. HMSO (1965)
19. ECONOMIC DEVELOPMENT COMMITTEE FOR BUILDING. Action on the Banwell Report. HMSO (1967)
20. NATIONAL JOINT CONSULTATIVE COMMITTEE OF ARCHITECTS, QUANTITY SURVEYORS AND BUILDERS. Code of procedure for selective tendering (1969)
21. MINISTRY OF PUBLIC BUILDING AND WORKS. Early selection of contractors and serial tendering. HMSO (1966)
22. J. B. CANNELL. Tendering procedures and contractual arrangements. *The Building Economist (Australian)* (February 1968)
23. I. H. SEELEY. *Civil Engineering Quantities*. Macmillan (1971)
24. H. C. MORRIS. Serial contracting. *The Chartered Surveyor*, **101.11** (1969)
25. L. W. MADDEN. Package dealing: shadow or spectre? *Building* (21 February 1969)
26. E. R. PARRINDER. Competitive selection for industrialised building. *The Chartered Surveyor* (August 1966)
27. ROYAL INSTITUTION OF CHARTERED SURVEYORS. The chartered quantity surveyor and package contracts (1963)
28. A. T. BRETT-JONES. Procedure, evaluation and financial control of contractor-designed projects: the quantity surveyor's contribution. *The Chartered Surveyor* (August 1965)
29. P. W. GRAFTON. Cost planning. *The Chartered Surveyor* (May 1966)
30. J. B. CANNELL. The history and future of the q.s. role as financial adviser to the construction industry. *The Chartered Surveyor* (May 1968)
31. A. W. KENYON. An architect's thoughts on the profession of quantity surveying. *The Quantity Surveyor* (July/August 1964)
32. G. HIGGIN. The future of quantity surveying: some reflections of an outsider. *The Chartered Surveyor* (October 1964)
33. DEPARTMENT OF EDUCATION AND SCIENCE. Building bulletin no. 4: Cost study. HMSO (1972)
34. ROYAL INSTITUTION OF CHARTERED SURVEYORS/BRIXTON SCHOOL OF BUILDING. Report of postgraduate cost planning course (1959)
35. A. V. WARD. Building economics: a bibliographical survey. *The Quantity Surveyor* (November/December 1966)
36. SCOTTISH HOUSING ADVISORY COMMITTEE (SIDWELL REPORT). The cost of private house building in Scotland. HMSO (1970)
37. INSTITUTE OF BUILDING. Building and the IOB (1971)
38. A. B. WATERS. Disciple of multi-discipline. *Building Design* (1 January 1971)
39. INSTITUTE OF QUANTITY SURVEYORS, MEMBERS ADVISORY BOARD. Cash flow within the construction industry. *The Quantity Surveyor*, **27.3** (1970)
40. P. DUNSTONE. Value analysis in building. *Building* (16 October 1970)
41. MINISTRY OF PUBLIC BUILDING AND WORKS. Serial tendering: a case study from Nottinghamshire County Council. HMSO (1970)

2 COST IMPLICATIONS OF DESIGN VARIABLES

THE COSTS OF buildings are influenced by a variety of factors, some of which are interrelated. It is essential that quantity surveyors should be fully aware of the cost consequences resulting from changes in shape, size, storey heights, total height, fenestration and other building characteristics. The cost effect of the main design variables will be examined and compared in this chapter.

It is important that the method used for expressing building costs permits alternative designs and forms of construction to be examined on a comparable basis. The most convenient unit is the square metre of floor area measured between the main enclosing walls (inner faces), and making no deductions for internal walls, staircases, lift shafts or other circulation space. The cost of the whole building or any part of it can be equally well expressed in this way. For instance, a six-storey office block in the Midlands might have cost about £70 per sq m of floor area in 1971, with internal finishings accounting for £8 per sq m. Considerable care has, however, to be taken in using costs expressed in this way to make allowance for widely differing conditions on different projects. For example, soil conditions can cause quite different foundation costs for otherwise similar buildings, the cost of finishings will be influenced appreciably by the requirements of the specification and lift costs will be affected by the height of the building and the area of each floor.

PLAN SHAPE

The shape of a building has an important effect on cost. As a general rule the simpler the shape of the building the lower will be its unit cost. As a building becomes longer and narrower or its outline is made more complicated and irregular, so the perimeter/floor area ratio will increase, accompanied by a higher unit cost. The significance of perimeter/floor area relationships will be considered in more detail later in the chapter. An irregular outline will also result in increased costs for other reasons; setting out, siteworks and drainage work are all likely to be more complicated and more expensive and this will be apparent by comparing the two buildings shown in figure 2.1, each of which have the same floor area. In building B where there is six per cent more external wall to enclose the same floor area, setting out costs are increased by about fifty per cent, excavation costs by about twenty per cent and drainage costs by approximately twenty-five per cent (two metres of additional 100 mm drain and two extra manholes). The additional costs do not finish there as brickwork and roofing will also be more costly due to the work being more complicated. It is important that both architect and client are aware of the probable additional costs arising from comparatively small changes in the shape of a building. They are then able to consider the advantages to be gained from variations in shape in the full knowledge of the additional costs involved and can adopt a rather rudimentary cost benefit approach. It does however involve a subjective judgment as far as the aesthetics are concerned.

Although the simplest plan shape, that is a square building, will be the most economical to construct it would not always be a practicable proposition. In dwellings, smaller offices, schools and hospitals considerable importance is attached to the desirability of securing adequate natural daylighting to most parts of the buildings. A large, square structure would contain areas in the centre of the building which would be deficient in natural lighting. Difficulties could also arise in the planning and internal layout of the accommodation. Hence, although a rectangular shaped building would be more expensive than a square one with the same floor area because of the smaller perimeter/floor area ratio, nevertheless practical or functional aspects, and possibly aesthetic ones in addition, may dictate the provision of a rectangular building. The following example illustrates the need to maintain a balance between various design criteria, in this case cost, function and appearance. It is pointless

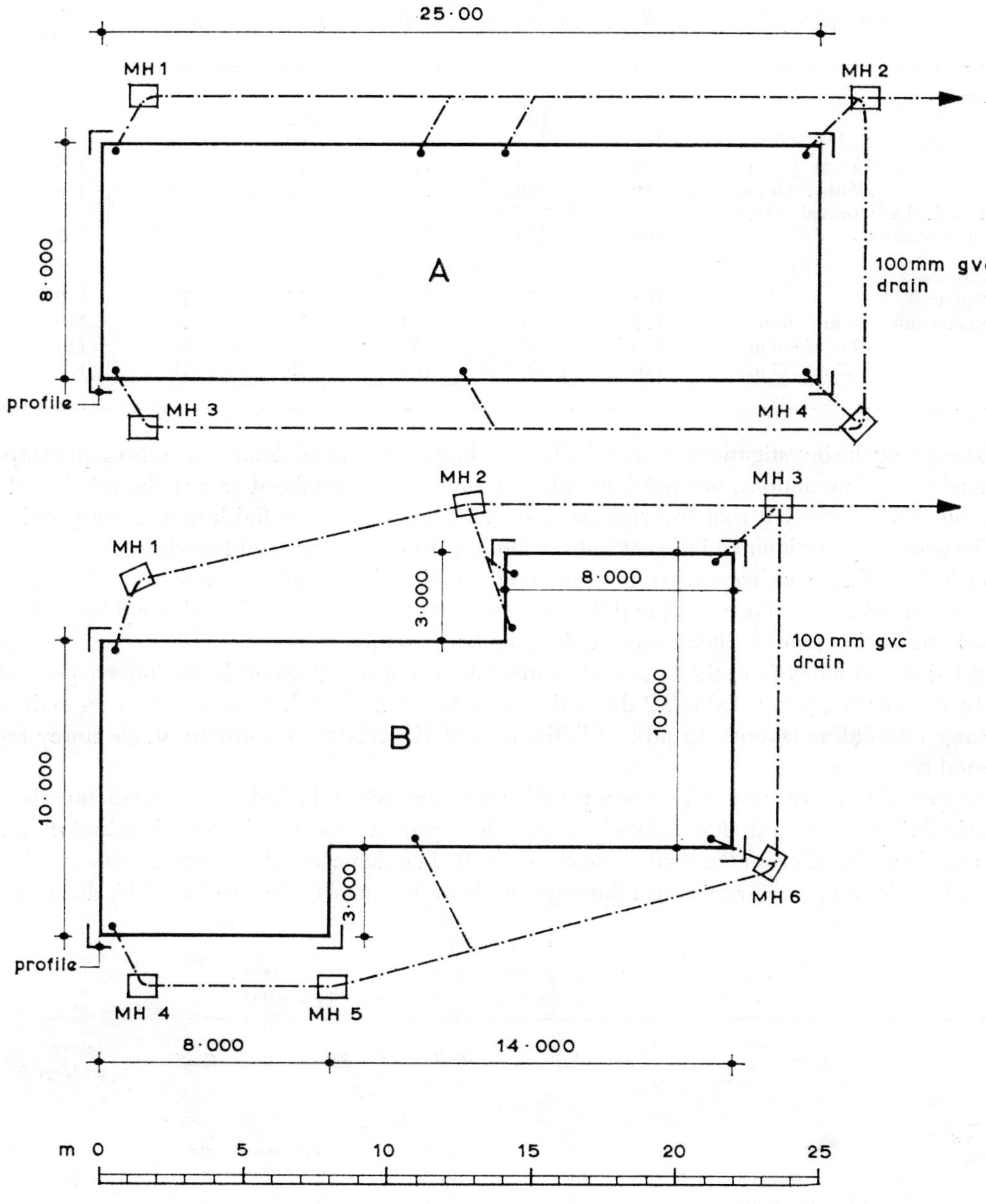

Figure 2.1 Higher cost of buildings of irregular shape

for the quantity surveyor to submit cost-saving alternatives which could not function satisfactorily or which would be aesthetically undesirable.

In the early sixties the Junior Organisation of the Royal Institution of Chartered Surveyors established a working party to consider the effect of shape and height on building costs.[1] The working party considered a number of buildings of varying shapes and heights but all had a total floor area of 930 sq m (1000 sq ft). In

TABLE 2.1

EFFECT OF SHAPE AND HEIGHT ON BUILDING COSTS: TOTAL AND ELEMENTAL COSTS

Form of building	*Total cost*	*Foundations*	*Walls*	*Elemental Costs* *Floors*	*Roof*	*Internal finishings*
Single-storey						
Square	100	100	100	100	100	100
Rectangular 12.2 m × 7.6 m	101	107	100	100	100	100
15.2 m × 6 m	106	113	110	100	100	105
20 m × 4.6 m	116	140	128	100	100	110
T or L shape formed of three equal sections	109	113	118	100	100	105
Two-storey						
Square	104	70	146	175	51	110
Rectangular 7.6 m × 6 m	105	70	148	175	51	110
10 m × 4.6 m	110	87	157	175	51	116
15.2 m × 3 m	126	103	193	175	51	127

addition to confining the investigations to small buildings, the constructional details were also kept comparatively simple, namely strip foundations, one-brick load-bearing external walls faced externally, reinforced concrete floors and roof, bituminous felt roof covering on a screed, granolithic floor finish and distempered walls and ceilings. The working party ignored internal walls, windows, doors, staircases and services.

The conclusions of the working party were twofold. Firstly, that the overall cost increases as the perimeter wall length increases in relation to floor area; and that this becomes more marked when the building is increased in height by adding further floors without altering the total floor area.

Table 2.1 shows in index form the total and elemental costs of buildings of the various shapes and heights examined by the working party. Table 2.2 shows the proportion which each element cost bears to the total cost over the range of building layouts. In table 2.1. the base of 100 relates to a square, single-storey building of 930 sq m total floor area.

Certain types of building present their own peculiar problems which in their turn may dictate the form and shape of the building. For instance hotels, for visual reasons, to provide guests with good views and the advertising effect of a prominent building on the skyline, need to be tall. The shape and floor area are closely related to the most economic bedroom per floor ratio and this is generally in the range of forty to fifty. This dictates a tall slab

TABLE 2.2

EFFECT OF SHAPE AND HEIGHT ON BUILDING COSTS:
PERCENTAGE OF ELEMENTAL COST TO TOTAL COST

Form of building	*Foundations*	*Walls*	*Floors*	*Roof*	*Internal finishings*
Single-storey					
Square	13	36	9	34	8
Rectangular 12.2 m × 7.6 m	14	36	9	33	8
15.2 m × 6 m	14	37	9	32	8
20 m × 4.6 m	16	40	7	29	8
T or L shape formed of three equal sections	14	39	8	31	8
Two-storey					
Square	9	51	15	17	8
Rectangular 7.6 m × 6 m	9	51	15	17	8
10 m × 4.6 m	10	52	14	16	8
15.2 m × 3 m	11	55	12	14	8

rather than a tower. Kaye[2] has described how slender towers are aesthetically very desirable but their relatively poor ratio of usable to gross floor area often renders them prohibitively expensive.

Taking another example, there are also functional limits on the depth of office blocks. Office buildings with depths of up to 18 m are acceptable in America and Australia, but the rental value obtainable per square metre of floor area reduces with the larger depth of building. Buildings of small depth, up to say 12 m, can be more readily split into small office units and will accommodate a greater variety of tenancies. Admittedly, an office building occupied by a large industrial concern will probably contain mainly large general offices and depths of up to 15 m may well be acceptable. Furthermore, an industrialist normally requires about 900 sq m of office floor area per floor.

There are occasions when the site itself will dictate the form or shape of the building. In some cases the designer may feel obliged to advise the building client to purchase additional land, where this is practicable, to make the development a more economical proposition. It may be worthwhile to underutilise an awkwardly shaped site in order to secure a regularly shaped and more economical building. Kaye[2] gives an instance where a strip form of development of possibly eight to ten storeys in height is involved, and means of escape considerations alone will dictate the optimum length of building to secure the maximum net to gross floor area relationship.

The shape of a building may also be influenced by the manner in which it is going to be used. For instance, in factory buildings the determining factors may be co-ordination of manufacturing processes and the form of the machines and finished products. In schools, dwellings and hospitals, and to a more limited extent in offices, shape is influenced considerably by the need to obtain natural lighting. Where the majority of rooms are to rely on natural lighting in daylight hours, the depth of the building is thereby restricted. Otherwise it is necessary to compensate for the increase in depth of building by installing taller windows which may compel increased storey heights. The aim in these circumstances should be to secure an ideal balanced solution which takes into account both the lighting factor and the constructional costs. Deeper rooms result in reduced perimeter/floor area ratios with a subsequent reduction in construction, maintenance and heating costs, but these savings may be offset by increased lighting costs. With taller rooms the conditions are reversed.

The Building Research Station,[11] in a survey of Midland factories, found that most factory buildings were simple rectangles with an average length/breadth ratio of about two to one, although there were a few which were almost square in plan. Most of the buildings with a high length/breadth ratio were used for engineering; the ratios in these cases sometimes being as much as five or six to one. On average single-storey factory buildings were around 38 m in breadth, whereas the multistorey blocks averaged about 23 m.

SIZE OF BUILDING

Increases in the size of buildings usually produce reductions in unit cost, such as the cost per square metre of floor area. The prime reason for this is that oncosts are likely to account for a smaller proportion of total costs with a larger project, or expressed in another way, they do not rise proportionately with increases in the plan size of a building. Certain fixed costs such as the transportation, erection and dismantling of site buildings and compounds for storage of materials and components, temporary water supply arrangements and the provision of temporary roads, may not vary appreciably with an extension of the size of job and will accordingly constitute a reduced proportion of total costs on a larger project. A larger project is often less costly to build as the wall/floor ratio reduces, rooms tend to be larger with a proportional reduction in the quantity of internal partitions, decorations, skirtings, etc., and there may also be a proportional reduction in the extra cost of windows and doors over walls. With high rise buildings a cost advantage may accrue due to lifts serving a larger floor area and greater number of occupants with an increased plan area.

Figure 2.2 shows the effect of doubling the length of a rectangular building on the ratio of enclosing wall to floor area. The length of external wall per square metre of floor area (one floor only) is reduced from 383 mm to 317 mm, a reduction of 17.25 per cent.

An example will serve to illustrate the cost advantage in lift provision by increasing the area on each floor of multistorey blocks of flats and offices. A six-storey block of offices built in the East Midlands in 1970 has 360 sq m of floor area on each floor and the six floors are served by two passenger lifts. The total cost of the project of £150 000 is equivalent to £68/sq m of floor area and the lifts cost £9000 and are equivalent to £3.80/sq m of floor

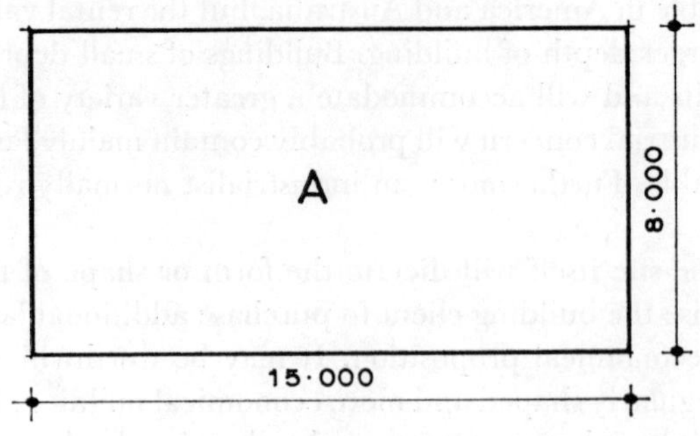

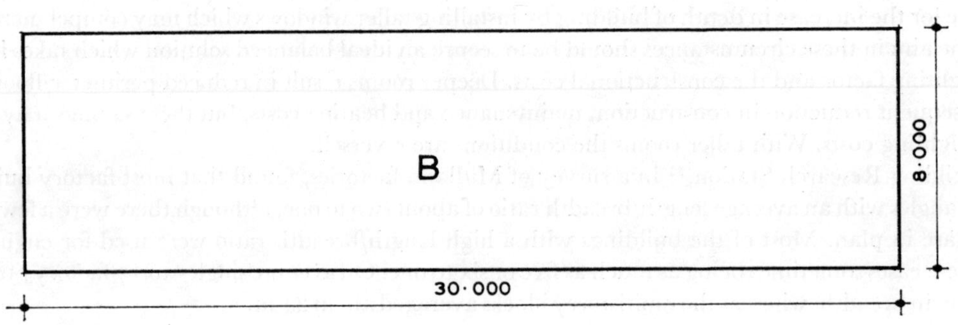

	BUILDING A	BUILDING B
FLOOR AREA	120 sq m	240 sq m
LENGTH OF ENCLOSING WALL	46 m	76 m
LENGTH OF WALL/ SQ M OF FLOOR	383 mm	317 mm

Figure 2.2 Effect of change in size of buildings

area. If the floor area was doubled on each storey the lift provision could remain the same and the cost of lifts would then be reduced to about £1.90/sq m of floor area, giving a saving of 2.8 per cent on total building costs.

Another interesting illustration of the effect of size on building costs emanates from a comparison of the costs of two, three and four-bedroom houses. Table 2.3 compares the costs of building two, three and four-bedroom

TABLE 2.3

COMPARATIVE COST OF TWO, THREE AND FOUR-BEDROOM LOCAL AUTHORITY HOUSES (two-storey)

	Two-bedroom four-person house	*Three-bedroom five-person house*	*Four-bedroom seven-person house*
Floor area (sq m)	75.3	85.1	106.5
Superstructure costs (£)	2097	2389	2878
Substructure costs (£)	386	446	515
Siteworks costs (£)	440	440	440
Total costs (£)	2923	3275	3833
Cost/sq m (£)	38.9	38.5	36.0
Cost/person (£)	731	655	548

houses in a local authority scheme in the West Midlands in 1970. The costs per sq m of floor area of the three-bedroom houses show a 1.3 per cent reduction on the cost of the two-bedroom houses, and the four-bedroom houses show a 3.9 per cent reduction on the three-bedroom, although the cost of siteworks is shown as remaining constant. When the costs are related to the number of occupiers they show much greater reductions, although it must be appreciated that only a small proportion of the larger houses are fully occupied.

These house prices do not support Reiners' findings at the Building Research Station that an increase in area of ten per cent in local authority houses and flats is accompanied by a change in total cost of about five per cent, as the unit cost equivalent to the additional area shows reductions in the order of 12.6 to 25.0 per cent. The unit cost of the additional floor area of a three-bedroom over a two-bedroom house is £34/sq m, that of a four-bedroom over a three-bedroom house is £26.8/sq m and that of a four-bedroom over a two-bedroom house is £29.2/sq m. Admittedly it would not be prudent to attach too much importance to the relationships drawn from the results of a single tender, nevertheless they do serve to indicate the broad pattern.

By way of comparison Table 2.4 shows the 1970 housing cost yardstick figures without regional variation allowances.

Investigations by Reiners[7] showed that tender prices, when expressed per square metre of net floor area, tend to fall as the area increases. For the schemes examined, an increase of ten per cent in the floor area of flats was accompanied by a decrease of 4.7 per cent in the price per square metre. Reiners then postulated that in general terms an increase in the area of flats of x per cent may be expected to reduce the price per square metre by $x/2$ per cent. A decrease in area of x per cent causes a corresponding increase of $x/2$ per cent in the price per square metre.

TABLE 2.4

TARGET COSTS FOR FOUR, FIVE AND SIX-PERSON TWO-STOREY HOUSES CALCULATED FROM HOUSING COST YARDSTICKS

Dwelling	4-person	5-person	6-person
Substructure (£)	244	275	306
Superstructure (£)	2092	2300	2508
External works (£)	288	325	360
Total (£)	2624	2900	3174

Source: *Metric house shells*[3]

This form of relationship is however limited to differences in area of up to about fifteen per cent, as over this limit rather more complex relationships operate.

PERIMETER/FLOOR AREA RATIOS

We have already seen that the plan shape directly conditions the external walls, windows and external doors which together constitute a composite element – the enclosing walls. Different plans can be compared by examining the ratio of enclosing walls to floor area in square metres (known as the wall/floor ratio). The lower the wall/floor ratio, the more economical will be the proposal. The best wall/floor ratio is produced by a circular building, but the saving in quantity of wall is usually more than offset by the much higher cost of circular work over straight, the increased cost varying between twenty and thirty per cent.

Figure 2.3 shows the outline of two buildings, one of which (building A) is L shaped and the other (building B) has a very irregular outline. Both buildings have an identical floor area on each floor of 244 sq m, and assuming that the buildings are each of two storeys, this gives a total floor area of 488 sq m for each building. Wall thicknesses have been ignored in this example to simplify the calculations. The length of enclosing wall in building A amounts to 70 m while that in building B totals 100 m – an increase of forty-three per cent. Assuming that the height of the walling is 6 m, the areas of enclosing walls are 420 sq m for building A and 600 sq m for building B and the wall/floor ratios are

$$\text{building A} = \frac{420}{488} = 0.86$$

$$\text{building B} = \frac{600}{488} = 1.23$$

Building B is very uneconomical with a much greater area of enclosing walls than A. It should be borne in mind that the perimeter cost of a building can be in the order of twenty to thirty per cent of total cost and an external wall can be two to four times as expensive as an internal partition. In this example building B is likely to be at least ten per cent more expensive than building A on account of the much increased perimeter costs.

The National Building Agency[3] found that with an economically designed five-person house, the cost of enclosing walls, including windows and doors, is likely to be twenty-five to thirty per cent of the total cost of the superstructure, services should account for about twenty per cent, and internal partition and associated doors fourteen to fifteen per cent. With traditional construction, the cost of the structure – enclosing walls, windows, roof and floors – increases steadily with an extension of the frontage. As a result wider frontage houses are more expensive than those of narrower frontage with similar floor area. The length of frontage also affects the costs of external services and external works. Consideration also needs to be given to the cost effects of varying the number of dwellings in a terrace where schemes include terraced blocks. The additional cost of an end-of-terrace-house superstructure over that of an intermediate house is likely to be in the order of £140 to £230, because of the higher cost of gable-end walls over party walls, but this will be partly offset by the increased cost of making provision for rear access to the intermediate dwellings.

Graphs produced by the National Building Agency[3] show that the narrow frontage house becomes relatively cheaper as the length of terrace increases, with costs falling more sharply, than with the medium and wide frontage houses. Where both narrow and wide frontage houses and short and long terraces are required, it is more economical to use wide frontage houses in the short terraces and the narrow frontage houses in the long ones.

Another aspect which has to be considered when investigating perimeter/floor area ratios is the adequacy of the natural lighting to the interior of the building and the practicability of the internal layout. By reducing the frontage and increasing the depth of a building the amount of natural light reaching the innermost parts will be reduced and may result in increased operating costs through higher artificial lighting charges. A deeper building may also result in wasteful and inconveniently shaped rooms such as long cubicles housing wcs. Thinner walls will also provide greater floor area for the same length of enclosing wall.

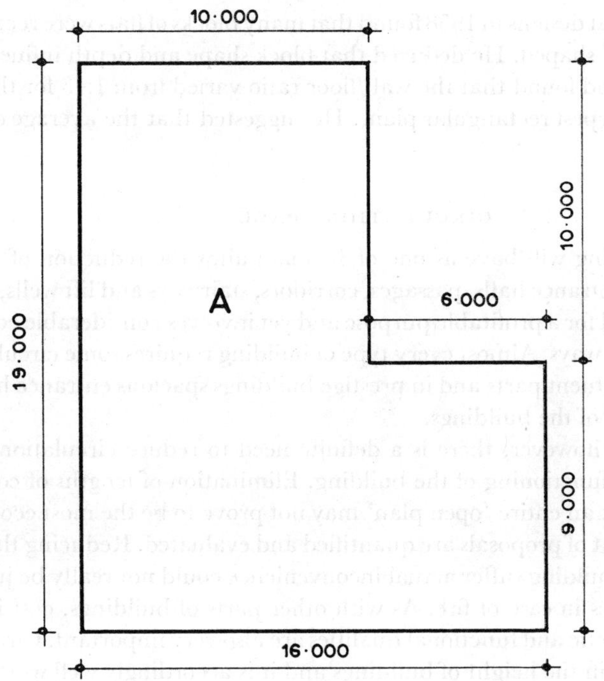

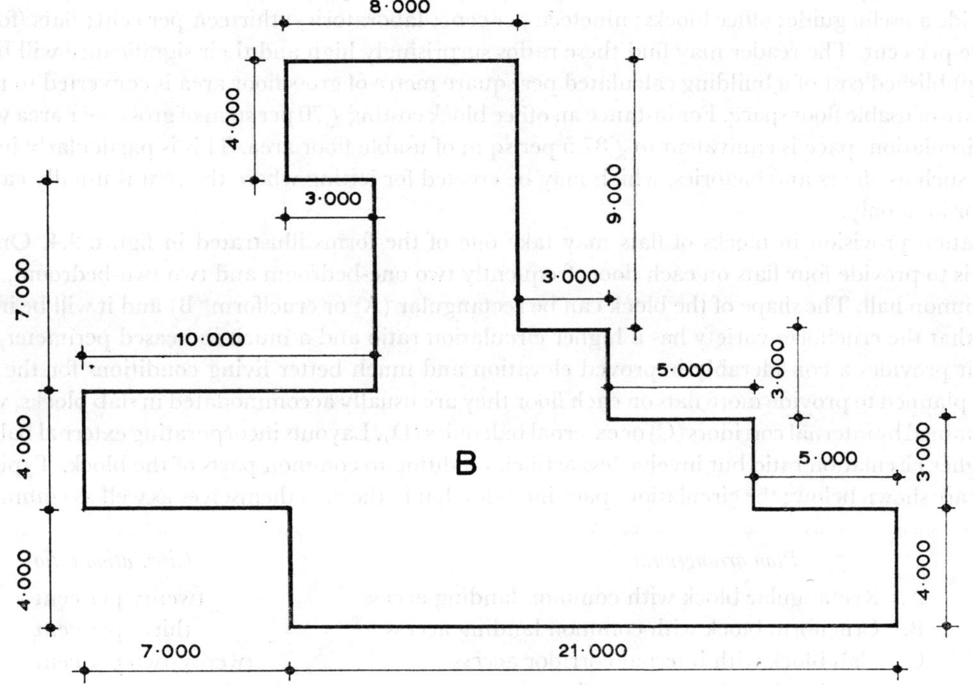

Figure 2.3 Perimeter/floor area ratios

Craig[8] in his investigation of flat designs in 1958 found that many blocks of flats were rectangular in plan shape, although some were U, L, Y or T shaped. He deduced that block shape and depth influences the proportion of external wall area to floor area and found that the wall/floor ratio varied from 1.43 for the most complex plans to 0.56 for the most simple and deepest rectangular plans. He suggested that the average of 1.00 might prove to be an acceptable target.

CIRCULATION SPACE

An economic layout for a building will have as one of its main aims the reduction of circulation space to a minimum. Circulation space in entrance halls, passages, corridors, stairways and lift wells, can all be regarded as 'dead space' which cannot be used for a profitable purpose and yet involves considerable cost in heating, lighting, cleaning, decorating and in other ways. Almost every type of building requires some circulation space to provide means of access between its constituent parts and in prestige buildings spacious entrance halls and corridors add to the impressiveness and dignity of the buildings.

In the majority of buildings, however, there is a definite need to reduce circulation space to a minimum compatible with the satisfactory functioning of the building. Elimination of lengths of corridor which result in communication through rooms or an entire 'open plan' may not prove to be the most economical proposition if all the costs and benefits of each set of proposals are quantified and evaluated. Reducing the width of corridors to an extent that persons using the building suffer actual inconvenience could not really be justified; corridors may also have to serve as escape routes in case of fire. As with other parts of buildings, cost is not the only criteria which has to be examined – aesthetic and functional qualities are also very important. Circulation space requirements tend to rise with increases in the height of buildings and it is accordingly well worth while to give special consideration to circulation aspects when designing high rise buildings.

Unfortunately, very few of the cost analyses issued in the past by the Building Cost Information Service give details of circulation space. This deficiency is now rectified in the form of cost analyses issued as from 1970. There can be little doubt that the proportion of floor space allocated to circulation purposes will vary considerably between different types of building. The following circulation ratios (proportion of circulation space to gross floor area) provide a useful guide: office blocks: nineteen per cent; laboratories: thirteen per cent; flats (four-storey): twenty-one per cent. The reader may find these ratios surprisingly high and their significance will be apparent when the published cost of a building calculated per square metre of gross floor area is converted to the cost of a square metre of usable floor space. For instance an office block costing £70 per sq m of gross floor area with twenty per cent circulation space is equivalent to £87.5 per sq m of usable floor area. This is particularly important in buildings, such as offices and factories, which may be erected for letting where the rent is usually calculated on usable floor area only.

Circulation provision in blocks of flats may take one of the forms illustrated in figure 2.4. One common approach is to provide four flats on each floor, frequently two one-bedroom and two two-bedroom, with access from a common hall. The shape of the block can be rectangular (A) or cruciform (B) and it will be immediately apparent that the cruciform variety has a higher circulation ratio and a much increased perimeter/floor ratio, although it provides a considerably improved elevation and much better living conditions for the occupants. When it is planned to provide more flats on each floor they are usually accommodated in slab blocks, when access may be obtained by internal corridors (C) or external balconies (D). Layouts incorporating external balcony access have a higher circulation ratio but involve less artificial lighting to common parts of the block. Typical circulation ratios are shown below; the circulation space includes that in the flats themselves as well as communal space.

Plan arrangement	*Circulation ratio*
A. Rectangular block with common landing access	twenty per cent
B. Cruciform block with common landing access	thirty per cent
C. Slab block with internal corridor access	twenty-two per cent
D. Slab block with external balcony access	thirty-two per cent

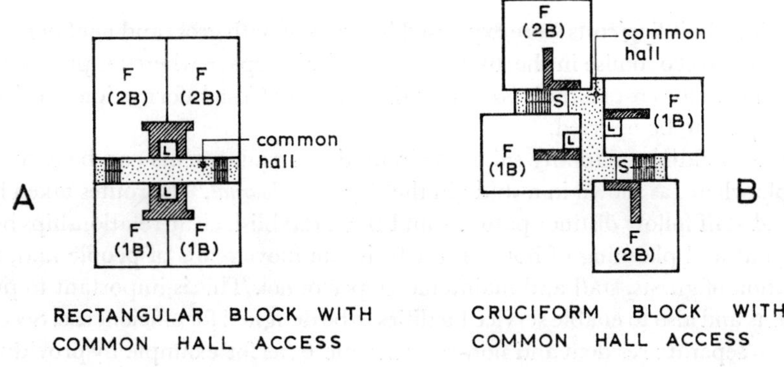

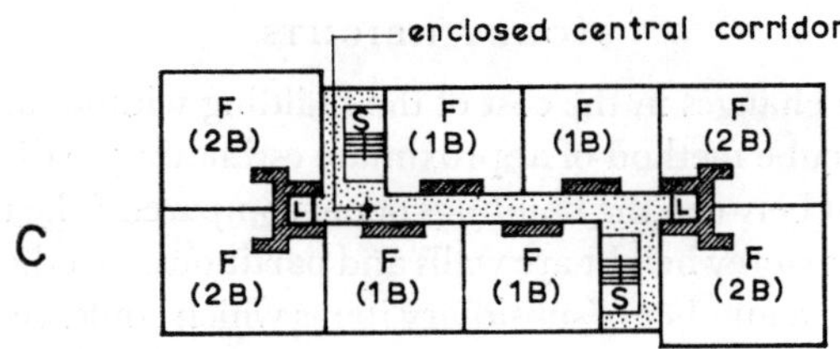

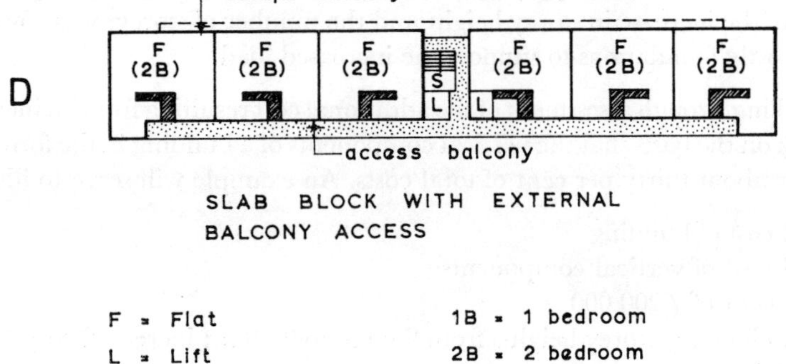

Figure 2.4 Means of access to flats

Craig[8] in his investigations in 1958 of seventy-two blocks of flats built in this country, found that over twenty per cent of the gross floor area of the blocks consisted of non-living area. The range was from eleven to thirty-five per cent.

It would be helpful if all building costs were expressed in terms of both gross and net floor areas as they would give emphasis for the need to economise in the amount of circulation space wherever practicable. As previously mentioned, the revised forms for cost analysis issued by the Building Cost Information Service in 1970[4] require the area of circulation space to be inserted.

With certain types of building, planning suitable circulation arrangements can be a complex task. Hotels provide a good example where, as shown in a study in the *Architects Journal*,[5] the routes taken by resident guests, non-resident diners, and staff follow distinct patterns and these establish clear relationships between the hotel's various parts. The layout and planning of hotels must facilitate movements of people and, as far as possible, provide for the separation of guests, staff and maintenance personnel. This is important to prevent annoyance and disturbance of guests and also to enable service facilities to be designed for efficient use. Secondary circulation may also be desirable to separate resident and non-resident guests, as for example by providing direct access to restaurants and banqueting halls. Figure 2.5 illustrates in diagrammatic form typical hotel circulation patterns and relationships.

STOREY HEIGHTS

Variations in storey heights cause changes in the cost of the building without altering the floor area, and this is one of the factors that makes the cube method of approximate estimating so difficult to operate when there are wide variations in the storey height between the buildings being compared. The main constructional items which would be affected by a variation in storey height are walls and partitions, together with their associated finishings and decorations. There will also be a number of subsidiary items which could be affected by an increase in storey height, as follows.

(1) Increased volume to be heated which could necessitate a larger heat source and longer lengths of pipes or cables.
(2) Longer service and waste pipes to supply sanitary appliances.
(3) Possibility of higher roof costs due to increased hoisting.
(4) Increased cost of constructing staircases, and lifts if provided.
(5) Possibility of additional cost in applying finishings and decorations to ceilings.
(6) If the impact of the increase in storey height and the number of storeys was considerable, it could result in the need for more costly foundations to support the increased load.

One method of making a rough assessment of the additional cost resulting from an increase in the storey height of a building is to work on the basis that the vertical components of a building in the form of walls, partitions and stanchions account for about thirty per cent of total costs. An example will serve to illustrate the approach.

Estimated cost of building	£200 000
Estimated cost of vertical components thirty per cent of £200 000	£60 000
Proposal to increase storey heights from 2.60 m to 2.80 m: increased cost	
would be $\frac{0.20}{2.60} \times 100 \times$ £60 000	£4620

It would, however, be necessary to consider the possible effect of some or all of the subsidiary items previously listed if the increase in storey height was substantial.

The average clearance height for single-storey factory buildings in a prewar sample taken by the Building Research Station[11] was about 4.5 m, with an average building height of about 5.8 m as few of the buildings had flat roofs. The only trade in which the average height of 4.5 m was much exceeded was the engineering group, where in some cases clearance heights were 6 m and 9 m. Such clearances were usually required for overhead

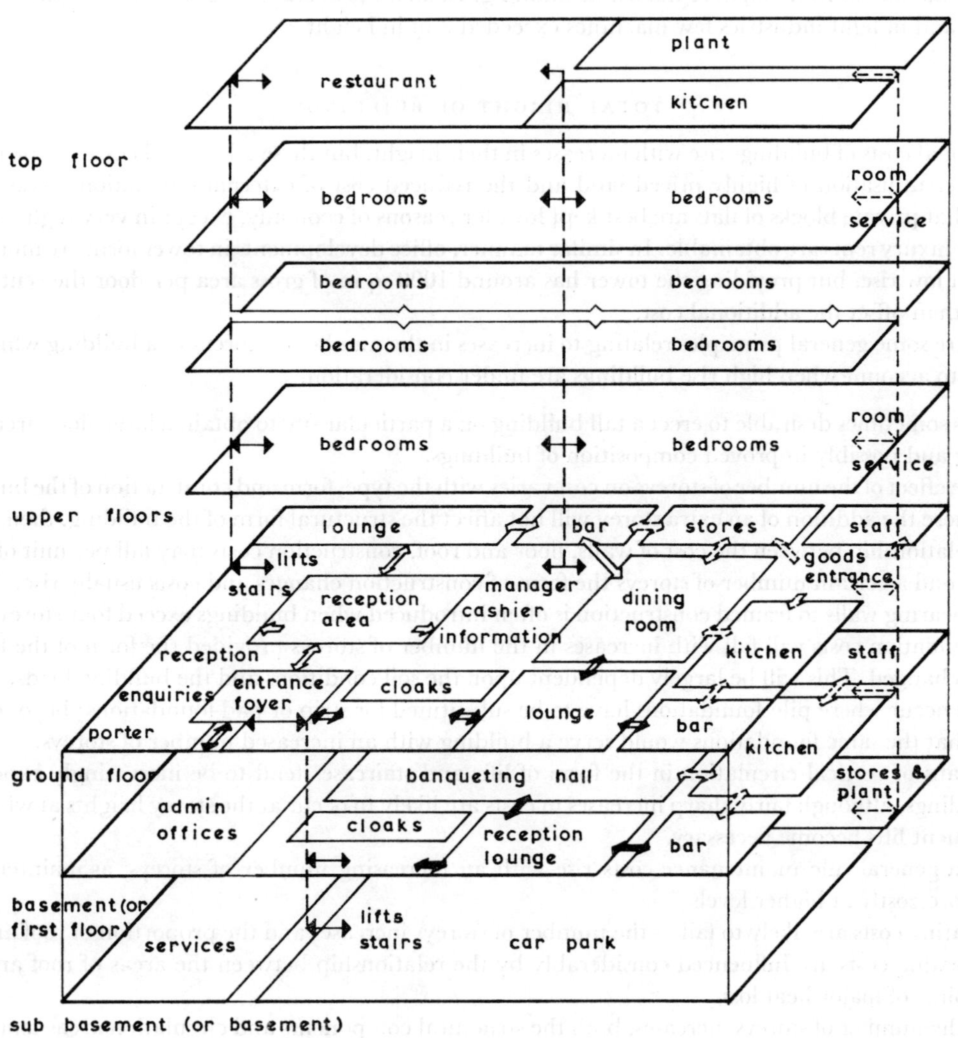

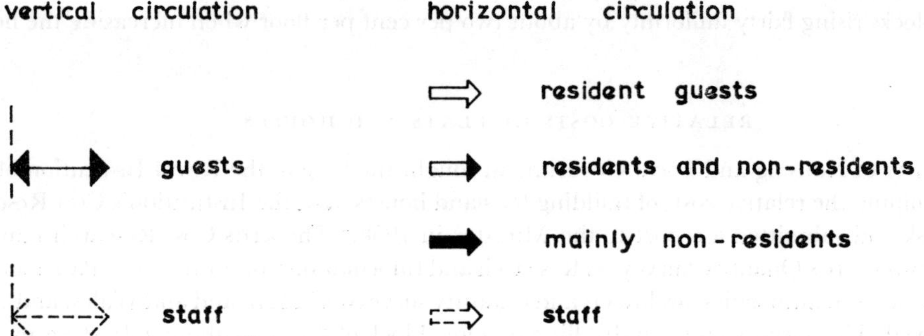

Figure 2.5 Hotel circulation patterns and relationships

cranes and did not often extend over the whole building. In fact, the average height for the engineering industry was 4.9 m, and in light industries few machines exceeded 3 m in height.

TOTAL HEIGHT OF BUILDINGS

Constructional costs of buildings rise with increases in their height, but these additional costs can be partly offset by the better utilisation of highly priced land and the reduced cost of external circulation works. Kaye[2] has suggested that private blocks of flats are best kept low, for reasons of economy, except in very high cost site locations where luxury rents are obtainable. In similar manner, office developments in tower form are more expensive in cost than low rise, but providing the tower has around 1000 sq m of gross area per floor the rent obtainable may more than offset the additional cost.

There are some general principles relating to increases in the number of storeys to a building which ought to be taken into account when high rise buildings are under consideration.

(1) It is sometimes desirable to erect a tall building on a particular site to obtain a large floor area with good daylighting and possibly improved composition of buildings.

(2) The effect of the number of storeys on cost varies with the type, form and construction of the building.

(3) Where the addition of an extra storey will not affect the structural form of the building, then, depending upon the relationship between the cost of walls, floor and roof, construction costs may fall per unit of floor area.

(4) Beyond a certain number of storeys the form of construction changes and costs usually rise. The change from load-bearing walls to framed construction is often introduced when buildings exceed four storeys in height.

(5) Foundation costs will fall with increases in the number of storeys provided the form of the foundations remains unchanged. This will be largely dependent upon the soil conditions and the building loads. A large rise in costs will occur where pile foundations have to be substituted for strip or pad foundations; beyond this point it is likely that the same foundations would serve a building with an increased number of storeys.

(6) Means of vertical circulation in the form of lifts and staircases tend to be increasingly expensive with higher buildings, although fairly sharp increases in costs are likely to occur at the storey heights at which the first and subsequent lifts become necessary.

(7) As a general rule maintenance costs rise with an increasing number of storeys, as maintenance work becomes more costly at higher levels.

(8) Heating costs are likely to fall as the number of storeys increases and the proportion of roof area to walls reduces. Heating costs are influenced considerably by the relationship between the areas of roof and walls, as roofs are points of major heat loss.

(9) As the number of storeys increases, both the structural components and circulation areas tend to occupy more space and the net floor area assumes a smaller proportion of the gross floor area, thus resulting in a higher cost per sq m of usable floor area. A study by the Department of the Environment[12] shows the cost of local authority office blocks rising fairly uniformly by about two per cent per floor when increasing the height above four storeys.

RELATIVE COSTS OF FLATS AND HOUSES

In 1956 the Minister of Housing and Local Government sought the help of the Royal Institution of Chartered Surveyors in examining the relative costs of building flats and houses, and the Institution's Cost Research Panel undertook the task and submitted a report to the Minister in 1958.[6] The RICS Cost Research Panel has since been superseded by the RICS Quantity Surveyors Research and Information Committee. The Panel sent out about 850 questionnaires to local authorities, architects and quantity surveyors in England and Wales, and 153 detailed replies were received. The average cost of a dwelling in a low block of flats was about £21.80/sq m in 1954, excluding site works. For the same period the average cost of a two-storey house of similar accommodation was about £15.35/sq m. A further study indicated that the costs of multistorey housing (six or more storeys) ranged from £24.30 to £48.0/sq m of net floor area during early 1955.

Multistorey design generally involves certain features which are not required in two-storey dwellings. These include costs arising from the height of the structure, the provision of common areas for access to the dwellings, additional Building Regulations and fireproofing requirements, and the installation of additional amenities needed to solve problems which only arise in multistorey housing. Such additional amenities include lifts in high blocks, clothes drying facilities, and refuse disposal and central heating installations. Table 2.5 shows a summary of relative costs of houses and flats in the mid-fifties broken down into four basic elements.

TABLE 2.5

RELATIVE COSTS OF LOCAL AUTHORITY HOUSES AND FLATS IN MID-FIFTIES

All costs are expressed in £/sq m of floor area

Component	*Two-storey house*	*Three-storey flat*	*Eight-storey flat*
Substructure	1.60	1.50	2.50
Superstructure	7.50	10.00	15.30
Internal finishings	2.70	5.70	3.70
Fittings and services	2.50	5.20	6.20
Total (£)	14.30	22.40	27.70

The figure of £15.30 for the superstructure of the eight-storey flat includes £8.60 for a structural frame, and the cost of £10.00 for the three-storey flat superstructure includes £2.90 for floors, staircases and balconies. It is interesting to note that the substructure costs of the three-storey flat are less per sq m of floor area than the two-storey house, as similar foundation costs are spread over a greater floor area, whilst the substructure costs of the eight-storey flat increase due to the more expensive foundations that are needed. The superstructure costs of the eight-storey flat are double those of the two-storey house on account of the structural frame and increased constructional costs associated with the taller building. Balcony facings and balustrades provide another source of additional expenditure with flats as distinct from houses.

Consideration will now be given to a comparison of costs of the various elements of flats and houses.

Substructures

The cost of substructures is influenced by site conditions as well as by the type of building to be erected. Normally, foundations of two-storey dwellings are capable of supporting greater loads, and some economy results from the substructure cost being shared by a larger number of dwelling units in taller blocks. There will, however, be limits on the savings that can be achieved in this way, particularly on poor load-bearing soils. The construction of multistorey blocks in slab form often involves building across the contours to secure satisfactory orientation, and this will result in additional costs in earthworks and provision of foundations. With multistorey blocks expensive piled foundations may be required unless the soil has a high load-bearing capacity, although the higher cost of piled foundations will be offset to some extent by the sharing of the substructure over a larger number of floors. Some typical substructure costs for different building types are given in chapter 9.

Superstructures

Superstructure costs vary considerably between blocks of different designs, and this particular element accounts for much of the extra cost of multistorey development over traditional two-storey housing. Brick load-bearing structures are generally the most economical for three to five-storey blocks, while reinforced concrete columns or walls, or a composite form using both, are frequently used for taller blocks.[8] The need for fire-resisting floors and staircases in multistorey blocks also results in increased costs. As described earlier, increased circulation ratios with multistorey blocks will also produce higher unit usable floor space costs.

Investigations into flat costs by Reiners[7] in 1958 indicated some rather surprising cost patterns related to blocks of flats with varying numbers of storeys. In general three-storey flats were about thirty per cent more expensive that two-storey houses, with costs related to a specific unit of floor area such as the square metre. Increasing the height of blocks of flats from three to five storeys raised costs by about twelve per cent (six per cent per storey). This trend continued when the total height was further increased to six to eight storeys, with a further rise in costs of about seventeen per cent. The rate of increase in costs appeared surprisingly to flatten above eight storeys in height to about a two per cent addition per floor. There were, however, substantial variations in the cost relationships between blocks of differing heights located in Inner London, Outer London and the provinces. The cost differences are greater than could readily be explained by variations in the cost of labour and materials between different parts of the country, and must presumably have their explanation in part in differences in design and specification and in the materials used. Craig[8] described how proportional cost differences between continental flats and houses were considerably less than in England in the late fifties, and he considered that lower costs could be secured by the preparation of simpler designs, keeping different materials to the minimum and incorporating a maximum of standardisation, with closer liaison between designers and producers and more detailed consideration of erection methods at design stage.

Roofs

Some reduction in the roof cost per dwelling unit is to be expected with multistorey design, where the total roof cost is shared by a larger number of dwellings. In the case of three and four-storey blocks the roof is often of similar construction to that used for two-storey houses. With taller blocks a more expensive flat roof is normally provided which partially offsets the savings from the shared roof.

Internal Finishings

The floor finishings to the upper floors of multistorey flats are often rather more expensive per unit area than the corresponding finishings in two-storey houses. Furthermore, it is necessary to provide additional areas of finishings to the floors, walls and ceilings of common access and circulation areas. The elimination of plaster in flats by the use of fair-faced concrete walls and ceilings combined with 'dry' partitions will help to offset these additional costs.

Joinery Fittings

The cost of joinery fittings per square metre of floor area in multistorey flats is generally higher than the equivalent provision in two-storey houses. There are two main reasons for this: (1) A higher standard of provision of cupboards and storage space is almost invariably expected by flat occupiers. (2) Flats are usually smaller in floor area than houses and so the cost of joinery fittings per unit of floor area is proportionately greater.

Plumbing

Plumbing costs are normally higher in multistorey blocks because of the need for larger pipes and additional valves. It is possible to reduce the extra costs by the careful grouping of sanitary appliances to restrict the number of vertical pipe runs and by the use of single stack plumbing. Craig[8] advocated storey-height prefabrication of pipework for ducts and pushfit jointing to help contain the extra costs.

Heating

Surprisingly enough all the flats considered by the RICS Cost Research Panel were heated by open fires involving high costs in flue construction. Today the majority of flats are centrally heated resulting in additional costs per dwelling in the order of £150 to £400 depending on the type and design of installation. Apart from the high installation costs there is also the disadvantage of the lack of individual control.

Electrical Installations

Electrical installations in two-storey houses are considerably cheaper than those in multistorey flats, as they are generally simpler in form with less fittings and are easier to install. Furthermore, in multistorey blocks costs are also increased by the need for heavier mains and switchgear and each flat will have to share part of the cost of the lighting installation required to serve access balconies, staircases and other common circulation areas.

Lifts

Lifts are necessary in buildings in which the entrance to any dwelling is on the fourth storey or above.[9] This means that blocks of flats exceeding three storeys and blocks of maisonettes exceeding four storeys must be provided with lifts. In buildings of more than six storeys, two lifts are needed to maintain continuity of service during periods of breakdown and maintenance. With two lifts it may be sufficient to have stops only at alternate floors. Lifts can satisfactorily serve up to about fifty dwellings each, with a lift car capacity of eight persons. Thus if maximum utilisation of lift capacity were the sole consideration, a building with six storeys and one lift would contain fifty dwellings and a block with more than six storeys and two lifts would contain 100 dwellings. In practice the economics of lift provision is only one of many matters to be considered, but it is important. It would, for instance, be very uneconomical to design a five-storey block which contains only two or four dwellings on each floor, and incorporate a lift.

In the investigations undertaken by Craig[8] lift provision in six-storey to twelve-storey blocks ranged from one lift per fourteen dwellings to one lift per sixty-two dwellings, with an average of one per thirty dwellings. He advocated a target provision of one lift per twenty to twenty-five dwellings, which is on the low side, and amounts to a cost of about £3.80 per square metre of floor area. Extra costs stemming from the introduction of lifts cannot readily be separated from superstructure costs in analyses, as they will include the costs of lift shaft, lift motor rooms and pits.

Additional Dwelling Amenities

Multistorey blocks of flats incorporate various facilities which are not provided in two-storey houses. Drying cabinets are often provided to compensate for loss of external drying areas and shared laundry rooms may be incorporated on the basis of about one room per twenty dwellings. Pram stores and on occasions other common facilities such as playrooms, are often provided at ground floor level. In blocks of flats exceeding four storeys in height it is customary to install a chute refuse disposal system with hoppers at each floor level discharging into a movable container in an enclosed chamber at ground level.

Site Works

With two-storey housing, each dwelling is complete within its own garden, except where common unfenced grassed areas are provided at the fronts of houses. Hence the site works for which the local authority is responsible are limited mainly to estate roads, sewers and other underground services. With multistorey design, the site preparation works usually include extensive landscaping, secondary roads to meet fire regulations, cycle and pram stores if not incorporated in the blocks of flats, and children's play areas.

Effect on Relative Costs of Contracting and Tendering Procedures

The building industry embraces a very large number of small firms and quite a high proportion of traditional house building is performed by these small firms with their limited resources and plant. The erection of multistorey blocks of flats is by comparison a complex task requiring considerable technical resources and skilled administrative ability. Only the largest firms are adequately equipped to undertake such schemes, particularly the very high blocks. Thus multistorey schemes attract a much narrower range of building firms entailing reduced competition. Tenders for building projects are often influenced by the amount of work available at any particular time and tenders for multistorey work are likely to be more sensitive to this condition.

In this, country houses have proved to be much more popular than flats and so it has not been possible to build up the vast amount of design and constructional knowledge and expertise concerning multistorey flats as exists for low rise residential buildings. Furthermore, multistorey design makes possible the introduction of a whole range of new constructional processes, materials and components. Estimators may accordingly be faced with many unfamiliar problems, with little experience of past costs to draw upon and may feel obliged to price high to cover the uncertainties.

Another essential difference with multistorey blocks is the much larger proportion of work allocated to nominated subcontractors. The former RICS Cost Research Panel found that subcontractors' work accounted for an average of thirty-two per cent of total cost rising to forty-three per cent for schemes incorporating blocks exceeding six storeys in height. In fact, lifts and central heating alone could account for much of this percentage. By way of contrast the proportion of nominated subcontractors' work in two-storey housing rarely exceeds fifteen per cent and the average was probably in the order of about eight per cent.

The Panel was very concerned at the wide range of costs which occurred on different schemes and was convinced that many schemes of multistorey development carried out in the fifties cost more than they should and that high prices often stemmed from uneconomic design. Table 2.6 illustrates the wide range of costs of multistorey housing blocks operating at the time the Panel was making its investigations.

TABLE 2.6

RANGE OF COSTS OF MULTISTOREY HOUSING BLOCKS IN MID-FIFTIES

Costs per square metre of floor area take the middle two-thirds of the sample.

Number of storeys	*Range in £ per sq m*	
	lowest	*highest*
3	16.00	24.00
4	19.70	28.00
5	25.00	29.70
6	27.50	36.50

Source: *Building Research Station* (with data metricated)

The Panel also found that with the earlier multistorey housing schemes there was a lack of clear budgetary direction from the inception of the schemes and that this resulted in high building costs, high rents and/or excessive deficiencies in the housing revenue account. Tenders were often invited on bills of quantities prepared from incomplete information leaving important details to be settled later by variation orders. Furthermore, architects complained that they were seldom allowed sufficient time by local authorities to plan schemes adequately, and the local authorities expressed concern at the inconvenience and delays stemming from the need to obtain planning and statutory approvals, rising costs and frequent changes in government policy. It is evident that contractors must price on the basis of the conditions which they expect to meet during the contract and must include an allowance for possible disorganisation arising from variations. Many building firms are improving efficiency by the introduction of management techniques and improved methods of programming and controlling building operations. Lack of precontract planning by the design team can nullify much of the benefit to be derived from these costly and time-consuming efforts of the contractor. It is evident that dwellings in high rise developments must cost considerably more than identical accommodation provided in low rise schemes. Yet, at the same time, it was equally apparent that in the fifties multistorey schemes were excessively expensive and that part at least of the excessive cost stemmed from poor design work and lack of effective cost control.

IMPLICATIONS OF VARIATIONS IN THE NUMBER OF STOREYS OF BUILDINGS

Two practical examples will serve to illustrate the probable cost implications on the various elements of alternative design solutions involving changes in the number of storeys to residential and office buildings.

Example (a)

Comparison of alternative proposals to provide a prescribed floor area of office space in a rectangular shaped three-storey block or a six-storey L shaped block.

The six-storey block will involve increased costs in respect of the majority of major elements for the reasons indicated:

Foundations. More expensive foundations will probably be needed in the six-storey block to take the increased load, assuming a soil with an average load-bearing capacity, although this will be partially offset by the reduced quantity of foundations. The irregular shape will however increase the amount of foundations relative to floor area.

Structure. It is probable that a structural frame will be required in place of load-bearing walls with consequent higher costs, and there will be an additional upper floor and flight of stairs.

Cladding. The constructional costs will increase due to the greater amount of hoisting and the larger area resulting from the more irregular shape of the block.

Roof. Constructional costs will be higher but these will be more than offset by the reduction in area of the roof.

Internal finishings. Increased area due to more irregular shape and slightly higher hoisting costs will result in increased expenditure.

Plumbing, heating and ventilating installations. Increased expenditure due to increased lengths of larger-sized pipework and ducting.

Passenger lifts. Might not be provided with a three-storey block but will be essential for the six-storey block.

Example (b)

Comparison of alternative proposals to provide a prescribed number of flats of identical floor area and specification in two five-storey blocks or one ten-storey block.

Element	*Two five-storey blocks*	*One ten-storey block*
Foundations	Double the quantity of column bases and concrete oversite. Possibility of less costly strip foundations if load-bearing walls.	Half the quantity of column bases but they will need to be larger and deeper. Possible need for more expensive piled foundations.
Structural frame	Possibility of load-bearing walls. Otherwise two sets of frames but some smaller column sizes and less hoisting, so likely to be cheapest proposition.	Larger column sizes to lower six storeys as will carry heavier loads and increased hoisting will make this the more expensive arrangement.
Upper floors and staircases	One less upper floor and flight of stairs.	One more upper floor and flight of stairs. Stairs may need to be wider to satisfy means of escape in case of fire requirements and there will also be increased hoisting costs.
Roof	Greater roof area.	Reduced roof area but savings in cost partially offset by higher constructional costs.

Element	*Two five-storey blocks*	*One ten-storey block*
Cladding	Less hoisting.	May require stronger cladding to withstand increased wind pressures, and extra hoisting will be involved.
Windows	Slight advantage.	Increased hoisting and possible need for thicker glass in windows on upper floors to withstand higher wind pressure.
External doors	Double the number of entrance doors.	Might involve more doors to balconies.
Internal partitions	Slight advantage.	Some increased hoisting costs.
Internal doors and joinery fittings	Much the same.	Much the same.
Wall, floor and ceiling finishings	Little difference.	Little difference except for possibly slightly increased hoisting costs.
External painting	Some advantage.	Rather more expensive.
Sanitary appliances	Much the same.	Much the same.
Soil and waste pipes.	Increased length of pipe.	May need larger-sized pipes on lower storeys.
Cold and hot water services	Double the number of cold water storage tanks and may need two boilers.	Larger cisterns, boilers, pumps, etc. and may need some larger-sized pipes and ducts.
Heating and ventilating installations	Two separate installations but some savings due to smaller-sized pipes or cables.	Cost advantage of single system but may be largely offset by larger pipes or cables and fittings.
Electrical installation	Two separate installations and intakes.	Cost advantage of single system but probably more than offset by increased size of cables.
Lifts	Two lift motor rooms but probably the same number of lift cars.	Saving from one lift motor room but may be necessary to install faster and more expensive lifts.
Sprinkler installation	Two separate sprinkler systems.	One system but some of pipework will need to be of larger size.
Drainage	More extensive and expensive system.	Some economies particularly in length of pipe runs and number of manholes.
Site works	Likely to be more expensive in paths and roads but reduced ground area.	Some savings likely.
Preliminaries and contingencies	May require two tower cranes if blocks are to be erected simultaneously.	Taller tower crane needed.

COLUMN SPACINGS

Single-storey framed structures almost invariably consist of a grid of stanchions or columns supporting roof trusses and/or beams or portal frames. Sheeting rails, purlins and windbracing all help to stiffen and strengthen the structure. By increasing the lengths or spans of the roof trusses, the number of columns can be reduced and this may be of considerable advantage in the use of the floor space below with less obstruction from columns. The trusses may need to be of heavier sections to cope with the greater loadings associated with larger spans, and will need to be of different design if the spans are lengthened sufficiently. In like manner the sizes and weights of columns will need to be increased to take the heavier loads transmitted through the longer trusses, and this will partially offset the reduction in number of columns. One method of assessing the probable cost effect of varying the column spacings and spans of trusses is to calculate the total weight of steelwork per square metre of floor space for the alternative designs, when the most economical arrangement will be readily apparent.

For instance if steel stanchions 4.5 m high were provided to support steel trusses 7.5 m long at 4.5 m centres, the weight of the stanchions would be approximately 7.7 kg/sq m of floor area. The weight of stanchions/sq m of floor area would reduce to 5 kg for trusses of 15 m span and to 3.7 kg for trusses of 24 m span. On the other hand, with riveted steel angle trusses to 1/5 pitch and spaced at 4.5 m centres, the weight of the trusses per square metre of floor area would increase with lengthening of the roof spans as indicated below

7.5 m long trusses – 5.3 kg/sq m
15 m long trusses – 8.2 kg/sq m
24 m long trusses – 11.9 kg/sq m

To the weight of stanchions and trusses must be added the weight of beams and purlins to arrive at the total weight of steelwork. Purlins spaced at 1.35 m centres and spanning 4.5 m between trusses would probably weigh about 8.5 kg/sq m of floor area.

The Wilderness study group[10] investigated the design cost relationships of a large number of hypothetical steelframed buildings of equal total floor area and similar specification but with the accommodation arranged on one or more storeys in buildings of varying shapes with varying bay sizes, column spacings, storey heights and superimposed floor loadings. The study group confined their investigations to the functional components of roofs, floor slabs, columns, beams, ties and column foundations, collectively termed *the core*. The group produced a set of charts designed to indicate cost relationships under varying conditions of numbers of storeys (one to eight), storey heights (mainly 3 to 4.5 m), loadings (2 to 10 kN/m^2) and column spacings (3 to 12 m), but all limited to steel-framed buildings of simple design with solid *in situ* reinforced concrete floor and roof slabs.

An examination of these charts show increasing costs with wider spacing of columns. Adopting a storey height of 3 m and floor loadings of 5 kN/m^2, a comparison is made of the effect of extreme column spacings of 4.5 m in each direction as against 12 m in each direction. The increased cost of the structure resulting from the wider spacing of columns with single-storey buildings is shown at sixty-seven per cent rising to over 100 per cent with eight-storey blocks. If the storey height is increased to 4.5 m, the increases in cost are less spectacular as they are partially offset by the extra material in the extended columns. The extra structural cost due to the more widely spaced columns in blocks with 4.5 m storey heights rises from about sixty per cent in single-storey buildings to ninety per cent in eight-storey blocks. It is interesting to note that the variations in the costs of structures due to different storey heights reduces as columns are spaced more widely apart, although the extra structural costs arising from increases in storey heights are relatively small compared with those stemming from wider spacing of columns and increased floor loadings. For example, increasing storey heights from 3 m to 4.5 m produces a six per cent addition to structural costs for single-storey buildings rising to sixteen per cent for eight-storey blocks. It must be recognised that these relationships cover only part of the costs of buildings and relate solely to steel-framed buildings with *in situ* reinforced concrete floors. Nevertheless within these limitations this study provides some extremely useful guidelines in assessing the probable cost relationships of different structural designs for a project.

An investigation by the Building Research Station[11] showed that the most usual column spacing in factory buildings was between 6 m and 9 m; spacings of less than 3 m were unusual, but those over 9 m occurred quite frequently and there were some cases of spacings of over 15 m. The design of factory frames is very flexible and while column spacing affects costs it is probably not the most important determinant of initial costs. The layout of the floor space depends on the size of machines, the production flow, the gangway space for internal circulation, and the space required for storing stock and work in progress. Most production line widths were quite small with a common width of between 1.5 m and 3 m. Many production lines with their gangways can be satisfactorily accommodated either singly or in multiples between columns spaced 6 m to 9 m apart. It is however often advantageous to have the column spacings wider in one direction than the other, and it appears that 6 m × 9 m spacings or multiples of these dimensions are often the most acceptable.

FLOOR SPANS

Floor spans deserve attention as suspended floor costs increase considerably with larger spans. Furthermore, the most expensive parts of a building structure are the floors and roof, namely the members which have to thrust upwards in the opposite direction to gravitational forces. As a very rough guide, horizontal structural members such as floors cost about twice as much as vertical structural members like walls.

Craig[8] has described how in the upper floors of blocks of flats stiffness is an essential quality and meeting sound insulation requirements dictates a minimum floor thickness of 125 mm. In this situation the most economical spans are likely to be in the order of 4.5 to 6 m. With crosswall construction floor spans are usually within a range of 3.6 to 5.2 m. Two-way spanning of *in situ* reinforced concrete floor slabs helps in keeping the slab thickness to a minimum, and one-way spanning is only economical for small spans.

FLOOR LOADINGS

The Wilderness study[10] has shown that variations in design of floor loadings can have an appreciable effect on structural costs. Adopting a 7.5 m grid of columns and a 3 m storey height, a comparison of structural costs for buildings with floor loadings of 2 and 10 kN/m^2 respectively, shows an increase in cost of about twenty per cent for two-storey buildings rising to about forty per cent for eight-storey buildings for the higher floor loadings. Further increases of two to four per cent occur if the storey height is increased to 4.5 m. Limited increases also arise from the wider spacing of columns when coupled with heavier floor loadings, and these increases become more pronounced in the taller blocks.

Heavy loads can be carried most economically by floors which rest on the ground, rather than by suspended upper floors. Where heavy loads have to be carried by suspended floors it is desirable to confine them, wherever practicable, to parts of the building where the columns can be positioned on a small dimensional grid. As indicated previously it is expensive to bridge large spans and it becomes quite a complex task to determine the point at which the unobstructed space stemming from larger spans equates the extra cost of providing it. Eccentric loading of vertical supports its always uneconomical and it may be worthwhile to increase a cantilever counterweight by moving the support nearer the centre of the load to reduce or eliminate the eccentricity. For this reason perimeter supports are less economical than those provided by crosswalls.

REFERENCES

1. ROYAL INSTITUTION OF CHARTERED SURVEYORS: JUNIOR ORGANISATION. The effect of shape and height on building costs. *The Chartered Surveyor* (May 1964)
2. S. KAYE. Narrow or wide, high or low, profit or loss? *The Chartered Surveyor* (June 1966)
3. NATIONAL BUILDING AGENCY. Metric house shells: Two storey plans – cost guide (1970)
4. ROYAL INSTITUTION OF CHARTERED SURVEYORS: BUILDING COST INFORMATION SERVICE. BCIS amplified and brief forms of cost analysis (1970)
5. THE ARCHITECTS' JOURNAL INFORMATION LIBRARY. Hotels. *Architects' Journal*, **151.25** (1970)
6. ROYAL INSTITUTION OF CHARTERED SURVEYORS: COST RESEARCH PANEL. Report on the costs of flats and houses. *The Chartered Surveyor* (July 1958)

7. W. J. REINERS. Studies in the cost of housing: the tender prices of local authority flats. *The Chartered Surveyor* (August 1958)
8. C. N. CRAIG. Factors affecting economy in multistorey flat design. *Journal of Royal Institute of British Architects*, **63.6** (1956); and Value for money in flats. Building Research Station (1958)
9. MINISTRY OF HOUSING AND LOCAL GOVERNMENT. Flats and houses 1958: design and economy. HMSO (1958)
10. THE WILDERNESS COST OF BUILDING STUDY GROUP. An investigation into building cost relationships of the following design variables: storey heights, floor loadings, column spacings, number of storeys. Royal Institution of Chartered Surveyors (1964)
11. BUILDING RESEARCH STATION. Factory building studies no. 12: The economics of factory buildings. HMSO (1962)
12. DEPARTMENT OF THE ENVIRONMENT. Local authority offices: areas and costs (1971)

3 FUNCTIONAL REQUIREMENTS AND COST IMPLICATIONS OF CONSTRUCTIONAL METHODS

THIS CHAPTER IS concerned with the functional requirements and cost implications of alternative constructional techniques for different types of building and of building elements. A comparison of maintenance problems and costs associated with different materials and components is made in chapter 11.

LOW AND HIGH RISE BUILDING

Residential Buildings

The relative costs of low and high rise residential buildings were compared in chapter 2 and reasons sought for the much higher cost of high rise developments over two-storey housing. Admittedly, the more intensive use of highly priced land will offset to some extent the increased costs resulting from multistorey development. In addition there may be social benefits to be gained by some occupiers through the erection of tall blocks of flats on central urban sites, such as ready access to town centre facilities and reduction in length of journey to work, although these benefits are often difficult to evaluate, as indicated in chapter 14. A RICS case study in 1961[1] showed high flats costing between sixty and eighty per cent more per square metre of floor area than houses, with superstructure costs, influenced particularly by the costs of the frame and floors, accounting for just over half of the increased cost, and the costs of services including lifts taking up most of the remainder. Furthermore, the high blocks of flats showed a greater range of costs than the low flats or houses.

Another case study undertaken by the RICS Cost Research Panel[2] relating to low flats, found that the main source of higher costs in this type of flat as compared with a house was in floors, landings, staircases, balconies and finishings. These costs could be as much as two to three times greater than corresponding items in two-storey housing, while the majority of other components averaged from twenty to seventy per cent higher in cost. Savings in roof and substructure costs were only partially reflected in the sample. Table 3.1 shows the relative costs of low flats to houses, and is subdivided between the main component parts.

Apart from higher constructional costs, problems have also been encountered through people not wanting to live so far from the ground, difficulties experienced by young married couples bringing up children in high flats, the government's decision to stop the payment of additional grants for multistorey flats, and general sociological and aesthetic problems. For these reasons Coventry Corporation decided in 1968 not to build any more multistorey flats, and most of the major provincial cities were at that time reducing the rate of construction of high rise residential blocks.

Industrial Buildings

Industrialists have a general preference for single-storey premises, but a variety of matters such as limitations on land available and increased demands for car parking space may justify a reappraisal.[3] A wide diversity of loadings occurs in factory buildings. In a specific case, the loads to be carried may affect the choice of site, influence the building design and, if very heavy, may have a considerable bearing on the choice between single and multistorey construction.

In a survey of Midland factories undertaken by the Building Research Station,[4] less than one-half of the

TABLE 3.1

RELATIVE COSTS OF HOUSES AND LOW FLATS

Component	*Items*	*Three-storey flats as percentage of two-storey houses (provinces)*	*Three to five-storey flats as percentage of two-storey houses (all regions)*
Substructure	Foundations and ground floor slab including all abnormals.	125	166
Brickwork	Structural brickwork, lintels and sundry columns, including internal partitions.	101	110
Facework	Extra over cost of facings.	100	155
External fittings	Windows, external doors, porches, balcony balustrades, including painting and glazing.	126	151
Floors	Floors, staircases, balconies and landings.	254	306
Roofs	Roof structure, covering and external plumbing.	79	85
Superstructure		119	138
Floor finishings	Floor, staircase and balcony finishings, including screeds.	209	204
Plastering	Including any wall tiling.	100	116
Decorations	Internal decorations only.	83	113
Joinery	Doors, door frames, cupboards, dressers, including ironmongery.	165	153
Internal finishings		140	146
Heating	Fires, chimneys, or other forms of heating including water heating.	100	204
Plumbing	All internal plumbing services, including sanitary appliances.	147	161
Gas and electrics	Installation only.	155	234
Sundry services	Lifts, refuse disposal, etc. in three to five-storey flats.	—	—
Services		137	208
Overall relative cost		126	152

Source: RICS *Cost Research Panel – the cost and design of low flats*[2]

factories had production areas on upper floors, although in some cases as many as six floors were used for production. As few as one-quarter of the postwar buildings had production areas on upper floors. This rate is probably higher than in the country as a whole, for other sources suggest that only ten per cent of the factories erected in England and Wales during 1937–46 were multistorey. It is further believed that the majority of factory buildings erected in central London since the war have been multistorey; outside the central areas of cities the proportion is lower, while in the new and expanding towns almost all the factory buildings are single-storey, although possibly fronted with a double-storey office block.

Other Buildings

The number of storeys incorporated into other types of building will often be influenced by the purpose for which the building is to be used and/or the value of the site. Shops are generally of single-storey construction for the convenience of users, whilst offices are often multistorey to make more intensive use of highly priced central sites and to enable the occupants to be as far removed as possible from traffic noises. An ideal approach is to build offices on top of shops in the centres of large towns and cities, and flats or maisonettes above shops in neighbourhood centres.

Schools are of varying heights with a predominance of single-storey buildings amongst primary schools;

secondary schools are often of two, three or four-storeys, and technical colleges and colleges of further education are often of several storeys except for single-storey workshops. A case study undertaken by the Ministry of Education in 1957[5] for a three-form entry secondary school of 3320 sq m floor area, showed that a two-storey building was the cheapest proposition, and that both three and four-storey buildings were significantly cheaper than a single-storey building. Factors contributing to the higher cost of single-storey buildings included the greater quantity of work below ground floor level and in roofs and drainage work, more external doors, which are relatively expensive, and more pipe and cable runs, which are also costly.

The unit costs of three and four-storey blocks exceeded those of two-storey blocks with the same total floor area for several reasons:

(1) With certain elements such as roofs, and work below ground floor level, the reduction in cost per square metre obtained between the single-storey and two-storey building diminishes as the floor area increases with the provision of additional storeys.

(2) The three and four-storey buildings must be provided with two staircases whereas one is sufficient for the two-storey building.

(3) As the height of a building increases there is a marked increase in the need for wind bracing.

SUBSTRUCTURES

Adequate information on subsoil conditions is vital before a decision can be made as to the most economical type of foundation. On the majority of smaller contracts, sufficient information can be obtained by excavating and examining trial holes, and an inspection will often reveal that strip foundations or simple concrete bases will be adequate. Large buildings and those on difficult sites will generally justify the cost of a borehole investigation, which may indicate the need for piled or raft foundations.

Building Research Station Digest 67[6] shows how the type of construction and subsequent use of the building are important determinants of foundation design. Strip foundations, sometimes termed footings, are usually adopted for buildings where the loads are carried mainly on walls. Pad foundations or piles are more appropriate when the structural loads are carried by columns. Where differential settlements are to be controlled within fine limits, raft foundations are often the best approach. Superimposed loadings are calculated in accordance with the Building Regulations 1965 and British Standard Code of Practice 3, whilst dead loads can be determined by reference to BS 684.[7]

A typical two-storey semidetached house of about 93 sq m floor area, in cavity brickwork with lightweight concrete or clay block partitions, timber floors and a tiled roof, weighs about 100 tonnes excluding the weight of foundations which has to be added after design. The weights at ground level in kN per lin m would be approximately: party wall – 22, gable-end wall – 17.5, front and back walls – 11.5 and internal partitions less than 7.5. The use of modern materials tends to reduce total loads, although some forms of construction, such as crosswalls, may result in greater loads on certain walls. For loads of not more than 16 kN/m, strip foundations would need to be 225 mm wide in compact gravel and 450 mm wide in soft silt or very soft clay, increasing to 300 mm and 900 mm respectively with loads of up to 32 kN/m.[6]

Smith and Bott[8] have identified four main types of foundation as follows.

Conventional strip foundations (or footings with mass or reinforced structural bases) are used for most residential, educational and industrial buildings which are fairly lightly loaded on subsoils with a reasonable load-bearing capacity, such as rock or sand and gravel containing only a small proportion of clay. Where heavier loadings occur, as with multistorey structures or warehouses, these may be met by introducing larger than normal bases.

Raft foundations (rigid or flexible) are on occasions used for lightly loaded structures where some degree of settlement may be expected on the site. Settlement could arise from mining subsidence, soft subsoils or presence of fill. Rigid rafts often comprise flat concrete slabs reinforced with concrete beams acting as stiffening ribs and are very expensive. Hence it is better to use flexible rafts which usually consist of thin reinforced concrete slabs on beds of sand.

Piled and ground beam foundations which are often needed for structures erected alongside rivers and estuaries and in other difficult locations. The piles are designed to support the loads either by friction from the various strata through which they pass or by bearing on a firm strata at a suitable level. *In situ* concrete piles are generally about fifteen to thirty per cent cheaper for relatively short piles, but as the length of pile increases the cost of *in situ* piles comes much closer to that of precast piles.

Vibrocompaction is a method of treating weak soils to improve their load-bearing characteristics. A tubular probe is lowered into the ground and contains an oscillating eccentric weight which imparts a transverse vibration to the tube and hence into the surrounding ground. The hole formed by the probe is backfilled with compacted coarse aggregate and conventional foundations are then cast on the treated ground. It often proves to be a quick and economic process and can be considerably cheaper than piled foundations or deepmass concrete blocks.

Economics of Foundation Construction

The availability of suitable plant can influence the relative costs of foundation types. Experience has shown that in areas of shrinkable clay, short piles bored with mechanical augers are competitive in cost with traditional strip foundations, even with quite small contracts, and could conceivably be about two-thirds the cost of a deep traditional strip foundation. For single house contracts, bored piles are usually slightly more costly than deep strip foundations (often 375 × 1000 mm) although the extra cost may be justified by the added safety factor.[6]

British Research Station Digest 67[6] emphasises that simple costing on a labour-plus-materials basis may be misleading. The construction of bored piles can often continue through the winter months when the trenches for strip foundations could be waterlogged or damaged by exposure to frost. In addition to use on clay sites, bored-pile foundations with precast ground beams may be useful for low rise industrialised systems, and in crosswall house construction it is rarely necessary to use more than ten piles per house. Orchard and Hill[54] have shown that concrete deep strip foundations for houses are often cheaper than traditional brick and concrete strip.

Removal of subsoil water from foundation excavations is always expensive, hence foundations on sands and gravels should be kept above the water table wherever possible. Some general rules with regard to relative excavation costs may also be helpful.

(1) It is almost invariably more economical to utilise the existing ground configuration rather than to dig it out or make it up.

(2) Removal of surplus excavated material from a site is far more expensive than depositing and levelling the soil elsewhere on the site.

(3) Over a certain minimum (probably about 2000 cu m), the cost of bulk excavation, but not its disposal, probably reduces with quantity more steeply than any other building operation price rate.

Lee,[9] in his evaluation of the comparative costs of different forms of municipal housing, calculated the percentage costs (table 3.2) for substructure work with various types of housing. It should be borne in mind that foundation prices may be 'weighted' to secure a quick return to the contractor.

Basement Construction

The South Wales and Monmouthshire branch (quantity surveying section) of the RICS undertook an investigation into basement construction costs in 1959/63 and the results were duly published.[10] The branch confined their study to basements with brick and reinforced concrete walls, and adopted the dual aims of providing general cost comparisons of various types of construction and waterproofing, and also of providing data for assessing approximate basement construction costs. In order to do this a hypothetical scheme was evolved incorporating basic data suitable for an office building or block of flats of crosswall construction. The principal details were length – 15 m; width – 7.50 m; number of storeys – four plus basement; superimposed floor loads – 5 kN/m^2 on upper floors and 10 kN/m^2 on the basement floor; and basement wall heights of 2.74 m, 3.00 m, 3.81 m, 4.27 m and 4.57 m.

The principal conclusions of the working party were

(1) the larger the basement the cheaper the cost/sq m;

TABLE 3.2

MUNICIPAL HOUSING – RELATIVE SUBSTRUCTURE COSTS

Type of housing	*Weighted percentage of total costs*
Traditional brick two-storey houses	
Average value	9
Lowest value for good sites and terrace blocks	8
Highest value for poor sites and development in pairs	10
(exceptionally difficult sites could involve extra costs of up to seven per cent)	
Brick flats and maisonettes	
Average value	8
Lowest value for four-storey blocks on good sites	6
Highest value for blocks involving piling and for those over 30 m in length	13
(exceptionally difficult sites or blocks with basements could involve extra costs of up to five per cent)	
Concrete structure flats and maisonettes	
Average value	10
Lowest value for high blocks with normal foundations	4
Highest value for low blocks with piled foundations	16
(exceptionally difficult sites or low blocks incorporating basements could involve extra costs of up to seven per cent)	

Source: M. Lee – *Comparative cost evaluation – municipal housing*[9]

(2) the nearer to a square shape the cheaper the cost, although the cost does not increase considerably until a ratio of one to three is exceeded;

(3) where basements are constructed in concrete, the cost/cu m does not increase with the depth up to 4.60 m, providing the damp-proofing requirements are constant;

(4) the use of brickwork for depths up to 3 m, providing the basement walls are loaded down by the superstructure, generally shows a distinct economy over the use of concrete, but beyond this depth, this trend either diminishes or reverses;

(5) the additional cost of waterproofing concrete is relatively small and therefore worthwhile when conditions do not justify asphalt tanking;

(6) the cost of asphalt tanking constitutes a high proportion of the total basement cost and underlines the need to ascertain the proposed method of waterproofing at the time of preparing preliminary estimates. Table 3.3 shows cost indices covering the total costs of basement construction for the different types of construction, under varying conditions.

STRUCTURAL COMPONENTS

A variety of structural forms are available from load-bearing brick walls to reinforced concrete, steel and timber frames. It is often necessary to prepare cost comparisons of different structural forms in order to determine the most economical constructional form for a particular job. The principal structural forms are now considered.

Load-bearing Brickwork

In recent years it has been found economical to use load-bearing brick crosswalls for buildings up to five storeys in height in place of structural frames of steel or reinforced concrete. The solid crosswalls carry all the floor and roof loads and have separating and insulating functions. These walls are able to provide structural support when the building permits the plan form to be repeated on each floor, or the load-bearing walls are so placed in relation to planning needs that they can, for the most part, continue uninterrupted from foundation to roof. This form of

TABLE 3.3

INDICES OF BASEMENT COSTS COMPLETE

Base 100 = bricks walls (A construction) in good ground, 2.74 m depth of basement (in fourth quarter 1963, approximately £20.50 per linear metre).	Base100 = floor (A, B or E construction) (in fourth quarter 1963, approximately £5.40 per square metre).

Depth of basement in metres	*Ground*	*BRICK WALLS* A	B	C	D	*CONCRETE WALLS* E	F	G	H	*FLOORS* A, B, E	F	C, G	D, H	*Depth in metres*
2.74	Good	100	128	186	197	166	173	225	236	100	105	130	143	2.74
3.00		113	142	206	218	181	190	247	259	106	110	136	148	3.00
3.81		204	227	305	319	245	256	325	339	118	123	148	161	3.81
4.27		261	289	376	393	269	280	356	372	127	131	157	169	4.27
4.57		321	351	444	462	287	300	381	398	132	136	162	174	4.57
2.74	Poor	120	139	197	208	163	171	223	233	107	114	137	149	2.74
3.00		132	153	217	229	178	187	245	257	113	120	143	155	3.00
3.81		260	268	346	361	243	254	322	337	125	133	155	168	3.81
4.27		319	327	414	431	267	279	355	371	133	141	163	176	4.27
4.57		381	390	483	501	286	298	378	395	138	146	168	181	4.57
2.74	Bad	321	342	400	411	362	370	423	434	122	129	152	164	2.74
3.00		352	375	439	451	396	405	465	477	128	136	158	171	3.00
3.81		567	576	654	669	512	523	593	607	145	153	175	188	3.81
4.27		657	666	753	770	567	578	656	672	155	163	185	198	4.27
4.57		737	747	840	858	603	615	697	714	162	169	192	204	4.57

Good ground	250 kN/m² load bearing.	E construction	Concrete walls.
Poor ground	190 kN/m² load bearing.	F ,,	Waterproofed concrete walls.
Bad ground	As last but waterlogged, necessitating sheet piling and pumping.	G ,,	Concrete walls with external tanking – asphalt to BS 1097.
A construction	Unprotected bricks walls.	H ,,	Concrete walls with external tanking – asphalt to BS 1418.
B ,,	Brick walls with external waterproof rendering.	Brickwork	Common bricks having a crushing strength of 21 N/mm² laid in cement lime mortar (1:1:6).
C ,,	Brick walls with external tanking – asphalt to BS 1097.	Concrete	Portland cement concrete (1:2:4, 19 mm ring).
D ,,	Brick walls with external tanking – asphalt to BS 1418.	Reinforcement	Mild steel rod.

Source: RICS *Design/cost research working party – basement construction*[10]

construction also permits flexibility in the choice of claddings to front and rear walls as they are no longer load-bearing. The optimum spacing of crosswalls is about 4.80 to 6.00 m centres and some claim that this inhibits planning by establishing too rigid a framework. The introduction of some column and beam construction would permit the provision of large openings in crosswalls, but with comparatively small and well-defined units such as flats the crosswalls are unlikely to present any real obstacle.

A five-storey block of maisonettes and flats over ground floor shops were erected at Allestree, Derby, using 380 mm brick cavity flank walls and 215 mm brick crosswalls and staircase walls.[11] The spacing of the crosswalls varied from 5.20 to 6.20 m. It is claimed that the brick crosswalls showed a saving of thirty-five per cent on the estimated cost of a reinforced concrete frame, and this is equivalent to an overall cost saving on the job of about six per cent. Building Research Station Digest 120[12] indicated that for two-storey house construction, crosswalls were not inherently more economical than more traditional forms of construction. The cheaper method will depend on many design features such as plan, roof and wall types and finishes, and on site conditions. Some of the factors to be considered include the supporting of timber joists on party walls and transference of roof loads to transverse walls by beams, possible buttressing of ends of walls and effect of discontinuity of operations on site.

An investigation of the relative costs of low flats by the former RICS Cost Research Panel[2] found that with three-storey blocks, the cost of the brick crosswall structure showed a similar order of costs to corresponding features of traditional blocks. Reinforced concrete crosswalls showed a cost higher than that of traditional brickwork. A study of high flats[1] showed a sixty per cent cost range for external infilling between the crosswalls, timber cladding being the cheapest and specialist curtain wall infilling the most expensive.

Reinforced Concrete Frames

Dunican[13] in 1960 found that cost analyses prepared by many surveyors and engineers showed that reinforced concrete was cheaper than structural steelwork for the main frameworks to most multistorey offices buildings and high flats. Dunican investigated the cost relationships of a variety of steel and concrete framework designs and found that the use of concrete produced a saving of between twenty and twenty-five per cent of the structural cost of a scheme in steel. Protagonists of structural steelwork postulated that in certain circumstances the cost differential in favour of reinforced concrete could disappear. The relative ease of modification of steelwork by cutting and welding, and the more ready adaption of steelframed buildings to meet changing user requirements favour structural steelwork as against reinforced concrete. Dunican's findings are supported by Creasy[14] who in 1959 analysed the costs of six similar multistorey buildings, three of which were constructed of cased steel with the rest made of reinforced concrete frameworks. The average cost of the structural items, relative to the overall cost of the building, was thirty-nine per cent for the reinforced concrete and forty-seven per cent for the steel frame. Reasons for using reinforced concrete frames to turbine houses of power stations included aesthetic considerations.

Continual developments in the design of steel structures and the manufacture of structural steel could alter the cost relationship with concrete, so it is necessary to keep the relationships under constant review. Aspects other than cost frequently deserve consideration, for instance, increased flexibility of concrete designs in the initial stage, both from the point of view of loading and of the opening of sections of the floor. It is also advisable to use structural materials in the best way to exploit their own particular potentialities – thus a job which is well-suited to be designed in concrete may not be viable as a steel building.

Leon and Wajda[50] assert that multirise dwellings, which have comparatively small room spans, are usually more economically constructed in concrete of boxshell construction, rather than with a structural steel framework. In general, buildings with short spans are particularly well suited for construction in reinforced concrete. With large spans, the weight of the concrete structure is high in relation to the load carried, the dead weight/imposed load ratio being approximately unity, whereas with steel structures it is about one-fifth.

Advantages of the use of reinforced concrete as compared with steel are as follows.

(1) Reinforced concrete structural sections can be precast to combine weather resistance, fire protection, sound and thermal insulation and attractive finish in a single operation, and permit more rapid completion of the building.

(2) Structural members can be economically formed by standardising shutters or moulds for beam and column sizes, with extra steel introduced where required to take heavier loads.

(3) Buildings based on steel frames are generally about ten to twenty per cent more expensive than similar buildings using reinforced concrete frames.

Smith and Bott[8] postulate that for multistorey warehouse buildings the most economic structure is likely to be an *in situ* reinforced concrete frame with a column grid of about 6 m and incorporating *in situ* concrete flat plate floor construction. With multistorey office and laboratory buildings the same authors believe that, depending on clear span requirements, the most economical arrangements will be either a completely precast concrete structure and flooring system, or an *in situ* reinforced concrete frame and precast floors.

Building Research Station Digest 48[15] outlines ways in which the cost of *in situ* concrete frame construction can be reduced. Economies in formwork stem from simplicity of design of multistorey buildings and too few designers appeared to take account of production difficulties at the design stage. The use of sliding formwork can show savings of around twenty per cent of the cost of traditional formwork. However the successful use of sliding formwork depends mainly upon:

(1) an appreciation that formwork, reinforcement and concreting become integrated and interdependent operations;

(2) the greater use of mixing plant and cranes, which enables high outputs of 5 to 10 cu m per hour of concrete to be placed; and

(3) the technique requires more overtime to be worked than traditional construction.[15]

It is further suggested that plate floors are more economical than slabs and beams in multistorey flats, the inclusive cost of the omitted beams being approximately twice the cost of the additional reinforcement in the plate floors. Further savings could often be secured by the use of designed concrete mixes.

George[51] has described how the structural cost of a multistorey building is usually in the order of twenty to thirty per cent of the total building cost. The formwork component of a structural element in normal concrete work may vary from as low as thirty-five per cent, in the case of thick suspended floors, to sixty-five per cent in the case of thin walls. George's analysis of the formwork costs of two *in situ* reinforced concrete framed multistorey office buildings in Brisbane in 1970 showed that the formwork was forty and forty-five per cent respectively of the cost of the reinforced concrete work, and ten and eight per cent of the total cost of the building.

Economies in formwork can be obtained mainly through simplicity of shape, and repetition of units to obtain the maximum number of uses; and other ways of reducing formwork costs are as follows.

(1) Formwork should be designed to reduce labour requirements in its assembly and to permit re-use without cutting.

(2) It is often more economical to use a little more concrete than is structurally necessary to secure repetition of formwork.

(3) Column sizes should be standardised as far as possible with differing structural loads accommodated by making variations in the quantity of steel reinforcement or concrete strengths and, where column sizes must be changed, reducing one dimension at a time will save cutting.

(4) Establish uniform spacing of columns and beams, and uniform beam dimensions wherever possible.

(5) Eliminate perimeter edge beams, in flat slabs wherever possible.

(6) Shapes of sections should preferably be designed to allow easy removal of forms, for example, sides of beams could advantageously slope slightly outwards from the bottom.

Precast Concrete Structures

Precast concrete has been defined as premoulded concrete units that fulfil both the architect's visual and functional requirements and the consulting engineer's structural requirements. Mansfield[16] has described how precast concrete offers a wide variety of finishes which rarely require any further treatment after erection. It offers a quick form of construction, although this may be adversely affected by the speed of construction of other elements of the building. Nevertheless, precast concrete structural members can prove expensive and it is vital that they be carefully designed to ensure maximum and efficient use of handling plant and moulds.

The Ronan Point tragedy highlighted some of the problems associated with the design of large panel structures, through the possibility of a consecutive failure of floor and wall components, that is 'progressive collapse'. Short and Miles[17] have described how an addendum to Code of Practice 116 (precast concrete) lays down new design rules for all large panel structures and, in addition, includes requirements for such structures where resistance against progressive collapse is an essential criterion for design. Tie forces needed for the security of large panel structures can be provided by the inclusion of reinforcement or prestressing tendons from floor to wall, peripherally, internally or as continuous vertical ties. The additional cost of these features is likely to be relatively small in most cases.

The size and shape of precast concrete units is generally governed by the limitations imposed by transportation and site erection. White[52] describes how these limitations are normally imposed by economic and legal requirements and not necessarily by technical problems; however, further limitations on design are imposed by the need to lift and stack precast components safely and at an early stage in the precaster's yard pending delivery to site.

Designers need to consider the problems of handling in the works and special care must be taken in respect of projecting steel, edge details, finishings, lifting methods and stacking arrangements, the problems and cost of transportation, and the cost of site erection, which is usually governed by the size of crane required. The cost of transportation is generally proportional to the distance covered and weight carried, but the rate per tonne/km reduces over longer distances. For exceptionally long and wide loads, the cost can double or treble as certain statutory precautions have to be taken to ensure the safety of the load and other road-users. Finally, to ensure smooth and quick erection, tolerances, joint details and fixings must be carefully chosen. Indeed, early design decisions can have a significant effect on the speed of erection and, hence, on the viability of the project.

Structural Steelwork

A survey of factories built between the wars in the Midlands, London and the Home Counties[4] indicated that about five-sixths of the structural frames to factories were of steel, and most of the remainder were of concrete. Since the last war some of the most successful steelbased systems in the United Kingdom have been developed by consortia of building clients. These include CLASP (Consortium of Local Authorities Special Programme) and SCOLA (Second Consortium of Local Authorities) – two groups of local education authorities who execute millions of pounds worth of school and other buildings each year; NENK, a system developed by the former Ministry of Public Building and Works for buildings of all types for the services; and 5M, a system for house building developed by the former Ministry of Housing and Local Government. Both Ministries now form part of the Department of the Environment which was established in 1970. In all these systems it has been stated that a steel frame was chosen as a result of user requirement studies and performance standards, from which steel was seen to emerge as the best solution. It was also pointed out that in these cases a lightweight structural steel frame offered greater flexibility in grid arrangements and allowed the easy threading through of services and other nonstructural components.[18]

CLASP's decision to adopt a light and flexible steel frame mainly resulted from the need to take into account ground movement due to mining subsidence. A further advantage was found to accrue from speedy erection with a steel frame for a primary school being erected within four to five days and occupying as little as one-and-a-half per cent of the total site labour involvement. SCOLA also saw the advantages associated with light steel frames of speedy erection, flexibility of internal layout and adaptability to meet changing future situations. Steel has also been used in the frames of a number of multistorey hospital buildings in recent years, and these have been selected after cost comparisons with other forms of construction, taking into account the effects on cost of piled foundations. A variety of steelframed systems have been developed for low, medium and high rise housing with the primary benefits of savings in time and labour and the use of constructional methods which can be carried out in all weathers.

Leon and Wajda[50] see the main economic advantages of structural steel frames for multistorey building as:

(1) overall reduction in size of structure, as the columns are smaller in section and obstruct less floor space than those required for reinforced concrete construction;
(2) precision and speed of erection;
(3) reduced foundation costs due to smaller dead weight of structure;
(4) flexibility of planning; and
(5) adaptability to requirements of possible future changes, including façade cladding and fenestration and internal layouts.

One of the principal objections to the use of structural steelwork in the past stemmed from the need to encase the steel for purposes of fire protection. Under conditions of intense heat unprotected steelwork could twist and buckle and result in the complete collapse of the superstructure. The original concrete and brick casings were both costly and very heavy. In recent years there has been a move towards lightweight casings, including prefabricated casings of plaster and asbestos board with a sheet steel outer finish. Another important development has been the production of a paint-like substance for direct application to the steel surfaces. Two products of the intumescent type are available for this purpose; one, a coating originally developed in Germany, can be applied by spray to a thickness of about 2 mm, and the other, available from the United States, can be used in greater thicknesses up to 20 mm. Both provide protection by intumescence (expansion under the effect of heat which forms an aerated mass acting as a heat shield). It is claimed that protection can be provided for up to three hours by applying coats of appropriate thickness. Yet another approach is to surround the steel members with a thin sheet of steel and to fill the intervening space with a lightweight insulating material.[19]

The relative economics of concrete and steel structural frameworks may also be influenced by local conditions. For instance, physical restrictions on a congested central area site may make it difficult, if not impossible, to use large steel erection cranes. Shortage of time coupled with the urgency of a particular job may not allow sufficient time for the curing of *in situ* concrete members. An abundance of sand and gravel workings in close proximity to a building site may enable concrete to be produced at an advantageous price. A good supply of skilled steel erectors in the area could result in cheaper steel erection costs. The design characteristics of a particular job may favour one particular type of construction. Hence it is difficult, if not dangerous, to generalise on the relative merits of different structural forms and it is usually necessary to cost up the alternative techniques for each job.

Timber

The greatest use of timber framing has been in the field of industrialised building for houses and flats, and the majority of systems make use of it in the construction of walls and floors. In place of the conventional joist and floor board assembly the trend is towards prefabricated floor sections, occasionally of the stressed skin type, which are brought to the site ready for assembly. Timber frame walls possess the dual advantages of flexibility of cladding design and speed of erection. Timber walls and floors are usually required to have a fire resistance of half-an-hour in one or two-storey houses, but the walls separating dwellings have to provide protection for one hour and this has necessitated further development work.[19]

The main impetus for the use of timber framing has come from Canada and the United States, both of whom possess an abundance of timber, and where an estimated ninety per cent of the 1.5 million annual output of dwellings are of timberframe construction. In this country there are about forty registered house building systems using timber frames and accounting for an annual output of about 10 000 houses. Levin[20] postulates that the main advantages of timberframe construction are that it requires no large capital investment, heavy plant or other exceptional site equipment, neither large contracts nor standardisation of house types, architectural treatment nor finishing materials. He further claims that with good management and use of conventional building skills, good houses can be provided at comparable cost to traditional construction, but using one-third to one-half of the site labour and with accelerated erection times, measured in weeks instead of months.

Timber possesses high bending strength and stiffness per unit of weight, high thermal insulation and relative ease of machining, cutting, assembling, fixing and transporting. Framed panel assembly is generally a relatively simple jig operation which can be carried out in the shop with rudimentary equipment, or even on a platform built on the site. A variety of cladding materials can be fixed with ease to timber framing.

WALLING

Walls and partitions with associated windows and doors constitute a major item of expenditure of a building. Lee,[9] in his investigation of the costs of municipal housing found that these components accounted for about twenty-four per cent of the total cost of brick four-storey maisonettes and about thirty-two per cent of the total cost of brick two-storey houses.

External Cladding

For low rise buildings, cavity walls are generally the lowest longterm cost solution, provided satisfactory detailing and workmanship is secured. Stone facings are very expensive and care is needed to select a stone which is suitable for the particular environment; for example close-textured stones withstand city atmospheres better than limestones, and dense cement mortars must be avoided. Precast concrete cladding and cast stone often provide economic alternatives to natural stone.

An investigation by the Building Research Station[15] found that 175 mm concrete walls plastered internally, and using traditional formwork for multistorey flats, cost approximately fifteen per cent less than brick and clinker block hollow walls, faced externally and plastered internally, including scaffolding costs. A further fifteen per cent reduction in cost was obtained by using sliding formwork for the concrete walls.

The natural resistance to decay of timbers such as teak and western red cedar has led to their increased use for external claddings, although their use in housing is often limited to spandrels under windows, gable ends and other small areas of external walling. Unfortunately, the weathering of these timbers if untreated, at least in an urban environment, frequently causes them to become unsightly. An annual application of a linseed oil/paraffin wax mixture containing a fungicide is the minimum necessary to preserve appearance, and when comparing alternative cladding costs the annual expenditure on treatment of timber cladding must be taken into account.

Similarly, when making cost comparisons of alternative metal claddings it is essential to consider also the cost of any periodic treatment that may be required. For instance, some metals such as copper oxidise attractively upon exposure, whereas others like aluminium alloys may become unsightly. Aluminium when used externally should preferably receive anodic treatment, followed by periodic washing, to maintain a good appearance. Where the aluminium has not been anodised, more frequent washing is necessary, the actual frequency varying with the degree of atmospheric pollution.[21]

It is often necessary to consider the use of newly introduced materials for use in external claddings and there may well be insufficient feedback of maintenance costs to make a realistic appraisal of probable total costs, when making cost comparisons with other forms of cladding. For example, reinforced polyester panels provide a useful material for the external cladding of buildings and, as a result of investigations conducted by the Fire Research Station, prefabricated external walls with such cladding and aerated concrete cores have been used in the construction of multistorey blocks of flats in the London area. A likely future development is the incorporation of plastics in concrete.[19] On industrial housing, interlocking plastic-faced sheet steel has been used coupled with an inner lining of insulated laminated plasterboard to give a lightweight but strong, fireproof, highly thermally insulated and sound resistant cavity wall.

Partitions

A Building Research Station investigation of the costs of multistorey flats[15] indicated that dry partitioning had no direct cost advantage over block partitioning with two-coat plastering on both sides. This was attributed to the greater detailing required for dry partitioning involving the preparation of drawings and cutting schedules by suppliers, the need for packing pieces, taking care to ensure that panels did not become wet and the additional cost of skim coat or 'dry' finishes. However, a more recent comprehensive study by Hill and Lazarus[22] found that the use of 50 mm laminated plasterboard partitions in lieu of traditional blockwork in domestic structures showed worthwhile savings in structural costs. In a cost comparison of a typical upper floor of a maisonette or two-storey house with timber joist floor construction, a 50 mm laminated plasterboard partition showed a twenty per cent reduction in cost compared with the use of 75 mm lightweight block partitions plastered on both sides. Another investigation covered a two-bedroom flat to Parker Morris standards using hollow pot floor construction, where the 50 mm laminated plasterboard partition showed a saving of 14.5 per cent over the use of 75 mm lightweight block partitions.

In addition, it was anticipated that other savings would also accrue from the use of 50 mm laminated partitions. The 50 mm laminated partition was selected as being the heaviest of the various plasterboard partition systems available in 1970 and therefore the most representative from the point of view of weight and design relationship. The other consequential savings are shown below.

(1) Reductions in structural costs of load-bearing brickwork, supporting frames and foundations due to a lower total dead weight of building.

(2) With the two-bedroom flats savings would accrue from the considerable simplification of construction and setting out of the floor ribs.

(3) Simplification of design procedure and consequent detailing arising from reduced load factors.

(4) Reduction in the expenditure usually necessary for the provision of temporary equipment, fuel and attendance for drying and controlling the humidity of the works.

(5) Reduction in the total cost of general conditions and preliminaries arising partly from reduced cost of building work and partly from reduced building contract period.

Newman[23] has described how the design team also has a large range of demountable partitions from which to choose. Selection is influenced by many factors including demountability, fire resistance, acoustic properties, appearance, cost, integration of services, weight of partitioning and feasibility of incorporating false ceilings. Where the building is to be air-conditioned, full research has to be conducted into air ventilation arrangements and where this is achieved by means of a false ceiling, often integrally illuminated, the layout of the partitioning needs careful thought.

Many large commercial and industrial organisations require a flexible layout when planning their office accommodation, particularly as departments can be expanded, contracted, integrated or fragmented at quite frequent intervals. In these situations a steel or aluminium framework enclosing a variety of infill materials may well offer an effective solution. With the larger offices, some firms are now favouring open plan 'landscaped' offices which can afford advantages in working environments, flexibility and communication. A complete appraisal of open plan and enclosed office proposals for any given situation could involve a cost-benefit study of the type described in chapter 14. Another recent innovation is the integration of office furniture and fittings into the partitioning system. Based on a fixed module, the partitioning panels are manufactured to the same standard dimensions as the office furniture or storage units, which clip onto the framework of the partitions.

Kilford[24] has shown how the quantity surveyor is often presented at the detailed design stage with quotations from specialist partitioning subcontractors, based on performance specifications, and on which he has to make observations and recommendations. He suggests that the quantity surveyor's task would be eased considerably if the partitioning specialists would:

(1) conform to the specification or state that their system will not comply with it;

(2) produce their terms and conditions in print of sufficient size to be clearly legible; and

(3) price the offer in detail so that variations can be adjusted in a fair and reasonable manner to the satisfaction of both parties to the contract.

Insulation

The majority of new buildings are designed to achieve reasonable standards of thermal insulation and the practical applications of this aspect will be considered in chapter 11. The U-value of a wall, roof or floor of a building is a measure of its thermal transmittance or ability to conduct heat out of the building; the greater the U-value, the greater the heat loss through the structure. The total heat loss through the building fabric is found by multiplying U-values and areas of the externally exposed parts of the building, and multiplying the result by the difference between internal and external temperatures. U-values are expressed in W/m^2 °C (heat flow in watts through one square metre of the construction for one degree Celsius difference in temperature between the inside and outside of building). Typical U-values are: 220 mm solid brick wall with 16 mm plaster – 2.0; 260 mm brick cavity wall (unventilated with 16 mm plaster on inside face – 1.4; 260 mm cavity wall (unventilated) with brick outer skin, lightweight concrete block inner skin and 16 mm plaster on inside face – 0.96; and, as last but with 13 mm expanded polystyrene board in cavity – 0.70.[25] The additional cost of the latter form of construction would be in the order of 25 to 30 p/sq m in 1971, and to inject urea formaldehyde foam into the cavity would be two-and-a-half to three times as expensive.

In certain situations sound insulation can also be an important consideration. Noise is becoming a greater problem as more mechanical equipment is used in buildings, and both road and air traffic noise is increasing in volume. An extensive study of postwar office buildings[26] showed that on average forty-five per cent of the occupants suffer annoyance from street noise, rising to sixty-five per cent with buildings located on main thoroughfares, and the nuisance is not mitigated by height. Surveys of noise in houses adjoining main roads in Greater London[27] enabled a correlation to be established between a measure of noise levels by traffic noise index (TNI) and dissatisfaction of occupants through difficulty in getting to sleep, being compelled to keep windows closed and/or being unable to entertain visitors in comfort.

Aircraft noise has been studied by the Wilson Committee[28] and it was found that by providing good double-glazing and a sound attenuating ventilator unit, a reduction in loudness to about one-sixteenth of the level outside the building could be obtained. This required the frame of the inner window to be sealed to the existing window surround with a minimum space of 180 mm between the panes of glass. This treatment when applied to three rooms per house cost about £250 per house in 1969.[29]

ROOFS

Roof Types

Craig,[30] in his investigation of multistorey flats, found that the majority had flat roofs. The need for lift motor rooms, tank rooms, ventilating plant on occasions, ducts and pipes, add to the complexity of roofs, and the economy of covering a number of dwellings with one roof is partially offset by the costlier construction. Buildings on roofs are both expensive and often aesthetically unsatisfactory, and their elimination would reduce costs and leave a worthwhile expanse of flat roof which might be put to other uses. By way of contrast, high blocks of flats on the Continent have been built with pitched roofs and although this raises aesthetic problems, they have the advantages of lower first cost, longer life and provision of storage space.

A study of postwar factory buildings[31] showed that about one-third had flat roofs, one-quarter north-light and there were also a considerable proportion of equal-pitch, monitor and barrel-vault designs. Most framed roofs were in steel, and flat roofs in reinforced concrete. The material most widely used as a roof covering to factory buildings was asbestos-cement sheeting, although felt and 'protected metal' were both used extensively. Asphalt was the most common covering to flat roofs. On average about one-fifth of the roof area was glazed, although few flat roofs had any glazing.

Smith and Bott[32] believe that the loading of the roofs of single-storey industrial buildings is often less than 1.2 kN/m^2, but that services suspended from the roof can often amount to an additional 0.25 kN/m^2. This twenty per cent increase in basic design load, unless suitably accommodated, can lead to overstressing of a slender roof structure.

Roof Finishes

The range of roof finishes available is immense; among materials available are clay, concrete, metals, timber, synthetics or a combination of various types. The main factors to be considered in the selection of finishes are:

(1) the provision of an impenetrable skin which does not change its characteristics on exposure;
(2) reasonable capital cost in relation to function;
(3) low cost and ease of maintenance and repair;
(4) speed and ease of application;
(5) long life expectancy;
(6) ready availability; and
(7) suitable visual qualities, such as colour, texture, scale and applicability to the required roof form.[33]

Concrete tiles account for ninety per cent of pitched roof coverings, presumably stemming from economy and suitability both technically and visually for low rise housing of medium to large frontage. For flat roofs there is no such clear-cut market situation, and the factors influencing choice are frequently more orientated towards the specialist skills of laying.

It is worth noting that the measured temperature range on a flat black insulated roof over a twelve-month period in this country was from 80°C to −25°C, and temperatures as high as 130°C have been recorded on south facing slopes. Surface reflection can make a worthwhile contribution to heat dispersion where temperatures of this order are involved. It seems evident that felts should be used on firm substrates, while single-layer plastic finishes are suitable for use in situations where building movement is likely. All flat roofs must have an effective vapour barrier and the insulation must be of sufficient thickness to prevent condensation occurring within the roof construction itself.

The Building Research Station[34] has drawn attention to the wide range of roof coverings now available, ranging from the more traditional type materials like bitumen felt and asphalt to less familiar coverings such as neoprene bonded asbestos sheet with PVF film, butyl rubber sheet with surface laminate of neoprene, and polyisobutylene sheets with or without glassfibre reinforcement. British experience of these coverings is limited although they would appear to be inherently durable materials. The choice of new materials does make for difficulties in costs-in-use calculations, because of the problem of assessing probable future costs. Corrugated transparent and translucent PVC sheeting has been used quite extensively as roof-lighting and cladding of small structures, but its effective life may be no more than ten years, whereas corrugated opaque sheets should remain mechanically sound for at least twenty years.[35]

Roofing Costs

In 1960 the RICS Lancashire and Cheshire Branch Cost Research Panel undertook an investigation of the cost of alternative types of roofs for local authority flats.[36] The method of approach adopted by the Panel is worth considering and an outline of it follows.

(1) The general outline and scope of the study was discussed.
(2) A branch member prepared details and a measured schedule for a basic design; this was discussed and general principles settled.
(3) Variations from the basic design were considered.
(4) Other branch members prepared details and measured schedules for the variations; these were discussed and co-ordinated.
(5) Two branch members prepared a schedule of rates based on an agreed priced list of materials.
(6) Schedules of measurement were priced.
(7) Cost relationships were examined.
(8) A report was prepared.

Parapet walls and perforations in roofs for dormers, stacks, etc. were excluded from the study. Measurements were taken to the outside face of external walls, although it was appreciated that wall thicknesses were a potential variable. The basic design for pitched roofs incorporated gable ends, and hipped ends were treated as variations, although it was subsequently found that hipped ends produced an overall saving stemming from the elimination of the cost of brickwork in gables. The rates for pricing measured schedules were built up in a scientific manner in preference to extracting prices from bills which varied widely, or from price books which were not relevant to the district under consideration.

Factory building studies[3] showed that north-light roofs of rolled-steel riveted lattice girder construction were about ten per cent more expensive in first cost than umbrella or equal-pitch roofs irrespective of span or storey height, while monitor roofs were about twenty-five per cent more expensive than umbrella roofs.

A cost study of low flats[2] found that flat roofs were consistently higher than comparable pitched roofs, the extra cost amounting on average to about thirty per cent. With pitched roofs, irregular shaped blocks resulted in considerably increased costs. Lee,[9] in his cost studies of municipal housing, found that roofing costs of traditional brick two-storey houses ranged from fourteen to eighteen per cent of total costs (the minimum being for a low pitched roof with low quality covering, and the maximum for a high pitched roof with high quality covering). With brick flats and maisonettes the proportion of roofing costs to total costs ranged from six to ten per cent (the minimum was for pitched roofs on high blocks, and the maximum for flat roofs on low blocks). With taller concrete framed flats and maisonettes the proportion dropped to about five per cent for blocks up to six storeys high and as low as two per cent for blocks of over ten storeys.

As a guide to comparative first costs of various coverings to pitched roofs, table 3.4 gives 1971 approximate average costs per square metre of roof covering to a gabled pair of houses, excluding perforations, hips and valleys, but including felt, ridge, eaves and verge.

TABLE 3.4

COMPARATIVE COSTS OF ALTERNATIVE COVERINGS TO PITCHED ROOFS

Type of covering	*Approximate average cost per square metre* (1971)
Concrete pantiles	£1.40
Concrete plain tiles	£2.00
Clay pantiles	£1.90
Machine-made clay plain tiles	£2.50
Hand-made clay plain tiles	£3.00
Best medium Welsh slates (500 × 250 mm)	£4.30

A Lancashire and Cheshire roof study[36] showed similar cost relationships when related to the enclosed area, and included the structural costs. Adopting an index of 100 for concrete interlocking tiles, plain concrete tiles were 120, machinemade plain clay tiles, 150 and handmade plain clay tiles, 185. By way of comparison a 17° pitched roof with copper coverings had an index of 300 and a flat reinforced concrete roof finished with a screed, insulation board and three layers of bituminous felt had an index of 165.

When comparing the cost of alternative insulating materials both the cost per square metre and the U-value must be taken into account. The unit cost multiplied by the U-value will give an index for comparison purposes. Adopting this approach, expanded polystyrene generally compares very favourably with insulating plasterboard, fibreboard or insulating screeds.

FLOORING

Craig,[30] in his study of multistorey flat design, emphasised the sound and fire resistance requirements of suspended floors to flats. In general, a concrete floor with a minimum thickness of 125 mm will meet these requirements. A Building Research Station study[15] showed how plate floors were often more economical than concrete slabs and beams for multistorey flats, where spans were not excessive. There is in fact a wide variety of constructional forms available for use in suspended floors incorporating *in situ* or precast reinforced concrete, with or without beams – pot slabs, prestressed construction, steel beams etc. It seems evident that there is a need for a detailed investigation of the comparative costs of different forms of construction to meet varying spans and floor loadings.

With a regular plan arrangement precast concrete units often show distinct advantages. Erection is speedy, costly shuttering is eliminated and the units provide an immediate working area. *In situ* slabs also have their own particular advantages which can make them the best solution in certain situations; they are adaptable to variations in plan shape and section thickness and give added rigidity to the structural frame through lateral support. It is also likely that some types of suspended floor are most economic at certain spans. One investigation at Nottingham indicated that precast beam and block floors showed advantages at small spans (800 to 2400 mm), hollow beams at medium spans (4800 to 5400 mm) and prestressed tee beams at large spans (8400 to 9000 mm).

An investigation of municipal housing costs by Lee[9] found that in traditional brick two-storey houses, floors, stairs and finishes accounted for eight to eleven per cent of total cost. With flats and maisonettes, floor finishes on average accounted for about six per cent of total costs. Craig[30] found that there was little difference in price between timber floorboards on battens and quilt or thermoplastic tiles on screed and quilt to suspended floors of multistorey flats. The timber boards did, however, give slightly better sound insulation and allowed electrical runs to be inserted at a relatively late stage of construction.

Floor finishes also vary considerably in unit costs and the thickness of the flooring can influence structural costs as a thick finish, like wood blocks, will produce a taller building than a thinner floor covering such as thermoplastic tiles. Cleaning and maintenance costs of floor finishes are other important considerations which should be taken into account in any cost appraisal.

Craig,[30] in his investigation of multistorey flat design found that the majority of stairs were of *in situ* concrete construction with a granolithic finish. With the steady improvement in lift reliability, staircases in high blocks are needed mainly for escape in case of fire and do not require a high standard of finish, and are in many cases isolated from landings by selfclosing doors. Furthermore, if storey heights could be standardised at two, or at most three, alternatives, it should be possible to produce precast flights for use in many types of block.

DOORS AND WINDOWS

Doors

The cost relationships of the principal types of external door are shown in table 3.5.

TABLE 3.5

COST RELATIONSHIPS OF EXTERNAL DOORS

Type of door (all 726 × 2040 mm)	*Cost index*
19 mm ledged and braced door	100
50 mm framed, ledged and braced door, with 19 mm boarding	140
50 mm standard flush door with skeleton core and plywood faces	130
50 mm standard flush door with glazed panel	150
50 mm standard panelled door with glazed panel	110

Windows

There is a wide range of choices available from timber to steel and aluminium, both single and double-glazed. Even with the same class and size of window, there can be wide variations in price according to the particular design of the window and the number of opening lights. For instance, with 1200 × 1100 mm steel windows the introduction of two opening lights over and above a single fixed light can increase the total cost of the window by twenty-five per cent, and weather-stripping can add a further ten per cent. The total cost of the window in this case includes the fixing of the window and its glazing and painting. The addition of a 75 × 75 mm painted softwood surround would increase total cost by about twenty per cent.

Rowland[37] has described how that until recently the use of stainless steel windows was confined almost entirely to banks, insurance offices and large department stores on account of their high initial cost. In 1967 the technique of adhesively bonding very thin stainless steel strip to aluminium alloy extrusions was introduced, and this resulted in rigid economical sections which combined the corrosion resistance of stainless steel with the flexibility of the extrusion process. At this juncture the stainless steel window became more competitive economically with those in other materials, particularly when savings on painting and other subsequent maintenance costs were taken into account. The latter aspect, which will be considered in more detail in chapter 11, prompted the use of 6000 stainless steel windows in a local authority housing project at Edmonton in North London.

In similar manner the higher initial costs of double-glazing have to be offset against savings in heating costs and the benefit of increased comfort, and this too will be further examined in chapter 11. The Building Research Station[38] has shown that a double window, openable but weather-stripped and incorporating a 150 to 200 mm airspace, will have double the sound reduction qualities of an ordinary single openable window which is closed.

Craig,[30] in his investigation of multistorey flats, found that about sixty per cent of the blocks had steel windows and most of the remainder used wood casements. The particular problems associated with windows at considerable heights are draught prevention, need for foolproof catches to prevent them swinging in high winds and facilities for cleaning, reglazing and general maintenance. At least one authority had developed a centre-hung window which could be reversed for cleaning. This represented a useful advance, but in general at the time of Craig's study these problems had not been satisfactorily solved. In addition there was a need to incorporate suitable extraction devices in kitchen windows to combat condensation.

A ten per cent reduction is considered possible if the variety of standard steel windows is reduced to one hundred models.[55]

FINISHINGS

The range of choices available for wall and ceiling finishings is probably greater than for any other component of a building, and the choice is influenced considerably by the class and use of the building. Lee,[9] when investigating municipal housing costs, found that finishings to walls and ceilings accounted for six to fifteen per cent of total building costs in varying situations. A RICS Cost Research Panel study[2] found very large regional differences in the cost of internal decorations; for example, the application of emulsion paint to plastered surfaces was on average forty-five per cent more expensive in Inner London than in the Provinces.

The use of plasterboard and similar sheet materials to form a dry lining to dwellings has become increasingly popular as it simplifies the work of the finishing trades and leads to earlier completion. If full advantage is to be gained, the implications of their use must be fully considered at the design stage and when planning the work on site. The Building Research Station[39] has indicated methods of fixing plasterboard whereby the walls of houses can be lined at costs generally comparable with that of two-coat plasterwork.

More recently a wide range of plastic-surfaced sheet materials have been introduced for use in a wide variety of situations from wall linings, cubicles and partitions to fitted furniture and sanitary fixtures. Decorative laminates offer a number of advantages: uniform coverage of large areas; attractive appearance; good durability; good resistance to wear, biological and chemical attack, heat and moisture; and low maintenance costs. They are particularly well suited for use in the communal parts of buildings, where an attractive, hardwearing and low maintenance cost surfacing is desirable.

The acoustic properties of finishings are becoming increasingly important and acoustic materials are manufactured in three main categories: porous materials for general sound absorption but with special reference to high frequencies; resonant panels for absorption at low frequencies; and cavity resonators which can be designed to provide maximum absorption at a particular frequency. The first category is the most commonly used and suitable porous absorbents include mineral wool, foamed plastics with interconnecting cells and proprietary tiles made of soft fibreboard or asbestos.

Proprietary acoustic tiles usually have a facing which is itself acoustically neutral-selected for ease of decoration and maintenance, but perforated to allow sound to pass through to the absorbent. From a design point of view there is a wide range of choice in panels, tiles, boards and strips. An investigation into costs of suspended ceilings in Nottingham in 1970 showed costs ranging from £2.50/sq m for 12.7 mm fibreboard and supporting framework to £5.50/sq m for aluminium strips enclosing 25 mm mineral wool, but the study also stressed the need to consider the ability to meet functional and performance requirements and maintenance and renewal costs when making comparisons of alternative techniques.

SERVICE INSTALLATIONS

Addleson[40] has described how buildings and their environmental services have become more complex and the range of choices continues to increase. Unfortunately, the wide choice, lack of experience of new techniques and the need for assistance from more specialist skills have tended to act as constraints. In particular, environmental requirements are often considered far too late in the design process for them to make a positive contribution to the final design. This is unfortunate when viewed against the high cost of service installations which may amount to as much as twenty-five per cent of total costs on a modern housing scheme and fifty per cent on a hospital project. There is a vital need for integrated design with all specialists contributing at each stage of the design process.

Addleson[90] has given an excellent example of how the means by which a multistoreyed office building is illuminated will have far-reaching implications on the structure generally, and how they must be considered together. The interior of a building can be lit in three basic ways by daylight alone for most of the working day; partly by daylight and partly by artificial light, more commonly known as permanent supplementary lighting of interiors (PSALI); and permanent artificial lighting (PAL).

In terms of building form illumination by daylight will mean an office block with a depth of about 14 m; if PSALI is adopted the depth may be increased to 22.5 m, while the use of PAL will enable the depth to be increased to at least 27 m.

Each alternative has different structural implications with varying forms of fenestration. Maximum window area is needed for daylight illumination, less window area with PSALI and vision strips only with PAL. The thermal consequences of each approach are also quite different as the daylight design will involve twenty per cent more exposed external wall area, than one using PAL for the same floor area. With the latter design, mechanical ventilation at the least and possibly air-conditioning will be essential. All the alternatives need costing very carefully to determine the most economical longterm solution.

Plumbing and Waste Disposal

It is important to group sanitary accommodation in the same plan positions on the various floors of a building and to pay special attention to the economics of the various drainage layouts that could be adopted. When designing pipework arrangements, it should be borne in mind that it is generally more economical to use a large pipe rather than a number of smaller pipes which together produce the same cross-sectional area.

Lillywhite and Wise[41] have described how a drainage system (including both vertical and horizontal pipework) should be designed to convey waste materials away quietly, avoiding blockages, and with a limitation on the air pressure fluctuations within pipework to ensure that an adequate water seal is retained at each appliance. By the use of single stack plumbing it is possible to reduce costs by thirty per cent over systems introducing vent piping. For example, field studies showed that all supplementary venting could be omitted in an office building eight storeys high with ranges of five WCs and basins and a drainage stack of 100 mm diameter. In large buildings it may be an advantage to increase the diameter of the bend and drain to 150 mm.

A shower installation is likely to be about thirty per cent more expensive than a bath installation. Yet advantages of showers, such as hygienic, refreshing action, smaller space requirement, reduced water needs, and safer and easier operation for elderly people, could well outweigh the cost disadvantage. A survey conducted at a teachers' training college showed that seventy-six per cent of the respondents favoured showers.

The designer must often be concerned with the economics of different materials for rainwater goods, although admittedly he must also be very much concerned with their appearance and other characteristics. Unplasticised PVC in black, grey and white, with durable supporting brackets of the same material, is commonly used for rainwater goods. There is a slight loss of resistance to impact, for example by ladders, as weathering proceeds but it is not serious. PVC rainwater goods are likely to last the life of the building.[35] A straight comparison of the first cost of supplying and fixing PVC rainwater goods with asbestos cement fittings shows increases of about thirty per cent for gutters and ten per cent for downpipes. On the other hand, PVC rainwater goods would show a considerable cost advantage if the asbestos cement goods were painted, or if colourglazed asbestos cement goods were used to improve the rather unsatisfactory appearance of natural asbestos cement.

A good example of aesthetic considerations outweighing economic advantages is illustrated by Thau[42] when he describes how more than 100 000 solar water heaters have been erected on the roofs of buildings in Israel to supply cheap hot water, but the disadvantage is that they disfigure the landscape, townscape and skyline of the country.

Heading

The need for central heating in residential as well as other types of building was highlighted in the Parker Morris report,[43] which proposed that the minimum standard for an installation should be 12.8°C in circulation areas and 18.3°C in living areas when the outside temperature is −1°C. Thomas[44] raises strong doubts as to the adequacy of these standards at the present time. A review of domestic heating in 1968[45] indicated that of central heating systems thirty-seven per cent were gas, thirty per cent electricity, twenty-six per cent solid fuel and six per cent oil. The majority of the gas-fired systems incorporated independent boilers while most of the electric installations used block storage heaters. By way of comparison, a study of factory buildings[4] found that solid fuel was the main fuel used for heating nearly half the buildings surveyed, but in a third of these it was supplemented by gas or electricity. Oil was the main fuel in a further third of the factories and in a quarter of these was supplemented by gas or electricity.

In cost planning heating services initial capital cost is rarely a sufficient criterion, as often the cheaper the heating system is to install the more it costs to operate. Furthermore, heating systems need to be compared on a comparable basis and Azzaro[46] has suggested the number of kJ/m^2 of gross floor area.

Installation costs of different heating systems to serve a house of about 90 sq m show a wide spread and table 3.6 shows the average 1971 prices.

TABLE 3.6

CAPITAL COSTS OF DOMESTIC HEATING SYSTEMS (1971 prices)

Hot water radiators and solid fuel boiler	£ 730
Hot water radiators with semi-automatic gasfired boiler	£1070
Ducted warm air with gas circulator and ducts	£ 700
Ducted warm air with oil-fired heating unit	£ 920
Electric under-floor heating (all floors)	£ 480

Electricity is generally an expensive method of space heating, although the use of off-peak electricity at about half the normal rate has improved its position relative to other fuels. Electric central heating is versatile, produces no fumes, can be automatic in operation and is 100 per cent efficient in that no heat escapes up flue. Off-peak electric floor heating costs about £2.20/sq m to install, whereas storage heaters cost about £30 to £40 each to install.

District heating permits the provision of heating services on a townwide scale and the Ministry of Power has stated that where district heating can be shown to meet the best socio-economic criteria, it should be adopted. Large-scale urban renewal, the building of New Towns and large expansions to existing towns provide ideal situations for heating services to be planned as an entity and so achieve economies of scale. The larger the scheme, the more significant will be the reduction in unit cost of the heat supplied. Not only can the heat be supplied to large numbers of dwellings and public, commercial and industrial buildings but it can also be used for ancillary purposes such as the heating of roads, shopping arcades and swimming pools. The burning of town refuse can provide a good source of heat to supplement the district heat supply as at Nottingham, whilst natural gas is to be used at Peterborough, and power station exhaust steam was used for heating residential and commercial buildings at Pimlico.

Pitman[46] has aptly described how illogical it is to improve heating standards without at the same time paying due regard to thermal insulation. If the correct standard of thermal insulation is built in at the design stage, there may well be an overall saving stemming from the cheaper heating installation; and even if the insulation is added later and at higher cost, the net outlay for a typical three-bedroom house will probably not exceed £100.

The central heating system needed for an uninsulated house required approximately 20 sq m of radiator surface and cost about £470 in 1970 (gas-fired installation), based on an inside temperature of 21°C, and the running costs were estimated at £90 for a heating season of 2250 hours. If the house was fully insulated after construction, as for instance by applying 50 mm of mineral wool to the first floor ceiling, filling the wall cavities with mineral wool, draughtproofing external doors with phosphor bronze weatherstrips and double-glazing all windows, then the required radiator surface is reduced to 11.5 sq m and the cost of the heating system is reduced to about £340, while the full cost of the insulation, as applied to an existing house, would be about £500. In addition to the saving on the first cost of the heating system of £130, there would also be a reduction in annual heating costs approaching £40.

Some parts of the insulation work showed a better return than others and if reductions in fuel costs were the sole criteria, then insulating cavity walls would have first priority, followed by roof insulation, then draughtproofing and finally double-glazing. It might be deemed advisable to confine the latter work to much used areas such as living rooms.

Air-Conditioning

A wall radiator and an open window are no longer adequate means of heating and ventilating larger buildings. Various means of heat distribution through ceilings are available from water-heated pipes to air circulation systems and use of lighting installations. For instance, with the development of higher levels of illumination and the better thermal insulation of walls, it is possible that most, if not all, the heating needs of a building can be satisfied by the lighting installations. Furthermore, an air-conditioned building generally gives greater freedom of planning, particularly in the width of the building and in the positioning of toilets, and thus partially offsetting the extra cost of air-conditioning.

It is claimed however that a fully air-conditioned office for 1000 employees with landscaped interior can be built for less than a multiroomed office block of similar capacity with central heating only.[47] An example is instanced where an open plan eight-storey office block with a plan shape tending more towards square than rectangular (to reduce perimeter heat gain and loss) and with only twenty per cent glazing compared with the more traditional fifty-five per cent, and almost identical lighting, could be erected for under £600 000 as against the more orthodox-type building at around £750 000. The cost of air-conditioning would probably amount to about £120 000 as against the cost of a normal central heating system at around £50 000, still showing an overall saving in cost coupled with improved working conditions.

A study by the Department of the Environment[53] of the costs of local authority office blocks with varying plan forms and heating systems showed the following order of total building costs: cellular plan, centrally heated: 100; cellular plan, mechanically ventilated: 116; open plan, air-conditioned: 131; cellular plan, air-conditioned: 139.

Lifts

Knight and Duck[48] carried out a comprehensive investigation of the costs of lifts in local authority multistorey flats showing how both the height of the block and the floor layout have a major influence on the costs of lifts per dwelling. A passenger lift is usually provided in blocks of flats exceeding three storeys and in blocks of maisonettes exceeding four storeys, while in blocks over six storeys in height a second lift is also provided. Considering the economics of lift provision alone would dictate layouts incorporating at least six dwellings on each floor of a block. In fact, the tender price for a lift installation varies with the height of the block, the number of floors served, the speed of the lift and the form of control.

These investigations showed wide variations in costs, and updating them to 1971 figures, gives £4000 for a single lift with a speed of 30 m/minute and simple automatic controls in a four-storey block, rising to £35 000 for a pair of 90 m/minute interconnected lifts with full up and down collective control in a twenty-four storey block. About one-half of the increased cost is attributable to the extra height and number of storeys served and the remainder equally to the cost of the increased speed and the more complex system of control. In blocks served by two lifts, an economy can be achieved by arranging for each lift to stop at alternate floors above the ground floor, so that each floor is served by one lift only.

Faster lifts require the use of either two-speed or variable voltage motors to provide smoother acceleration and deceleration, thereby substantially increasing capital, maintenance and running costs. A 60 m/minute lift costs about £1600 more than a 30 m/minute lift. Down collective control, whereby a downward travelling car responds to landing calls in floor sequence, entails increased costs of about £40 to £50 per lift per floor served. Full collective control would involve another addition in cost of about the same amount. Increasing the car capacity from eight to ten persons would probably increase costs by about £250 per lift. For a pair of lifts, the pit and motor room would cost about £1100, while the lift well could amount to approximately £200 to £250 per storey.

Knight and Duck[48] found that mechanical installation accounts for fifty to sixty per cent of the total cost, the building work for ten to fifteen per cent, and the running and maintenance costs for thirty to thirty-five per cent.

Refuse Disposal

The following range of refuse disposal systems is currently available.

(1) Individual storage containers usually in the form of metal, rubber or plastics dustbins, with loose or captive lids, or alternatively plastic or paper sacks, each with a capacity of up to 0.9 cu m. This is the cheapest method with total annual equivalent costs of provision and maintenance of containers and refuse collection and disposal at about £4.80 (1971 prices).

(2) Communal storage containers without chutes, with an individual capacity of up to 1 cu m, and emptied mechanically by special collecting vehicles. This offers some improvement over the first method but is still not entirely satisfactory, as the containers need to be housed in close proximity to the dwellings they serve. Total annual equivalent costs are likely to be about £5.30.

(3) Communal storage containers with chutes within 30 m of each dwelling which enable occupiers to drop refuse from upper levels into movable containers. This is a satisfactory method for use in blocks of flats and maisonettes and the comparable annual equivalent costs are likely to be about £7.10.

(4) Waterborne disposal system whereby part of the refuse is passed through a special sink unit into the sewerage system. Apart from the cost of renewal of fittings, the total annual equivalent cost would probably be in excess of £17 making this the most expensive arrangement.

(5) Other and more recently introduced systems include sack compression, container compression, onsite incineration and pneumatic conveying.[49]

EXTERNAL WORKS

It is interesting to break down the cost of siteworks on a housing contract to see the relative values of the various works, although it is appreciated that their distribution will vary from site to site. The average 1971 sitework costs for a five-person house in the west Midlands were as follows:

Site preparation	£90
Retaining walls	15
Screen walls and fencing	130
Paved areas	55
Drainage	150
External services	45
Landscaping	5
	£490

External works do therefore form a significant part of total building costs and justify taking steps to reduce costs by reducing the amount of earthwork and retaining walls, restricting paved areas to a minimum, reducing pipe runs of drains and other services and seeking materials and components which perform their functions satisfactorily and at the same time show favourable costs-in-use figures.

On drainage work, cast-iron drains are twice as costly as glazed vitrified drain pipes, whilst pitch-fibre will often show a saving on glazed vitrified clay. Savings in cost can often be obtained by building shallower manholes, not exceeding 1 m deep, in halfbrick walls in class B engineering bricks, and by omitting the spread of concrete bases beyond the manhole walls and thereby reducing the amount of concrete, excavation and fill. On occasions it may be possible to reduce drainage costs by laying both foul and surface water drains in the same trench but with 300 mm between their inverts, and by using combined manholes. Another alternative is to lay the sewers alongside houses and so dispense with long lengths of house drain. The use of drop manholes can effect considerable reductions in drain trench excavation costs on a sloping site.

There are wide variations in the cost of different forms of paving. Probably the cheapest in first cost is gravel paving (about 50 mm thick) on a bed of hardcore (usually about 100 mm in thickness). Tarmacadam, 50 mm thick, on a 100 mm bed of hardcore would be about sixty per cent more expensive, whilst 50 mm precast concrete paving slabs on a 100 mm bed of ashes would be approximately 120 per cent more expensive than the gravel paving. A concrete carriageway is likely to be about twenty-five per cent more expensive than a tarmacadam one. Once again it is important to consider maintenance costs in addition to first costs in order to make a really meaningful comparison, and this aspect will be further investigated in chapter 11.

Similarly with fencing there are wide cost ranges. One of the cheapest forms of fencing is post and wire, with cleft chestnut paling at nearly double the cost of post and wire. Chain-link fencing in its turn is about twice as expensive as chestnut paling in first cost and close-boarded fencing is almost twice as expensive as chain-link. Apart from maintenance costs, other aspects need to be considered including the purpose which the fencing is to serve and its appearance, and these may have more influence than cost on the choice of fencing.

REFERENCES

1. ROYAL INSTITUTION OF CHARTERED SURVEYORS: COST RESEARCH PANEL. The cost and design of high flats – a case study. *The Chartered Surveyor* (May 1961)
2. ROYAL INSTITUTION OF CHARTERED SURVEYORS: COST RESEARCH PANEL. The cost and design of low flats – a case study. *The Chartered Surveyor* (June 1960)
3. BUILDING RESEARCH STATION. *Building Research Station Digest 28 (Second Series)*. Factory building studies (1962)
4. BUILDING RESEARCH STATION. Factory building studies no. 12: The economics of factory buildings. HMSO (1962)
5. DEPARTMENT OF EDUCATION AND SCIENCE. Building Bulletin No. 4: Cost study. HMSO (1972)
6. BUILDING RESEARCH STATION. *Building Research Station Digest 67 (Second Series)*. Soils and foundations: 3 (1966)
7. BRITISH STANDARDS INSTITUTION. British Standard 648: Schedule of the weight of building materials (1969)
8. R. A. SMITH and C. S. BOTT. The structural engineer. *Architecture East Midlands*, **34** (1971)
9. M. LEE. Comparative cost evaluation – municipal housing. *The Chartered Surveyor* (July 1961)
10. ROYAL INSTITUTION OF CHARTERED SURVEYORS, DESIGN/COST RESEARCH WORKING PARTY. Cost research paper: basement construction. *The Chartered Surveyor* (March 1965)
11. CLAY PRODUCTS TECHNICAL BUREAU. Technical note 1.8: Multistorey housing over shops in load-bearing brickwork (1966)
12. BUILDING RESEARCH STATION. *Building Research Station Digest 120 (First Series)*. Questions and answers: cross-wall construction costs (1959)
13. P. DUNICAN. Structural steelwork and reinforced concrete for framed buildings: some notes on comparative economics. *The Chartered Surveyor* (August 1960)
14. L. R. CREASY. Economics of framed structures. *Paper 6323, Proceedings of Institution of Civil Engineers* (March 1959)
15. BUILDING RESEARCH STATION. *Building Research Station Digest 48 (Second Series)*. Multistorey flats – design, building methods and costs (1964)
16. C. E. MANSFIELD. Why precast concrete? A contractor's view. *Building Technology and Management*, **8.1** (1970)
17. A. SHORT and J. R. MILES. Large panel structures: notes on draft addendum 1 to CP116 (1965). *Building Research Station Current Paper*, **30/69** (1969)
18. NEW BUILDING. Building in steel. *New Building* (August 1969)
19. H. L. MALHOTRA. Recent developments in structural fire protection for buildings. *Building Technology and Management*, **8.10** (1970)
20. E. LEVIN. Timber frame housing forges trends for the '70s. *The Illustrated Carpenter and Builder* (19 July 1968)
21. BUILDING RESEARCH STATION. *Building Research Station Digest 46 (Second Series)*. Design and appearance – 2 (1964)
22. P. H. HILL and S. LAZARUS. Laminated plasterboard partitions: research into cost effectiveness in domestic structures. *Building* (12 February 1971)
23. S. A. NEWMAN. Architect's brief. *Building Design* (26 March 1971)
24. E. KILFORD. Surveyor's report. *Building Design* (26 March 1971)
25. BUILDING RESEARCH STATION. *Building Research Station Digest 108 (Second Series)*. Standardised U-values (1969)
26. F. J. LANGDON. Modern offices: a user study – National Building Study 41. HMSO (1966)
27. I. D. GRIFFITHS and F. J. LANGDON. Subjective response to road traffic noise. *Journal of Sound and Vibration*, **8.1** (1968)
28. CMND. 2056. Final report on noise. HMSO (1963)

29. A. F. E. WISE. *Building Research Station Current Paper*, **43/69.** Buildings in noisy areas – interaction of acoustic and thermal design (1969)
30. C. N. CRAIG. Factors affecting economy in multistorey flat design. *Journal of Royal Institute of British Architects* (April 1956)
31. P. A. STONE. The economics of factory buildings – Factory building studies 12. HMSO (1962)
32. R. A. SMITH and C. S. BOTT. The structural engineer. *Architecture East Midlands*, **34** (1971)
33. J. MUSGROVE. Roof finishes. *Building Technology and Management*, **8.7** (1970)
34. BUILDING RESEARCH STATION. *Building Research Station Digest 51* (*Second Series*). Developments in roofing (1964)
35. BUILDING RESEARCH STATION. *Building Research Station Digest 69* (*Second Series*). Applications and durability of plastics (1966)
36. LANCASHIRE, CHESHIRE AND ISLE OF MAN BRANCH OF RICS: COST RESEARCH PANEL. Cost comparison of alternative roofs suitable for local authority type flats. *The Chartered Surveyor* (March 1960)
37. K. T. ROWLAND. Stainless steel windows. *Building with Steel* (February 1970)
38. BUILDING RESEARCH STATION. *Building Research Station Digest 128*. Insulation against external noise – 1 (1971)
39. BUILDING RESEARCH STATION. *Building Research Station Digest 9* (*Second Series*). Dry-lined interiors to dwellings (1961)
40. L. ADDLESON. Challenge of complexity, growth and change. *Building Design* (27 February 1970)
41. M. S. T. LILLYWHITE and A. F. E. WISE. Towards a general method for the design of drainage systems in large buildings. *Building Research Station, Current Paper*, **27/69** (1969)
42. I. A. THAU. Solar water heating: useful but unsightly. *Build International*, **2.10** (1969)
43. MINISTRY OF HOUSING AND LOCAL GOVERNMENT. Home for today and tomorrow. HMSO (1961)
44. I. O. THOMAS. The builder and domestic heating standards. *Building Technology and Management*, **8.5** (1970)
45. THE ILLUSTRATED CARPENTER AND BUILDER. Domestic heating. *The Illustrated Carpenter and Builder*, **13** (September 1968)
46. R. G. PITMAN. Economics of insulation. *Official Architect and Planner* (November 1969)
47. BUILDING DESIGN. Cold comfort. *Building Design* (12 March 1971)
48. T. L. KNIGHT and A. E. DUCK. The costs of lifts in multistorey flats for local authorities. *The Chartered Surveyor* (January 1962)
49. BUILDING RESEARCH STATION. *Building Research Station Digest 116*. Domestic refuse (1970)
50. G. LEON and R. L. WAJDA. Economic principles of multistorey industrialised buildings. *The Quantity Surveyor* (September/October and November/December 1967)
51. W. H. GEORGE. Cost planning in formwork design. *The Building Economist* (*Australia*), **9.3** (1970)
52. J. WHITE. Transportation and erection of precast concrete units. *Precast Concrete* (September 1970)
53. DEPARTMENT OF THE ENVIRONMENT. Local authority offices: areas and costs (1971)
54. G. F. A. ORCHARD and P. H. HILL. Concrete trench fill for house foundations. *Building* (21 April 1972)
55. D. W. PEDDAR-SMITH. Cost benefits from variety reduction in standard steel windows. *Building Research Station, Current Paper*, 39/71 (December 1971)

4 INFLUENCE OF SITE AND MARKET CONDITIONS AND ECONOMICS OF PREFABRICATION AND INDUSTRIALISATION

THIS CHAPTER EXPLORES the effect of site and market conditions on building costs and the way in which they account for variations in the price of similar type buildings erected in different locations. The origins and forms of prefabrication and industrialised building are examined and the economics of these processes critically investigated.

EFFECT OF SITE CONDITIONS ON BUILDING COSTS

Each site has its own peculiar characteristics which can have a considerable influence on the total cost of development. Some of the more important site factors are now examined.

Location of Site

The cost of building on a site in London could be as much as twenty per cent more expensive than erecting a similar building on a provincial site, due to higher wages, materials and other costs. Some parts of the country are subject to higher rainfall than others and this can lead to greater loss of working time. Even within the same region the costs of operating on different sites can vary tremendously. For instance, a project on a remote country site may involve long lengths of temporary access road and of temporary power cable for electricity supplies and increased costs of transporting operatives and materials to the site. By way of comparison a site in a congested central area of a city will give rise to major problems in delivery and storage of materials and components, protection of adjoining buildings and the public, and restrictions on the use of mechanical plant. Overcoming these problems involves considerable additional costs. Furthermore, a very exposed site may make working conditions more difficult and costly and some locations may be more vulnerable to vandalism and theft and so require more costly protective measures.

Demolition and Site Clearance

One site may be clear of all obstructions, whilst another may contain substantial buildings requiring demolition, extensive paved areas which need breaking up and a number of large trees which require felling, together with the grubbing up and disposal of their roots.

Contours

Few sites are entirely level and the more steeply sloping the site, the greater will be the cost of foundations and earthworks generally. The stepping of strip foundations increases their costs. Most buildings require constant floor levels and this will involve considerable excavation and fill on a sloping site. It is cheaper to form a sloping bank than to construct a retaining wall, but this may have to be balanced against space considerations. A basement boilerhouse ought ideally to be located in an open area to reduce excavation and tanking costs.

Ground Conditions

Where the strata is of low load-bearing capacity, it may be necessary to introduce piled or other more expensive types of foundations. Raft foundations on made-up ground or in areas liable to mining subsidence may be three times as expensive as normal strip foundations, whereas piled foundations used to convey loads to a deeper load-bearing strata could be as much as five times as expensive. In these circumstances it might be advisable to consider increasing the number of storeys in the building to take fuller advantage of the higher load-bearing capacity of the piles. The cost of excavation in rock could be five to eight times as expensive as working in normal ground. Indeed bad ground conditions could conceivably increase overall building costs by as much as five per cent. The probable length of haul in the disposal of surplus soil also needs consideration.

Where ground water level is close to the surface of the site, costly pumping operations may be needed throughout the substructural work. A wet site may also involve raising temporary sheds and offices on brick bases and more costly temporary roads. The Institute of Building Estimators[1] has drawn attention to the need to examine trial holes and to note the depth of vegetable soil, nature of subsoil, and evidence of water table or standing surface water. The type of strata will also influence the form and extent of timbering or other means of support that will be needed to the sides of excavations.

Planning and Building Regulations

The ratio of site value to cost of building should always be considered. In expensive central locations the aim must generally be to secure the most profitable permitted use, and coupled with this is the desirability of obtaining maximum site utilisation. The operation of planning controls through plot ratios and floor space indices are very relevant and will be further considered in chapter 12. The shape and size of the building, as well as probable cost, are affected by height restrictions, building and road improvement lines, parking requirements, light and air restrictions, landscaping conditions, access requirements, etc. In like manner the probable impact of Building Regulations or the London Building Acts must always be borne in mind when working from sketch or preliminary drawings, as they can have far-reaching consequences on building costs.

Services

The position and capacity of existing services such as main drainage, water supply, electricity, gas and telephones are other important influences on site costs. Connection to a public sewer 6 m deep on the far side of a busy dual carriageway can be a costly item of work and the need to install pumping plant to drain a low-lying site has repercussions in both initial and future costs. The cost implications of combined, separate and partially separate systems of sewerage must be appreciated. On occasions it is necessary to divert existing services which cross sites, to improve and regrade watercourses which adjoin sites, and to obtain pipe easements to lay essential services across land in other ownerships.

Availability of Labour, Materials and Plant

The Institute of Building[2] has drawn attention to the need for builders to consider the labour situation in the area of the job and the availability of materials when preparing tenders. Investigations into the availability of materials would include local sand, gravel and ballast pits and brickworks. Contractors will have to make several important decisions at the tendering stage, each of which will influence costs.

(1) Whether all labour requirements can be met from within the organisation or whether it will be necessary to recruit for the project.

(2) Whether mechanical and nonmechanical plant already owned by the organisation is suitable and likely to be available or whether it will be necessary to purchase or hire for the project.

(3) Whether it will be desirable to sublet to specialists certain aspects of the job, such as excavation, formwork or scaffolding.

Other Factors influencing Cost

Miller[3] has described how a contractor can influence the cost of a job by his selection of constructional methods and by adjusting these methods to increase the effectiveness of the resources used. The free choice of method is however constrained by the design of the building, the availability of the numbers and types of resources needed for each method and by the relative cost of employing each one of these sets of resources. Another important cost aspect is the quantity of materials wasted on building sites, and Abbott[4] believes that the cost of wastage of materials is generally in excess of the allowances made by estimators and, in many cases, the total wastage of a material could have been halved by effective supervision.

USE OF PLANT

With the continual rise in labour costs, both direct and indirect, many contractors are making greater use of plant. This requires an intimate knowledge of all types of plant, when each can be used most profitably and of the fullest possible utilisation of plant. The latter point is significant as the use of machinery on building sites rarely exceeds seventy per cent of its working capacity.

A contractor has sometimes to choose between hiring or purchasing plant, and the decision will be influenced considerably by the likely future demand for the particular item of plant. Owning plant offers a number of advantages:

(1) plant is readily available at all times;
(2) plant may be retained on a particular site if circumstances make this desirable;
(3) plant can be transferred from one site to another without great difficulty; and
(4) in emergency situations machines can be taken off less important work as the contractor has complete control of the plant.

Nevertheless, plant hire also serves a valuable function in that it offers a wide variety of plant types to the contractor free from the liabilities attached to the purchase of plant, particularly when a contractor is short of capital. Hiring plant often ensures maximum economy with full plant utilisation and is an aid to quicker building.

The main factors to be considered when evaluating the economics of buying or hiring plant are:

(1) forecasts of commitments to assess plant requirements;
(2) availability of workshop facilities for servicing plant;
(3) length of time for which plant will be required; if it is only required for a short period of time with little prospect of use in the foreseeable future then it it is better to hire than purchase;
(4) adequacy of capital available for purchase;
(5) availability of personnel for controlling and operating plant holdings; and
(6) cost of transporting plant to sites.

Costs of owned plant are made up of a wide range of items – capital costs, interest on capital, depreciation, licences and insurance, overheads, maintenance, repairs and replacements, haulage to sites, fuel and operating costs. The total annual costs have to be spread over the period of effective use to give an inclusive hourly or daily cost.

There are three principal methods of charging for plant.

(1) Percentage of the contract price, whereby total plant costs are allocated to each contract in proportion to the overall cost of the job. Its main advantage is one of simplicity of calculation but its disadvantages include lack of incentive to obtain maximum use of plant, lack of comparative information for preparing tenders and costing, and lack of information to use as a check on plant operations as a whole.

(2) Direct cost to contract, where the size and nature of the job and expected use of the plant justifies charging an item of plant completely to the job. This method is also easy to apply but runs into difficulties if the plant is used on other contracts.

(3) Hire charge is the most usual method whereby the plant is charged to the site by the unit of time employed. The primary advantages are ease of control over economic use of plant, adequate information available for estimating purposes and for providing a check on plant operations.

In all cases an efficient costing system is necessary to ensure that realistic rates are charged for use of plant.

MARKET CONSIDERATIONS

The level of prices submitted for a particular job may reflect the market situation at that point in time. This is governed by the volume of work in progress in the area and the relative keenness of building prices. The situation can change quite dramatically over a comparatively short period of time.

In addition many unforeseeable and external factors can influence building costs, such as national and local shortages of labour and/or materials, a credit squeeze, abnormal rainfall resulting in a sharp rise in ground water levels, and sudden increases in the price of building materials or components or fuel. Other factors bearing upon costs include time for completion and special requirements of the building owner, such as phased completion of various sections of the work. Preliminary items may on occasions account for more than ten per cent of the tender sum and this may indicate that contractors are somewhat uncertain about the requirements of the job at the tendering stage. Tender prices can also be influenced by the way in which the contract documents are prepared and the amount and adequacy of the information which is supplied to the contractor at the tendering stage. One contractor described to the author how he put a price on some architects to meet the extra cost arising from their general lack of attention to detail and the consequent delays and disruption of the work. The method of tendering can also affect the price of a job and negotiated contracts are often more expensive than competitive tenders, but factors other than cost may influence a decision on the method to be adopted.

Many cost aspects may need considering in the feasibility studies of projects. A common problem is to assess the relative merits of converting and modernising existing buildings as against building new ones. Probable trends in prices can be an important factor. Cost limits of development projects may be set by the rents obtainable and an important aim may be to maximise the areas available for use by occupants. The periods needed for both precontract and construction stages may be of considerable importance to the building client, as for example where a new factory is urgently required to enable large export orders to be met on a tight time schedule.

COST IMPLICATIONS OF PREFABRICATION AND STANDARDISATION

Increased mechanisation of building work can speed up production and frequently results in reduced costs of construction. Power tools such as power floats and power saws and the shot-firing of components are suitable for use on small sites, whilst cranes and forklift trucks are better suited to larger sites. On all types of site the extensive use of prefabricated components eliminates much of the cutting on site and many of the time-consuming wet trades. Hence manufactured joinery has largely superseded site-produced joinery, plasterboard often replaces wet plaster, and there are few building projects which do not incorporate some precast concrete units as part of the structure. These units offer distinct advantages through economy in site work and independence from weather conditions, as well as the high quality and often superior finish of concrete units produced under factory conditions. Dense and lightweight, solid and hollow concrete load-bearing blocks, some with thermal insulating properties, are now being mass produced in fully automated plants with an output of 20 000 units per eight-hour shift.[5]

Advantages of Prefabrication

Most building processes can be better accomplished in the workshop under superior working conditions to those on the site. The ideal solution would be to produce the complete building with all its ancillary services in the workshop, independent of weather and time of year, under the best possible environmental and working conditions, and subsequently to transport the work complete to the site. Unfortunately physical and practical difficulties generally prevent this ideal from being achieved. Hence it is usually necessary to subdivide the work into construction components or elements which can be prefabricated in the workshop and then transported to the site.[6]

Maximum efficiency and economy can be achieved by mass production methods in factories aimed at producing large outputs of selected, standardised, dimensionally co-ordinated and interchangeable components suited to a range of building types.[7]

Standardisation

Leon[7] has postulated that optimum standardisation is essential for component manufacturers to obtain larger series runs to balance turnover against capital, minimise time taken in changing over machines, reduce production costs by the bulk purchase of materials, and attain better quality production of fewer varieties of component types with fast-operating cycles related to a large and sustained volume of demand. Standardisation of storey height permits the production of ranges of standardised units for staircases, refuse chutes, wall units and other components.

Redpath[8] has described how standardisation is a time-consuming and costly process. It is particularly expensive in the research and development effort required to achieve solutions worthy of standardising and then to obtain agreement to the results. Where the final product is a component, it can also be expensive in capital investment for the machinery requirements. The prime factor in standardisation is repetition. Yet the normal mode of working on building sites militates against repetition where each project involves establishing a new team of men to undertake a different job to the last, and the team is subsequently disbanded when it is achieving high productivity. Common standards of communication, components, construction and procedures in design offices, in factories and on sites could lead to the same high increases in productivity as have been secured in the manufacturing industries.

The Department of the Environment has found that one contractor was able to reduce the time required to build a house from over 2000 man-hours down to 1300, when he was engaged on a series of large contracts each of about 500 houses. Nevertheless, the savings resulting from standardisation are not always as extensive as might be expected. For instance a Building Research Station study of the effects of variety reduction in doormaking[9] found that maximum feasible price reduction was eleven per cent on the cheapest door resulting from drastic variety reduction, increased efficiency and a massive increase in the scale of production. Redpath[8] points out, however, that reductions in the variety of door set types (doors and frames) and painting doors on the site can yield significant price reductions. Variety reduction and the repetition which flows from it can save large sums of money in the design office, in the factory and on the site. Where it is applied intelligently, it does not restrict the designer unduly and leads to better value for money.

A study by the Building Research Station[10] stressed the need for a uniform approach in the production and coding of information, and highlighted the problems encountered in retrieving and processing information in the construction industry. It showed the need for a common language and code for ease of communication and improved efficiency.

Economies of Rationalisation

As a general rule the cost per unit of a manufactured product will fall as the length of the production run is extended. After a certain level of production the rate of reduction in cost falls. Stone[11] has described how in Finland the cost of doors fell by fourteen per cent when the production run increased from twenty to fifty but only by five per cent when from 200 to 500. Similarly, sanitary goods showed an eighteen per cent saving when the run was increased from ten to 100 but only six per cent from 100 to 400. There is a possibility that savings resulting from large extensions to production runs could be offset by increased costs resulting from larger stocks and longer delivery distances.

The logical sequence to standardisation of components is a method of production whereby all components are related to a unified set of dimensions, preferably operating on an international basis. A good approach is through modular co-ordination whereby all components are designed in terms of a common dimension, such as 100 mm, to permit all components to be interchangeable without the need for cutting or packing. This approach should produce savings in labour and materials, a reduction in ranges of sizes, smaller stocks, longer production runs and lower production costs. Changes of this type involve expensive retooling and problems arise from the lack of dimensional accuracy in traditional components and the tendency to be not too precise in setting out building projects.

Whittle[12] described the need to rationalise the dimensions of buildings and their parts and to secure precision in the manufacture of those parts, if the techniques of mass production were to be fully exploited in the building industry. He argued that the cutting away and making good, and the scribing and fiddling to make things fit, all add considerably to the cost of the building. A United Nations Organisation publication[13] stated 'that large scale production with its resulting economies depends on the reduction in the number of sizes and shapes of products and to achieve this the process of typification and standardisation are essential'.

INDUSTRIALISED BUILDING METHODS

Nature of Industrialised Building

The aim of industrialised building is to apply the best available methods and techniques to an integrated process of demand, research, design, manufacture and construction. Basically industrialised building systems should aim to combine aesthetic value and user satisfaction with economy of materials, production methods and erection techniques. Industrialised building has been defined as the application of power and machinery and quantity production to those building processes which can effectively be undertaken in this way. This latter definition fails however to give sufficient emphasis to the important integration aspect of the process. A basic concept of industrialisation must be the organisation of the whole construction process in an integrated way, so that materials, components, plant and labour are available at the appropriate times to secure continuity both in the factory and on the site.

Historical Background

Large panel systems of construction were introduced into America and Europe at the beginning of the present century. Approximately one per cent of the dwellings erected between the wars in Great Britain were of prefabricated construction but they failed to offer any real advantage over traditional dwellings.[11] In 1945 a Government Interdepartmental Committee approved 101 systems for use by local authorities in the construction of houses to assist in meeting the urgent postwar housing needs. They were based mainly on precast or *in situ* concrete walls, or steel or timber frames. Special grants were made available to help in meeting the cost of converting and tooling factories for the prefabrication of components. These systems proved very costly and the grants were withdrawn at the end of 1947. Typical housing costs in 1947 were: temporary bungalow – £1170; aluminium bungalow – £1600; and traditional house – £1250. During the period 1945 to 1955, twenty per cent of the houses built in England and Wales were of nontraditional construction but by 1961/62 this had dropped to six per cent.

In the early nineteen-sixties industrialised methods secured increased support, particularly for use in high rise buildings, when the Government accepted the objective of substantially increasing the output of new dwellings, and industrialised building seemed to offer the only method of achieving it, faced as they were with a shortage of skilled labour. Over a five-year period hundreds of proprietary systems were brought into existence, but few had any real chance of achieving the size or continuity of orders that were essential to make them viable. The use of industrialised systems was not confined to housing; a proportion of offices, factories, hospitals, stores and schools were constructed by industrialised methods.

Classification of Systems

Industrialised building systems have often been classified into two categories: closed systems and open systems.

Closed systems (contractor-based) are operated by sponsoring contractors and are often subject to the payment of royalties for production under licence covering a limited range of building types. A restricted number of structural elements, which are not interchangeable with other systems, are individually designed and manufactured by each separate sponsor of a closed system, and supplemented with standardised nonproprietary components obtainable in the open market from outside manufacturers.[14]

Another variant is the manufacturer-sponsored system in which manufacturers or distributors of building materials or components develop them to form the essential features of building systems. The items concerned commonly form the structural elements and in some cases, such as steel frames for schools and low rise housing, may not account for more than ten per cent of the total cost of the building.

Open systems (client-based) are not subject to the payment of royalties, and are based on components designed by government departments, local authorities and other clients for buildings with similar basic requirements and performance standards, such as schools, hospitals and government office buildings. A system of standardised structural elements is dimensionally co-ordinated with a subsystem of standard nonproprietary components which are designed and selected for assembly in a variety of ways.[14]

Examples of open systems include the *public building frame*, of precast concrete structural members designed by the Department of the Environment in conjunction with the Cement and Concrete Association; the Compendium of Hospital Building Assemblies, issued by the Ministry of Health for the industrialised hospital building programme; and the SCOLA and CLASP systems, particularly suited for schools and developed and controlled by consortia representing groups of local education authorities. Many open systems are of 'light and dry' construction often using light steel or concrete frames.

Leon[15] has also show how industrialised systems can be classified according to structural type; materials, weight and method of assembling structural elements; and building type. Structural types include the following.

(1) Load-bearing crosswall construction of concrete, bricks or concrete blocks; floors and roofs of precast or *in situ* concrete, timber or steel framing; and cladding of various types, all of which are well suited for housing.

(2) Storey height plank construction of 400 to 600 mm wide aerated concrete planks forming load-bearing walls, supporting floors and roofs of various types.

(3) Framed structures of concrete, steel or timber; prefabricated floors and roofs, with cladding produced in a factory or on the site. These systems are well-suited for buildings with large rooms, such as schools, fire stations and factories.

(4) Load-bearing large panel construction for high rise housing formed of medium ($2\frac{1}{2}$ tonnes) or heavy weight (6 tonnes) precast concrete panels for full-size, or parts of, walls and floors. *In situ* concrete staircase walls or lift shafts give rigidity to the structure and cladding can be formed from a wide variety of materials.

(5) Box construction of monolithic or composite prefabricated units, finished and equipped to form complete dwellings, incorporating plumbing and electrical services, doors, windows and fittings, and decorated in the factory.

(6) Composite systems with steel or concrete frames; load-bearing precast or *in situ* concrete walls and crosswalls; and cladding of precast concrete, brick, aluminium sheeting or other materials.

Advantages of Industrialised Building

There are certain advantages claimed for industrialised building which include increased productivity; lessening the adverse effects of inclement weather; a reduction in amount of nonproductive work, such as scaffolding; securing better working conditions for site operatives; and the elimination of waste in both labour and materials.

There are also supplementary benefits some of which have not yet been fully realised. In a system like CLASP the components have been carefully and thoroughly designed and the designs have been subsequently modified as a result of feedback of information from sites. By using a system which has been tried and tested, considerable savings in time accrue at the working drawing stage. This permits the designer to spend more time on clarifying his brief, carrying out user research, achieving improved planning solutions and possibly saving space and improving quality within the same cost limit and overall time scale. These developments should lead to better value for money in terms of appearance, durability and performance in the design of components.

Honey[16] criticises the use of the terms open and closed to describe building systems, because they lead to confusion due to lack of general agreement as to their main characteristics, and the fact that there is no absolute condition of complete limitation or complete freedom in many of the fields of choice. He prefers classifying them according to whether or not they are component-based, their type of sponsorship and whether or not their development is directed towards interchangeability with other systems.

TABLE 4.1
BUILDING SYSTEMS BY TYPE OF SPONSORSHIP IN RELATION TO BUILDING TYPES AND TECHNIQUES

	Type of sponsorship				
	Contractor	*Manufacturer*	*Client*	*Designer*	*Outsider*
Principal building type for which system developed					
Housing, high rise	24	4	2	–	–
Housing, low rise	42	28	4	4	2
Schools	5	7	4	2	1
Other	2	1	1	1	–
	73	40	11	7	3
Techniques or major materials					
Precast concrete	36	9	1	2	–
In situ concrete	9	1	1	–	–
Aerated concrete	1	–	–	–	–
Rationalised timber	8	3	2	–	–
Timber	9	19	1	1	–
Composite (for example frame and infill)	10	8	6	4	3
	73	40	11	7	3

Source: C. R. Honey, *System building, sponsorship and disciplines. Building Research Design Series 52.*

Effect on Contractual Arrangements

In the development and operation of a building system, the traditional relationships between the client, architect and contractor are modified and the role of the component manufacturer assumes greater significance. Other parties may be involved such as the corporate client, whose needs the system must satisfy, and the system designer, who may be quite separate and distinct from the designer of the individual building. Honey[16] has described how the sponsor of a building system undertakes a number of duties which do not occur in traditional building. These include the rationalisation of user requirements for the building type as a whole, the commissioning of design and prototype work, the co-ordination of the manufacture of components with the execution of the whole work, the advising of the designers and contractors for individual buildings on the proper use of the system, the promotion of the system and the appraisal of its performance as a basis for further development. The essence of the sponsor's duties is the provision of capital and co-ordination.

System building imposes on the design of an individual building a number of disciplines which do not occur with traditional methods of construction. These may restrict design flexibility, the choice of components and contractors, and the extent to which components can be interchanged with those outside the system. When interchangeability is provided, the design of the components will still be subject to dimension and jointing restrictions. Hence designers who wish to make use of prefabricating techniques must be prepared to submit to a working discipline based on an appreciation of operational factors.[16]

INDUSTRIALISED HOUSING

The Labour Government in the mid-nineteen-sixties provided a great impetus to the use of industrialised methods in order to secure a substantially higher output of dwellings from the available labour resources. There was considerable capital investment in plant and equipment on the basis of Government promises. Riley[17] postulates that one of the most serious miscalculations was to believe that systems of construction producing standard house types with standard exterior finishes and allround standard architecture could compete satisfactorily with more flexible forms of construction. Others argue that rationalisation and variety reduction can produce more attractive dwellings and can with every justification refer to most delightful and successful examples of standardisation in previous centuries at Bath, Cheltenham, Edinburgh and Bloomsbury (London).

High rise construction lends itself to component standardisation and repetition of elements, and Riley[17] has drawn attention to the greater opportunity for sponsors of high rise systems to persuade architects to accept system buildings in these situations than in 'close to ground' human scale development where aesthetic values are more local, personal, human and intimate. Hence a considerable swing soon developed towards the industrialisation of high rise flat construction. In the late nineteen-sixties the popularity of high rise flats declined, not so much due to a failure of industrialised construction techniques to secure economy and speed in construction, as to a social reaction against the compartmentalisation of human beings. The decline was also influenced by the higher constructional costs of high rise over low rise dwellings and the abolition of the additional Government grant for high rise dwellings.

For high rise development it was found generally that precast concrete systems were erected more quickly and cheaply than *in situ* concrete systems. In contrast, *in situ* concrete systems often proved the cheaper method for low rise dwellings although not the quickest. This results mainly from reduction in repetition and standardisation in low rise housing compared with high rise construction. Furthermore, the high initial cost of a precast concrete factory and its high overheads tend to make precast concrete units expensive. Timber and steel frames have both been used extensively for low rise housing and have the benefit of fast erection times. Both materials are, however, relatively expensive.

It is interesting to note the criteria against which the National Building Agency judged systems for use in local authority housing contracts and on the basis of which the Agency issued certificates of suitability:

(1) system can produce dwellings at reasonable initial and maintenance costs;
(2) system is capable of producing dwellings which comply with the necessary technical and space requirements;
(3) system is sufficiently flexible to be capable of producing a satisfactory environment;
(4) condensation problems are no worse than with traditional construction;
(5) adequate productivity;
(6) inspection of prototype revealed no serious defect.

PROBLEMS WITH INDUSTRIALISED BUILDING

General Problems

There are some disadvantages inherent in factory-produced building systems. They are almost certain to give rise to high initial unit costs unless very large contracts have been secured; factory employees usually receive higher wages that site operatives; these systems generally result in a reduction in flexibility as they are usually based on modular units, and there is lack of interchangeability between the systems. Too many projects in the housing field fall below the minimum economic threshold of fifty dwellings which is generally recognised as being necessary to achieve the full benefits of industrialisation.

Brouwer[18] has described how assembly operations occupy a dominant position in the industrialised building process. There is a concentration of capital-intensive machines and of highly qualified personnel. In order to obtain maximum utilisation of crane capacity it is necessary to synchronise all preceding and subsequent phases to the speed of assembly, and this requires thorough and skilled programming. Reiners[19] has shown that some of the advantages claimed for industrialisation were based on an oversimplified analysis and also pointed out that many of the potential advantages can be lost by poor design details which leave small but awkward pieces of work to be carried out by traditional means. The site construction may be made very much more complicated with a greater number of breaks in continuity, by the transfer of some of the work to a factory.

Hole[20] has emphasised that industrialised building methods, coupled with long production runs, militate against modifying a design in the course of a building programme, so that user needs, as well as other aspects of the design brief, require clear definition at the outset. To achieve the essential consistency of standards, the client will have to accept a more restricted range of choices than with traditional building.

Burgess and Morris[21] have described how industrialised building brings its own demands upon accuracy as it approaches ever nearer to the technique of mass production. These techniques, with their attendant economies, are achieved through the standardisation of components and a greater degree of component interchangeability. If dimensional standardisation is to be achieved, dimensional accuracy is essential in both the overall manufacturing dimensions and in the correct positioning of the units during erection. The degree of accuracy is controlled through tolerances but these are not entirely effective in practice as they are difficult to specify, measure and communicate.

Problems with Housing

The Government was prepared initially to sponsor schemes of industrialised building even to the extent of a production run of 20 000 houses, but it was not prepared to enter into obligations for the continuous supply of houses over a number of years. Promoters claim that they never had a real opportunity to substantiate the economy and viability of a system because of the uncertainty over future contracts. Furthermore, many local authorities possessed comparatively small housing sites and they only need a few houses each of various types. There were far too many firms trying to share the market, each of whom were faced with high labour costs in design and production, the need for expensive equipment and a noticeable lack of flexibility in operation.

Industrialised building calls for a systematic and disciplined approach from the building client. He will be buying a building which is largely composed of a range of mass-produced components, which can only be produced economically as part of a large, continuous manufacturing programme. In these circumstances, he may have to combine his requirements with those of others, to accept a place in a phased manufacturing and building programme and to adopt contractual methods to which he is unaccustomed.

To meet these restrictions, the larger housing authorities may find it necessary to combine in consortia, to secure a total programme adequate in size for the application of industrialised systems. The use of industrialised building posed the greatest problems for the smaller housing authorities. The nature of the problem becomes immediately apparent on consideration of housing administration arrangements in England and Wales. In 1971 there were almost 1500 housing authorities made up of boroughs, and urban and rural district councils. The average local authority housing contract was for less than fifty units and well over eighty per cent of the local authorities had annual programmes of less than 100 dwellings. It would be difficult for the smaller authorities to form consortia as they lack the necessary professional staff and it would need too many constituent members to produce a viable programme. Hence the National Building Agency came on to the scene to enable a large number of small demands to be collated into satisfactory programmes, and to give guidance on the selection of systems and contract arrangements.

Another major problem stems from the financing arrangements on an annual basis which operate in the local government sector. Local authorities are seldom prepared to formulate their building programmes for more than a year ahead, although a strong case can be made for rolling programmes extending over much longer periods. Yet another problem is associated with tendering arrangements; local authorities are often under pressure to invite tenders from a number of builders on a competitive basis, even although the Banwell Report favoured selective tendering. Changes are however taking place and some local authorities are using negotiated contracts and other modified tendering methods, permitting tenders to be obtained for the supply of components in advance of the main building contracts, or the selection of a system and a contractor at an early stage in the design process with the tender being negotiated later. Finally, the architect also has to adjust to a new discipline and to face up to the limitations on design imposed by the use of prefabricated components and a relatively limited range of claddings and finishes.

The recent decline in the use of industrialised building systems has meant that the majority of systems have not been fully developed. The cost of industrialised dwellings for medium and most low rise housing has been higher than that of comparable traditionally built dwellings. Furthermore, savings in total construction time have not been as great as anticipated. Although the shell is erected more quickly, the finishing trades and fittings often have not been sufficiently standardised and have taken as long to complete as those in traditional dwellings. Nevertheless, it would be wrong to write off system building as a complete failure. Many of the troubles experienced stem from the timing and scale of the system building revolution in the housing sphere.

ECONOMICS OF INDUSTRIALISED BUILDING

Comparative Costs of Different Constructional Methods

Leon[7] has laid down some general principles relating to relative costs of constructional techniques in different situations.

(1) Load-bearing wall construction is best suited to building types with small spans, such as dwellings.

(2) Systems for buildings up to five storeys in height are more economically designed as column and beam structures.

(3) Structures over five storeys in height with increases in vertical loading are well suited to panel construction.

(4) Savings in cost stem from a reduction in the number of different materials used for walls, floors and façades.

Figure 4.1 shows diagrammatically the comparative costs of different systems for various building ranges all related to the common unit of the square metre of floor area. There is a noticeable consistency in the average costs of the different systems for each type of building. However, precast concrete systems show a distinct cost advantage for high rise buildings whilst timber and steel systems seem particularly well suited to low rise buildings.

Leon and Wajda[22] have shown how multirise dwellings, with comparatively small room spans, are usually more economically constructed in concrete of boxshell construction. When using steel and reinforced concrete frames in structures subject to light loadings, it is usually preferable for the floor members to span in the opposite direction so that each bay can secure more evenly distributed loads on beams. With heavier floor loading, the units are more economically designed to span in two directions.

Large panel construction is often the most economical form of construction for high rise buildings and reduces the number of vulnerable external joints. They are, however, expensive in casting, fixing of reinforcement, provision of moulds and in handling. These disadvantages are usually more than offset by the high quality of finish, more rapid construction and improved working conditions which flow from their use. Bishop[23] has also shown how the selection of the type of mould is conditioned by the scale of the activity and by the expected economic life of the development. Given the expectation of a reasonable economic life and supply of accurate and well-finished panels, it is possible to obtain considerable savings in the labour costs of erection, jointing and finishings which are likely to more than offset the higher costs stemming from the use of more accurate or more sophisticated moulds.

Labour Implications

Bishop[24] has referred to cases where factory-made components have replaced only part of the corresponding conventional operations, resulting in an increase in the total number of operations and greater complexity. Furthermore, in these situations it is not uncommon for the work remaining to be performed by traditional processes to be the most difficult and awkward part, and therefore very expensive. Hence, the aim at the design stage should be to replace the whole of the traditional work and to reduce the number of operations required. Furthermore, the use of industrialised methods creates a demand for new skills. In this country the wage rates in the building industry are slightly below those in manufacturing industries, hence the transfer of work from site to factory is less financially attractive than in the United States where construction workers receive about thirty per cent more than workers in manufacturing industries.

Stone[11] has described how material costs are frequently higher in industrialised building as compared with traditional work and that it is therefore necessary to reduce labour costs. This highlights the need for effective preplanning and organisation of all phases of system building by integrating production with deliveries to sites, keeping erection teams fully occupied and minimising nonproductive time and site costs. A productivity study by the National Building Agency[27] found that an industrialised house takes about thirty weeks to build compared with a national average for local authority housing of about fifty-three weeks.

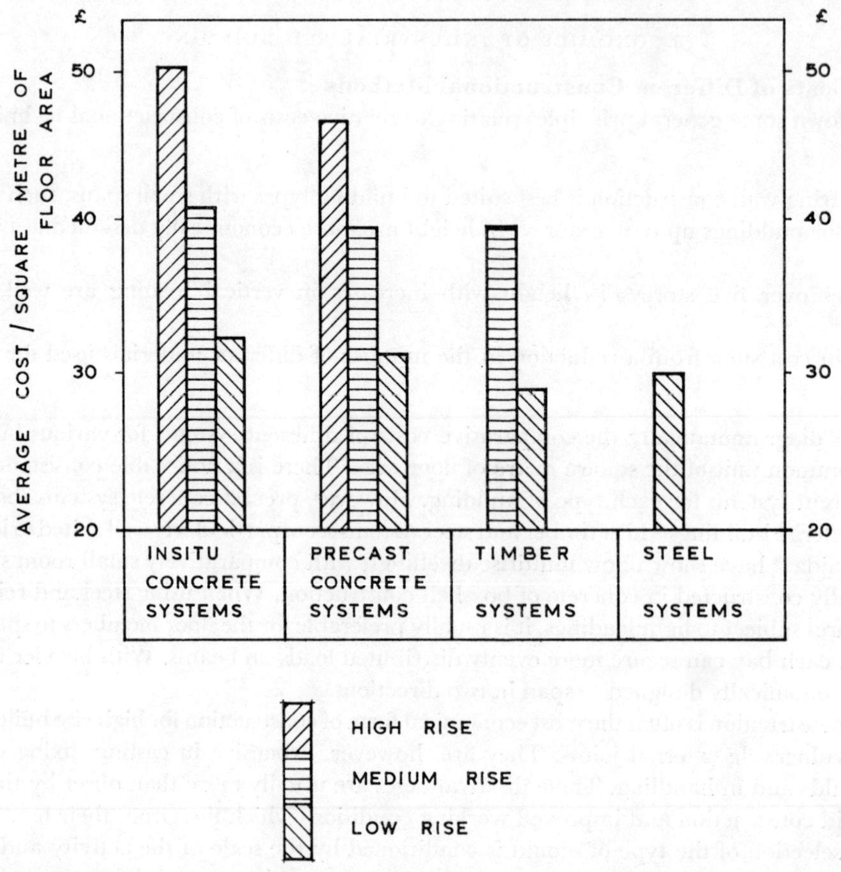

Figure 4.1 Comparative cost of industrialised systems for high, medium and low rise buildings

Capital Costs

Industrialised building entails much greater capital investment than for traditional building. In 1966 Bishop[24] estimated that in conventional building the investment per dwelling per day capacity would be about £50 000 to £60 000, whereas the investment in system building using large panel construction could be up to twice as much. Amortisation costs are likely to be increased since investors will probably anticipate a shorter economic life from their assets invested in industrialised building, stemming from particularity and possibility of early obsolescence. Interest charges may also be higher owing to the need to attract risk capital in the face of an uncertain market. Bishop has used a twenty per cent rate of interest where a short economic life is expected, and fifteen per cent and ten per cent rates for medium and long economic lives respectively.

Capital costs must include that of development work in designing and testing the system and the production process, including prototypes where required. Fixed assets include factory buildings and plant, site equipment, vehicles, services and temporary works generally. Sufficient working capital is required to finance work and bridge the gap between receipt of income and the making of payments to labour and creditors, and could amount to as much as two to three per cent of turnover.

Another approach is to compare the investment per operative. In 1966, Bishop[25] estimated that the investment per operative in conventional building was approximately £300 to £400, whereas the corresponding figure for system building could have been £800 to £1200. In 1963 it was estimated that the capital costs of a plant to produce dwellings by industrialised systems varied from £300 000 for an annual production of 600 dwellings up to £750 000 for 1500 dwellings.[12] Leon[26] found that the cost of a factory producing precast concrete panels could vary between £60 000 and £600 000, according to the scale of production and the degree of mechanisation and automation. It should also be borne in mind that up to thirty per cent of the cost of the factory may be spent on transport and trailers to carry components to sites, particularly in the case of large panels.

Building Demand

A permanent offsite factory to produce precast concrete structural elements is based on mechanised handling methods and semiautomated production line principles under controlled conditions to secure a high output ratio of accurately prefinished units. The viability of such a project is largely dependent on the maintenance of an adequate and sustained volume of demand, in order to justify the highly intensive capitalisation required for production. Indeed, some sponsors of industrialised systems have been forced out of business through failure to secure sufficient continuity of orders for economic production. Yet the demand for building, besides being dispersed on sites, is variable in volume, bespoke in character and uncertain in timing.[24]

Admittedly offsite factories carry with them the disadvantages of the need for long production runs, high overheads and high transportation and handling costs. In some situations a better approach might be to establish an onsite factory geared to the production needs of the site, with consequent reduction of transportation and handling costs, and being capable of transfer to another site on completion of work on the original site.

Economics of Production

It is evident that industrialisation almost invariably involves increased capital investment and other indirect costs and may, at the same time, result in decreased utilisation because of particularity. Hence, if industrialisation is to be economically worthwhile, these increased costs must be offset by reduced labour costs through higher productivity. Higher productivity should be feasible through increased technical aids, improved organisation and superior working conditions. Stone[11] has described one good way of reducing costs by eliminating operations, such as by accommodating services in wall and floor units or by producing units which are self-finished.

Too often, in the past, even the sponsors of systems have not been in possession of realistic production costs, and yet a full cost appraisal and feasibility study ought to have been undertaken before the component factory was erected. Another very real danger consists of available orders being spread over too many firms and systems, as each manufacturer views the market in isolation, and this can easily result in few systems achieving optimum production. It has been calculated that efficient traditional construction requires about 2000 man-hours to build a flat in a tall block, whilst an efficient concrete panel system of industrialised building would require 500 man-hours on site, 300 in the factory for concrete components and a further 300 for other factory work; a total of 1100 man-hours.[28]

Leon[29] postulates that production costs of concrete components are influenced by the following factors.

(1) Installation and running costs of factory, plant and equipment, and dismantling costs in the case of temporary factories.

(2) Mould costs, which can also be affected by increased use resulting from accelerated maturing of concrete and the introduction of shift work to increase output.

(3) Material costs, which are affected by the surface treatment and final colour of the units.

(4) Labour costs, which can be affected by the integration of the casting and erection cycles.
(5) Type, size, finish and number of units.

Plain and relatively simple units can be cast economically on site. Those of intricate design are better suited to offsite production under more closely controlled conditions. Unless units of the same shape and size are used on every floor, additional costs will be incurred in varying the moulds to suit the units for use at different floor levels.

Leon[30] has also described how an effective computer system can assist with rapid and yet flexible control over preplanning and costing by taped programmes for:

(1) ordering materials to phased deliveries;
(2) comparing estimated costs with the current costs of all jobs in progress by a detailed breakdown which forms a pre-expenditure control budget;
(3) rapidly determining payments and oustanding balances due to subcontractors on the basis of coded budget programmes; and
(4) readily obtaining monthly efficiency reports which provide detailed information about differences in costs and quantities of materials used on site, labour, plant and subcontract works.

Obviously long contracts are needed for system building because of the way in which costs rise if factories operate at a low level of output. A high proportion of factory costs are fixed, since neither the labour force nor overheads can be reduced easily when output falls below optimum. The establishment of consortia and the grouping of orders, coupled with a restriction in the number of systems, should help to secure viability of production. The concentration of system building factories in the vicinity of new towns and large redevelopment areas has much to commend it.

Comparison of Industrialised and Traditional Building

The labour requirements per dwelling with traditional building cover a wide range from 700 to 2400 man-hours with an average of around 1800 man-hours per dwelling. The performance with industrialised building also shows an extremely wide range with site labour requirements per dwelling varying between 700 and 1300 man-hours. The difference between the labour requirements for conventional and system building is likely to increase in tall buildings because these are inherently more complicated and the work involved is better suited to industrialisation. The work directly affected by industrialisation often approaches seventy per cent of the total, the remainder being siteworks, foundations and work of specialists. It would seem rather pointless using a system which does not extend beyond the structure, as the savings realised in the finishings are proportionately greater than those obtained in the structure.[24]

Comparisons of the alternative methods of house building can conveniently be made under four main headings.

Relative labour costs. The average hourly earnings in factories in the United Kingdom producing engineering products are about 106 per cent of construction industry earnings.

Indirect labour costs. These are generally increased in system building because there is a greater demand for management skills and services, and for technical expertise in design and production. Bishop[25] has estimated that they may be two-thirds higher than for conventional building.

Capital charges. These are made up of the capital sum, amortisation and maintenance charges and the interest rates necessary to raise the capital. Capital investment is nearly always increased with system building, although ideally the investment should be related to the capacity of the production process. This is possible with closed systems, but with open systems it is difficult to establish because of the wide range of building types and products and the widespread use of hiring and subcontracting in the industry.

A rather more assessable index is the investment per operative as described earlier in the chapter. Capital costs may range from £1.25 to £1.75 per day per £1000 invested for conventional building, and from £1.25 to £2.60 for system building.

Utilisation. In conventional building problems of utilisation rarely arise as most firms can undertake a wide range of building work, and the general employment of subcontractors gives considerable flexibility to the industry. System building radically alters this situation in that resources are committed to the production of specific components (in open systems) or to the production of specific building types (in closed systems).

With industrialised systems contracts are usually negotiated and hence tend to be rather less competitive than traditional dwellings which are normally the subject of tenders. This could be an oversimplification, as by concentrating on design and development work the sponsor of an industrialised system could provide a very efficient and economical building. On the other hand industrialised buildings rarely seem to function any more satisfactorily than their traditional counterparts and they are generally more expensive in maintenance. Charges of monotony and poor design of industrialised buildings are rarely well founded and open systems provide kits of parts which can be assembled in a large variety of ways and so provide the designer with a reasonable range of choices. System-built structures normally make fuller use of the inherent properties of the materials than traditional buildings, although the satisfactory jointing of components may prove expensive.

THE FUTURE IN INDUSTRIALISED BUILDING

It is believed that local authorities in England and Wales placed orders for no more than 15 000 system-built dwellings in 1970, representing a sixty per cent drop on the previous year; and on these figures it seems unlikely that any more than forty systems can remain viable. This presents a very different picture to the Government White Paper of 1965 and Ministry of Housing and Local Government circulars 21/65 and 76/65 to local authorities, setting a target of 500 000 new dwellings per year by 1970, half to be built by the private sector and half by local authorities, with forty per cent of the local authority houses to be system built. Nearly 300 systems appeared during the last decade, although a proportion of them never developed beyond the illustrated brochure stage.

It is evident that there were too many systems for the available work and that much of the anticipated workload in local authority housing did not materialise. Changes in the membership of local authorities can also have significant effects on housing policies. In the longterm, system building can only remain healthy provided users are satisfied as to its benefits and there is sufficient flexibility to respond to variations in demand. There is also a need for the private sector to show much more interest in industrialised methods by accepting some limitation in variety in terms of basic shape and layout, in order to secure cheaper and quicker manufacture of the factory components.

Manufacturers of industrialised buildings need relatively long periods, certainly not less than ten years, to amortise their costs; local authorities must be able to implement rolling programmes uninterrupted by financial restrictions; and the Government must give directions to local authorities on the use of system building if it is to be really successful. Until local authority orders for industrialised housing are handled on a regional basis and large longterm orders are given, it is bound to remain a risky capital venture. Another drawback to the present development of industrialised systems is the fact that they are largely restricted to the housing field. While the housing market is large, it still does not represent more than about one-third of the total output of the building industry.

REFERENCES

1. INSTITUTE OF BUILDING ESTIMATORS. Notes of guidance for the teaching of estimating. (Undated)
2. INSTITUTE OF BUILDING. Code of estimating practice (1966)
3. P. F. MILLER. The production unit – the basis of design and control. *The Quantity Surveyor* (September/October 1969)
4. W. W. ABBOTT. Reducing materials wastage on site. *Building Technology and Management*, **8.11** (1970)
5. K. M. WOOD. Early precast problems have been beaten. *Building Industry News* (3 June 1966)
6. BUILD INTERNATIONAL. Why more quality in building? *Build International* (July/August 1970)
7. G. LEON. Component production. *Illustrated Carpenter and Builder* (5 July 1968)
8. J. REDPATH. Standardisation – for what? *Illustrated Carpenter and Builder* (17 October 1969)
9. A. J. LOCKWOOD and D. W. PEDDER-SMITH. Variety reduction in door making. *Building* (13 June 1969)
10. BUILDING RESEARCH STATION. Coding and data co-ordination for the construction industry (1969)

11. P. A. STONE. *Building Economy: Design, Production and Organisation*. Pergamon Press (1966)
12. J. WHITTLE. The problems of mass production. *The Chartered Surveyor* (April 1963)
13. UNITED NATIONS ORGANISATION, THE ECONOMIC COMMISSION FOR EUROPE. Government policies and the cost of building (1959)
14. G. LEON. Industrialised system building. *Illustrated Carpenter and Builder* (21 June 1968)
15. G. LEON. Introduction to industrialised building. *The Quantity Surveyor*, **23.4** (1967)
16. C. R. HONEY. System building, sponsorship and disciplines. *Building Research Station Current Papers Design Series*, **52** (1967)
17. J. RILEY. All-round flexibility of system building. *Building Technology and Management* (April 1971)
18. I. F. BROUWER. Organising the building process. *Industrialised Building Systems and Components* (November 1969)
19. W. J. REINERS. The study of operations and economics at the Building Research Station. *The Chartered Surveyor* (April 1962)
20. W. V. HOLE. User needs and the design of houses. *Building Research Station, Current Paper*, **51/68** (1968)
21. R. A. BURGESS and P. W. G. MORRIS. Accuracy and productivity in industrialised building. *Building* (15 November 1968)
22. G. LEON and R. L. WAJDA. Economic principles of multistorey industrialised buildings. *The Quantity Surveyor* (September/October and November/December 1967).
23. D. BISHOP. Industrialised building – with special reference to formwork. *Building Research Station Current Papers*, **45/68** (1968)
24. D. BISHOP. The economics of industrialised building. *The Chartered Surveyor*, **99.4** (1966)
25. D. BISHOP. Traditional building costs – the target for system building. *Building Research Current Paper, Design Series*, **42** (1966)
26. G. LEON. 800 man-hours per dwelling. *Illustrated Carpenter and Builder* (13 December 1968)
27. NATIONAL BUILDING AGENCY. Industrialised two-storey housing: a productivity study (1970)
28. J. A. DENTON. Planning and design for industrialisation. *The Chartered Surveyor* (November 1965)
29. G. LEON. Analysing production costs. *Illustrated Carpenter and Builder* (27 December 1968)
30. G. LEON. Computer-production control: five basic schedules. *Illustrated Carpenter and Builder* (7 February 1969)

5 ECONOMICS OF RESIDENTIAL DEVELOPMENT

THIS CHAPTER IS concerned with the problems associated with housing provision; the alternative forms of layout that can be employed to meet varying housing requirements and their comparative costs; the methods and economics of different forms of car-parking provision; and the considerations involved in a comparison of the relative merits of redeveloping or rehabilitating twilight areas.

BACKGROUND TO PUBLIC HOUSING PROVISION

In the period between the passing of the Housing of the Working Classes Act in 1890 and the outbreak of the first world war, local authorities collectively provided no more than 600 houses per annum on average. By way of comparison private housing provision throughout the period 1900–05 averaged 150 000 dwellings per annum.[1] At the end of the first world war rent restrictions and high labour rates meant that the required dwellings to rent could not be provided profitably by the private sector. Hence the Addison Act of 1919 enabled local authorities to bear the cost of providing houses for the working class to the extent of the product of a penny rate (5/12p) and the Government bore the remainder. Sharply rising prices generally and inefficient administration caused average tender prices to rise by fifteen per cent within a year, and 175 000 dwellings cost the Government £200 million.[2]

In the interwar years a succession of Housing Acts varied the subsidy arrangements introduced by the Addison Act and the housing shortage remained sufficiently serious for rent restriction to continue. Local authorities built one million houses which was equivalent to about one-third of total provision. After the second world war the Labour Government introduced a number of additional restrictions on development, including licensing restrictions and rationing of building materials. Eighty per cent of the dwellings built between 1945 and 1951 were provided by the public sector. Orders were placed for 500 000 'prefabs' (prefabricated bungalows), but these were subsequently reduced to 170 000 due to problems of rising cost and balance of payments difficulties, followed by devaluation. The Conservative Governments of the fifties progressively reduced the public housing programme, licensing was abandoned and the private sector provided an increasing proportion of new houses. The Rent Act 1957 raised controlled rents to a level closer to that of the free market, and allowed controls to lapse from properties falling vacant. Between 1957 and 1965, over two million houses had been removed from control, mainly as a result of the movement of tenants. The Rent Acts of 1965 and 1968 subsequently provided machinery for determining rents on a basis that sought to express fairly the real value of the tenancy. Local authorities borrowing activities were restricted and subsidies reduced, with the result that public sector annual completions fell from 220 000 in 1954 to an average of 100 000 in the period 1959–63.

In the mid-fifties many private landlords sold their residential properties as they had long since ceased to be profitable. The number of households living in privately rented property dropped from sixty-one per cent in 1947 to thirty-one per cent in 1961[3] and fourteen per cent in 1971.[41] It has been estimated that sixty per cent of the population of this country can afford to house itself with the aid of tax rebates, option mortgage schemes and other hidden subsidies. The remainder are compelled to rely on Government intervention or legal protection of one kind or another.

The Parker Morris report[4] established minimum space standards for local authority dwellings and this resulted in a general increase in the cost of publicly provided houses. Furthermore, heavy increases in land prices, stemming in part from the operation of betterment levy introduced by the Land Commission Act and a general scarcity of building land, resulted in speculatively provided urban dwellings costing so much to build that there was little possibility of them being rented. In the public sector the introduction of housing yardsticks in 1967 caused too much emphasis to be directed towards first costs as opposed to longterm costs. Already loan payments in some local authority areas are matched by maintenance costs. Some local authorities have adopted a policy of selling some of their houses in an effort to correct adverse balances on housing accounts.

USE OF LAND FOR HOUSING PURPOSES

Planning schemes prepared by local planning authorities restrict the land that can be developed for residential purposes. Residential areas should desirably be conveniently located in relation to workplaces, shops, schools and other essential facilities. Sites for residential development should be physically suitable – not excessively steep, although some undulation gives character, be above flood plains and be reasonably healthy. Adequate public utility services and good transport facilities are other important needs. The nature of adjoining development needs consideration as it must be compatible.

The area of land required to house a specified number of people will depend on the operative residential density. Town maps and other plans showing planning proposals usually indicate the permitted density for each area of land zoned for residential use. Housing densities are expressed in a number of different ways. In prewar planning schemes residential densities were usually expressed in houses per acre, but in more recent times it has become customary to use persons per acre or hectare (population density) or habitable rooms per acre or hectare (accommodation density), as these give the developer a greater degree of flexibility in his choice of dwelling types. A habitable room is a living room or bedroom, but not a kitchen or bathroom. Population density can be converted to accommodation density by deciding on the average number of persons per habitable room. A house with five habitable rooms and occupied by five persons would have an occupancy rate of one, but in practice dwellings are frequently underutilised and the average family size is about three persons. Hence there is a significant variation between design and actual population densities.

The density most commonly used is that of net residential density. This is the population (or accommodation) divided by the area in hectares, including dwellings, gardens, any incidental open space (for example children's play spaces or parking spaces for visitors' cars) and half the width of boundary roads (up to a maximum of 6 m). Shops, schools and most open spaces are excluded. Gross residential density applies to a complete neighbourhood and includes all the land uses within it, and James[5] has argued that this is the more meaningful indicator of density. He postulates that the fixing of net residential density standards by local planning authorities over wide areas can be deceptive and misleading; particular conditions of the site, its shape, and its relation to surroundings all influence the intensity to which it should be used. The people who are to reside on the site will also be concerned with the ancillary features that are available in the neighbourhood.

In the garden cities, in the first generation new towns and in most of the earlier postwar local authority housing schemes, net residential densities of about thirty houses to the hectare (twelve to the acre) were used. It was often argued that this was a good practical density, which could be applied to varying orientations without overcrowding and overshadowing. It permitted reasonable distances between houses, giving satisfactory street pictures with suitable depths of front garden and ample back gardens. Space was available for kitchen gardens, which are generally cultivated far more intensively than farmland by sparetime labour, which would not otherwise contribute to food production.[6]

In 1962, the Ministry of Housing and Local Government called for higher residential densities.[7] It was pointed out that the population of England and Wales was increasing by about 250 000 persons per annum and all available data indicated that this rate of increase would be maintained. People now marry younger and live longer, and greater prosperity enables more families to have a home of their own, thus increasing the number of households. Furthermore, the occupancy rate had fallen from 3.3 to 3.0 persons per house between 1951 and 1961.

The relief of overcrowding, new roads, schools and open spaces, all require more land. The plea for higher residential densities was made to reduce the total demand for land, to help preserve the countryside and protect good agricultural land. Substantial savings in land could be secured by increasing the net density to 100 persons per hectare, particularly when some development was proceeding at about sixty persons per hectare. It was further shown that net densities of 150 persons per hectare are attainable with two-storey houses and the introduction of a proportion of three or four-storey flats can produce net densities up to 225 persons per hectare. Higher densities still can be obtained by using tall blocks of flats, but these need not predominate until net densities of 350 persons per hectare are reached.

On the financial side it is pointed out that as the permitted density rises so the cost of land increases, although the cost of land per dwelling decreases. Building costs decrease as densities rise up to about 150 persons per hectare, due to the greater use of terraced blocks, but above this density building costs rise sharply. Service costs of roads and engineering services decrease with more compact development.

With many housing schemes in the public sector the aim is to secure densities in the 100 to 150 persons per hectare range. The task of designing the layouts economically revolves around the choice of types of buildings and the proportions of each type. The director of housing or housing manager will advise on the proportions of the various sizes of dwelling that are needed. These requirements will be influenced by the range and numbers of existing dwellings; the family composition of housing applicants; the extent to which the authority's houses are under-occupied; the numbers and sizes of other dwellings being built by the authority; and the possibility of exchanges with families in privately owned accommodation.

Given the site area, the density standard and the proportions in which the various sizes of dwellings are needed, the designer can calculate the number of dwellings of each size to be provided. It is customary to build in a tolerance of about five per cent in respect of numbers and proportions. With small sites, the range of dwelling sizes should be kept small to produce economical layouts.[8]

ASSESSMENT OF HOUSING NEED

Crude housing statistics might seem to indicate that the overall shortage of houses has been remedied: in 1970 there was a stock of 18.6 million dwellings in England and Wales[9] and a total of 18.3 million households.[10] These figures are however deceptive in that they include an appreciable amount of housing that may be deemed unfit, and dwellings that lie vacant through changes of occupancy or to permit repairs and conversions, and they also mask wide regional variations. Indeed 4.6 million dwellings, or about one-quarter of the total housing stock in 1967, could justifiably have been classified as uninhabitable on grounds either of poor condition or of lack of basic amenities.[11] The pressures of shortage are felt more intensely in some areas than others: Scotland and the north of England in general have a surplus of dwellings over households whereas the south-east has a deficiency. The housing waiting list of the Greater London Council increased from 150 000 families in 1965 to over a quarter of a million in 1970.[1]

As to future needs, the Registrar General's forecasts of future population indicate a total population for England and Wales of 64.6 million for the year 2000, or an increase of almost one-quarter on the present population. Household needs and a reasonable allowance for a vacancy reserve may together create the need for about twenty-four million dwellings by the year 2000, involving an addition of about five million dwellings to our present housing stock, to which must be added several million replacements at present running at the rate of about 70 000 dwellings per annum.

Increasing life expectancy coupled with the large number of births at the end of the last century produced in the nineteen-fifties a population containing large numbers of elderly persons. The proportion of elderly people will increase slightly in the next twenty years, but will then decline as the smaller number of persons born in the nineteen-twenties and nineteen-thirties reach old age. Hole[12] has described how longer life expectancy, a long-term trend towards a higher proportion of married persons in the population, and more recently towards youthful marriages, has resulted in a sharp increase in the number of households and this has exerted a continuous pressure on the housing market.

It is reasonable to assume that the proportion of married women in employment is likely to continue to rise with various implications in relation to housing. Their earnings may in the early stages of marriage assist with house purchase, and later may permit the family to move to larger or more expensive accommodation and increase the demand for such features as fully automatic heating and more labour-saving appliances.

The assessment of the adequacy of housing provision is further complicated by the consistent trend towards small households over the last fifty years, which has been accompanied by a concentration on the building of five-room family houses. Figure 5.1 serves to illustrate the significant disparity between the range of dwelling sizes available in the housing stock and the range of household sizes which has resulted from demographic changes.

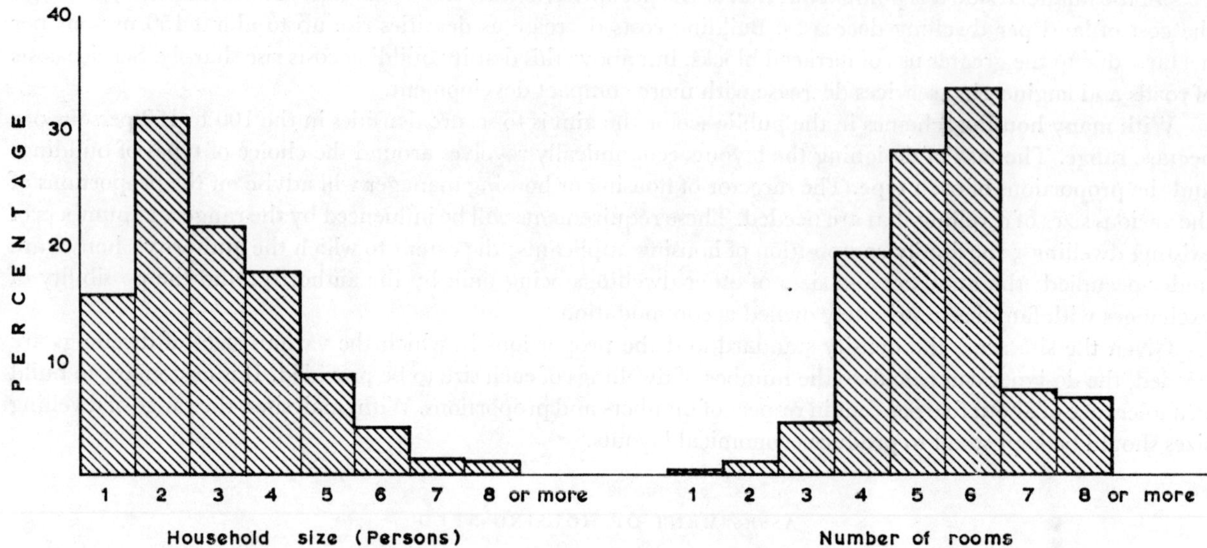

Figure 5.1 Distribution of household and dwelling sizes in England and Wales in 1966

DWELLING TYPES

There are numerous types of dwelling built by both the public and private sectors, varying both in size and in physical characteristics. The majority of dwellings are still provided in houses, which can be detached, semi-detached or built in terraced blocks, and the number of bedrooms can vary from one to four to suit different sizes of family. Special accommodation is needed to cater for aged persons, disabled persons, single workers, students and apprentices. In recent years rather more emphasis has been placed on the construction of flats and maisonettes, generally with a view to securing higher densities and the fuller use of highly priced land. Some of the best planned layouts frequently contain a mixture of dwelling types.

House Designs

Local authorities have over the years received considerable guidance from the Ministry of Housing and Local Government on the design and layout of dwellings. The guidance has been largely aimed at achieving good housing standards yet, at the same time, securing economical designs and layouts. Four-persons houses generally have two bedrooms and five-person houses have three bedrooms. A variation to this general theme is the three-bedroom house which has one double and two single bedrooms, to house a family of four with two older children of opposite sex. There are three basic designs:

(1) living room, small dining room and a working kitchen;
(2) large living room including a dining area, which may be in the form of a recess, and a working kitchen;
(3) living room and large dining kitchen.

With terraced houses, access to the rear of the house can take one of several different forms. The more orthodox arrangement consists of a common covered passageway between adjacent houses, with storage accommodation provided in outbuildings at the rear of the block. Another form of layout provides access through an enclosed store built within the walls of the house, which eliminates the draughty covered passage and provides a sound barrier between the ground floors of adjacent houses. Other designs incorporate a store at the front of the house with access through the kitchen to the rear, or alternatively access can be provided through the hall at the front and the store at the rear of the house.[6]

Parker Morris Standards

In more recent years there has been a change of approach to the design of housing accommodation stemming from the Parker Morris report.[4] In this report accommodation requirements are not based on minimum room sizes, but on functional requirements and levels of performance, with minimum overall sizes for the dwelling related to the size of family. There should be space for activities demanding privacy and quiet, for satisfactory circulation, for adequate storage and to accommodate new household equipment, in addition to a kitchen arranged for easy housework and with sufficient room in which to take at least some meals. The suggested minimum net floor area, enclosed by the walls of a dwelling, of a five-person house ranges from 97.50 m^2 for a three-storey house, 92.00 m^2 for a two-storey house and 86.40 m^2 for a flat, to 83.60 m^2 for a single-storey dwelling. For four-person houses the net floor areas range from 74.30 m^2 for two-storey houses to 66.00 m^2 for single-storey dwellings.

Three-storey Houses

Three-storey houses in terraced blocks form a convenient and economical way of housing large families. Where densities are in the order of 250 to 290 habitable rooms per hectare, and this type of block is used in conjunction with three-storey corner flats and four-storey blocks of maisonettes, high blocks can be avoided. Typical arrangements of three-storey houses are:

Floor	*Scheme A*	*Scheme B*
Ground floor	Hall, store and dining/kitchen.	Hall, garage and all-purpose room.
First floor	Living room, bedroom and bathroom.	Living room and working kitchen.
Second floor	Three bedrooms	Three bedrooms and bathroom.

Flats

Tower blocks of eleven storeys or more in height are mainly provided to accommodate smaller families. A common arrangement is approximately one-half one-bedroom and one-half two-bedroom flats. Access is obtained in a variety of ways – daylighted common hall, enclosed common hall, cross-ventilated common hall or enclosed central corridor, and these are illustrated in the Ministry of Housing and Local Government's 1958 report on flats and houses.[8] Typical arrangements would be four flats per floor with the first three forms of access and eight flats per floor with corridor access. There would probably be two staircases and two lifts in all cases.

High slab blocks have been used where it was not practicable to house all the larger families in houses or four-storey maisonettes. The slab blocks of rectangular plan may contain primarily four or five-person living units arranged as flats or maisonettes. Access may be by balconies, corridors or staircases. The maisonettes are often arranged as three-floor units, whereby access to two maisonettes is obtained from the intermediate floor via halls and staircases, with the larger bedroom to each dwelling also located on this floor. The lower floor contains the living and dining areas, kitchen, bathroom and smaller bedroom for one dwelling, and the upper floor contains similar accommodation for the other dwelling.

Four-storey maisonette blocks usually consist of three-bedroom units spread over two floors, with access to upper dwellings by a balcony served by a common staircase.

Three-storey flats usually contain two and three-bedroom flats (two on each floor) served by a common staircase. All the rooms may be entered from an entrance hall or access to bedrooms may be secured through a separate inner lobby. Other variations are working-kitchens and dining-kitchens.

Elderly Persons' Dwellings

Many elderly people who are still fit prefer to live in fully selfcontained dwellings. The most popular form of dwelling is the single-bedroom bungalow with a living room, small kitchen and combined bathroom and WC. There are also a large number of elderly persons who, by reason of age or infirmity, are unable to manage entirely on their own in normal type bungalows. Where these persons can be relieved of some of the burdens of normal tenancy and, given friendly oversight and a limited amount of assistance by a warden, they can continue to lead independent lives and remain an integral part of the community.

In recent years many schemes of grouped flatlets have been implemented. Basically these schemes usually provide a number of centrally heated flatlets, shared baths and WCs, some communal facilities and a warden's residence. The flatlets can be designed in a number of ways: with bed-sitting rooms and kitchen fitments in cupboards, with separate kitchenettes, or with separate bedrooms. Another alternative is to locate beds in bed recesses in living rooms which can be closed off with curtains.

Baths are generally shared in the ratio of about one bath to three to five persons, and WCs in the ratio of one appliance to two or three persons. Communal facilities often include a common room, laundry, heated drying room(s) and guest room(s). The warden's quarters adjoin the flatlets and generally consist of a two-bedroom dwelling. Useful information on the operation of grouped flatlets for elderly persons and a number of plans of operative schemes are shown in the Ministry of Housing and Local Government's pamphlet *Grouped Flatlets for Old People.*[13]

HOUSING REQUIREMENTS OF OCCUPANTS

A Building Research Station report[14] has described how as a result of longterm trends in demographic structure and house building the range of house sizes in this country's current stock of dwellings is out of alignment with the range of household sizes. It is sometimes asserted that this situation can be remedied by the movement of households from larger dwellings to smaller ones, when families have reached the later stage of the family cycle; but this can only be a partial remedy as older households become less mobile. It is likely that about one-fifth of future housing demand will be of the nonfamily type, catering for single, widowed and divorced persons. Of the remainder a proportion of the dwellings will cater for small households, such as newlyweds and elderly couples.

Hole[15] has described how the architect concerned with mass housing has little opportunity for direct contact with the people who will inhabit his houses. Industrialised building methods which presuppose long production runs militate against modifying the design during a building programme, so that user needs require clear definition at the outset. Furthermore it is often argued that people do not know what they want or that the needs they express are only in terms of the types of house with which they are familiar, or of the social norms to which they subscribe. A natural extension of this line of reasoning is that housing needs would be better defined by a panel of experts, whose views would be less influenced by their immediate social environment. The more recent findings of anthropology, sociology and psychology have shown this view of 'natural man' to be untenable. Hole[15] postulates that if the pattern of user activities is related to the characteristics of the house plan on the one hand, and to characteristics of the user such as age and stage in the family cycle, education and social class on the other, it is possible to determine whether the user's pattern of activity is due to constraints imposed by the physical environment, or whether they reflect the needs of different categories of user. Roberts[16] has described how in assessing housing requirements of occupants in Hackney, an analysis was made of the features that most families want in a new home, such as privacy, safety, convenience, attractive setting, etc. The mounting pressure to limit costs and secure maximum value is likely to lead to a measure of standardisation in constructional methods.

The majority of persons in this country favour low rise dwellings and the reasons most commonly advanced for not wanting high rise dwellings are:

(1) inconvenience stemming from use of lifts and stairs;
(2) lack of garden;
(3) lack of privacy;
(4) sharing of some facilities;
(5) noise from adjoining tenants, despite sound insulation;
(6) unsuitability for young children; and
(7) height factor and possible lack of safety.

Indeed, as long ago as 1875 Octavia Hill[17] expressed the view that even the poorest would like to have their own homes to themselves, and she found from her work in Deptford that the smallest cottages were the most popular, because in them a family was more likely to achieve its ambition of having a whole dwelling to itself.

A Brick Development Association survey[18] showed that of the 1 845 000 persons obtaining mortgages through building societies during the period 1965–8, approximately forty per cent were for semidetached houses, and twenty per cent for each of detached houses, bungalows and terraced houses. This does not necessarily prove that semidetached houses are the most popular, as the numbers available would be greater than for the other categories and might therefore be merely an indicator of relative availability. A survey which asked persons what type of residence they would like showed thirty-six per cent wanting bungalows, thirty-nine per cent detached houses and fifteen per cent semidetached houses.

Changing occupational status, car ownership and educational attainments of a considerable proportion of the population must influence the types of accommodation required. In quite recent times there has been emphasis on the separation of the laundry from the kitchen and, in some cases, a demand for personal bathrooms, playrooms and 'dens', to provide separate space for individual activities. There has also been a trend toward open plan interiors and more living space and it has been suggested that the average family house with a present day floor area of about 84 sq m could increase to around 100 sq m by the end of the century.[11] Increased car ownership will accentuate present access problems to houses and may point the way to more housing layouts incorporating arrangements for the separation of vehicles and pedestrians.

As well as requiring larger homes, occupants of the future will in all probability also expect better homes providing greater comfort in terms of warmth, ventilation, noise levels and privacy.[19] Differing surveys put the number of centrally heated houses in the United Kingdom at between three and three-and-a-half million in 1969 (eighteen to twenty-one per cent of the total housing stock). The position is however changing quite rapidly as a high proportion of new houses have central heating installations (ninety per cent of the dwellings in the public sector for which tenders were approved in 1969 had some form of central heating[9]). Different patterns in the use of space within the home may develop in response to a more even distribution of warmth.[12]

PATTERNS OF DEVELOPMENT TO MEET VARYING DENSITY REQUIREMENTS

Higher Residential Densities

The protagonists of high density developments base their philosophy on the maximum use of highly priced land, restricting the loss of agricultural land and reducing the extent of urban sprawl. Wibberley[20] has shown that the present average increase in agricultural productivity of 1.3 per cent per annum will more than offset the additional land likely to be needed for all urban uses. The Town and Country Planning Association[21] abhors high density residential development on both cost and social grounds. Hole[12] draws attention to the evidence that mothers experience real difficulties in coping with small children in high flats, and certain values associated with the privacy of the family appear to be threatened by flat life. Much has still to be learnt about the design, management and maintenance of communal spaces in and around flats. Hole and Allen[22] have demonstrated that certain types of persons such as elderly persons or families whose occupation necessitates living near the job can be reasonably well suited in high rise development. Economic and social objections to high rise blocks have however precipitated interesting experiments in low rise high density development which will be described later in this chapter. Nevertheless, the large proportion of childless households and the deficiency of small dwellings suggest that flats, including some in high blocks, have a useful part to play in the housing stock.[12]

Table 5.1 shows the proportion of flats built in England and Wales since the last war, and how they assumed a significant proportion of all dwellings in the public sector in the last decade.

TABLE 5.1

PROPORTION OF FLATS BUILT IN ENGLAND AND WALES

Type of flat	*Period*	*Total dwellings completed*	*Flats as percentage of total dwellings*
Local authorities and new towns	1945 to 1960	2 017 215	23
Local authorities and new towns	1961 to 1969	1 137 365	49
Private owners	1961 to 1969	1 699 394	9
Public and private	1961 to 1969	2 836 759	25

Source: *Great Britain Housing Statistics, No. 17, Table 7* (1970)

Lord Llewellyn-Davies[23] has examined the premise that the urban quality of the community depends on density. New towns have been attacked by architects and planners for wasting land on low density systems with huge interspaces – windy, draughty and miserable places where human beings are unhappy because of lack of urban feeling. Yet the urban densities of new towns are almost identical with, or higher than, those of older towns. Hence urban design character depends on the local, small conurbations of densities and not on average density across the area as a whole. James,[5] however, postulates that gross density offers a better measure of residential development and asserts that raising gross densities to high levels produces relatively small savings of land and in so doing involves disproportionate increase in net density. At gross densities of 150 to 160 persons per hectare, dwellings take up about half the gross residential envelope, the remainder being needed for ancillary uses; hence any increase in gross density involves double that increase in net density. James[5] believes it to be of first rate importance to our land economy that housing and local facilities should be planned together at medium gross densities of seventy-five to 100 persons per hectare and should not often fall below these figures.

Forms of Development

The form or pattern of development will be very much influenced by the density standard prescribed for the site, although diverse variations in the form and disposition of the building blocks are also possible to achieve a specific density. The primary need is to translate the terms of the brief into a total volume or bulk of building per unit area (hectare), and then to assess the cost effects of distributing this amount of building on the site in blocks of varying height, plan area and shape, with due regard to daylighting, sunlight, access, privacy and other factors. The main objective is to minimise the proportion of the more expensive types of block and to maximise the less costly ones, and this entails a thorough knowledge of the relative costs and capacities of low, medium and tall blocks, of various plan shapes and sizes and a variety of structural forms. Furthermore, consideration must be given to running and maintenance costs in addition to initial capital costs, as dwellings in tall blocks are more expensive in both first and future costs than those in low rise developments.

In the preparation of a layout a designer is faced with a number of relatively fixed factors which have considerable influence on his design, such as nature of site, residential density to be achieved, approximate proportions of dwellings of varying types (for example, one, two, three and four-bedrooms), minimum spacing of blocks to secure satisfactory daylighting and adequate privacy, and in public housing space standards of dwellings may also be specified. Daylight indicators are used at the planning stage to ensure that the new blocks respect the light of others and permit the recommended standards of light to be attained within them, and their use is well described and illustrated in Planning Bulletin 5.[24]

Apart from these fixed factors there are many other matters upon which decisions have to be made by the client or designer or both. This second category of factors includes methods of accommodating large and small families (in what type of block), proportion of low rise development and whether two or three-storey, acceptability of four-storey maisonettes, height limitations on blocks, number of dwellings per floor of tall blocks, extent of provision of children's play spaces and other ancillary facilities, ratio of garages and/or car spaces to dwellings, extent of landscaping and road pattern.

TABLE 5.2

THE EFFECT OF PROVIDING VARIOUS PROPORTIONS OF ROOMS IN HOUSES

Overall net density for the whole site: hrh.	*Percentage of rooms in the scheme provided in houses laid out at 150 hrh.*	*Density which has to be achieved over remainder of site (in flats and maisonettes): hrh.*	*Percentage of rooms in the scheme which could be provided in houses for the same density over remainder of site, if houses were laid out at 200 hrh.*
250	20	300	40
	30	350	53
	40	450	64
350	10	410	16
	15	458	24
	20	525	31
400	5	437	8
	10	490	16
	15	567	23

Source: *Flats and houses* (1958)

It is a generally established principle that large families are best housed in low blocks and only small families accommodated in high blocks. The proportion of houses that can be incorporated in a development will be largely dependent on the density adopted. In medium density developments a maximum allocation of houses is a primary aim, but with higher densities it would be unwise to include a large number of houses as this would result in the provision of many excessively tall expensive blocks to achieve the required density. Three-storey houses and flats and four-storey maisonettes have a useful role to play in medium density schemes. The introduction of one or two high blocks enables a high density to be achieved for that part of the site and so releases a larger area for cheaper and more popular low rise buildings. Where very high densities are required, it is more economical to use tall blocks with a greater number of flats per floor, possibly increasing the provision from four to eight.

Table 5.2 shows the effect of providing various proportions of rooms in houses at overall net residential densities of 250, 350 and 400 hrh (habitable rooms per hectare). At an overall density of about 250 hrh, it is desirable to provide a proportion of houses which does not require a density for the remainder of the site in excess of 275 to 300 hrh; this density can probably be obtained with four-storey maisonettes. A high proportion of houses entails the use of expensive high blocks. At high densities a proportion of dwellings in high blocks is generally necessary;

and this can be kept to a minimum only by limiting the proportion of rooms in houses. Major advantages stemming from the use of four-storey maisonettes include increased numbers of dwellings with private gardens and with entrances at ground level.

The following general criteria warrant consideration in the design of housing schemes.

(1) Wherever the required density can be achieved without the use of high blocks, this should be done on grounds of economy alone. The ideal solution is to incorporate the maximum number of two-storey houses with the balance of three or four-storey blocks.

(2) Where some high blocks are needed to secure the required density the number of high blocks should be kept to a minimum and use can advantageously be made of cheaper four-storey maisonettes.

(3) At higher densities it is often good practice to use a few very high blocks of up to twenty storeys rather than a larger number of medium rise blocks.

(4) Compact layouts assist in keeping the quantity of high rise building to a minimum and in securing the greatest possible amount of low rise development to obtain the required density. It also helps to provide a more urban character to the development.

(5) More economical layouts can often be secured by the use of three-storey houses.

The National Building Agency[25] designed some minimum frontage, three-storey blocks of flats at Harlow New Town for use on restricted infill sites and these are illustrated in figure 5.2. Each block contains six flats and four garages with two-person (one-bedroom) flats and garages on the ground floor and four-person (two-bedroom) flats on the first and second floors. The frontage of each block was kept to a minimum and was determined by the sum of the widths of four bedrooms, plus wall thicknesses. The minimum bedroom width of 2550 mm conforms to the recommendations in DB6 (*Space in the Home*) and this is compatible with the width of a garage below, and the flat frontage of 6300 mm proved to be a economical span for precast concrete floor units supported by crosswalls. All controlling dimensions are in multiples of 300 mm.

Housing Layouts

There are three main forms of housing estate layout: the conventional or corridor street layout; use of culs-de-sac and courts; and finally various types of Radburn layout. Each will now be considered in turn.

Conventional layouts. In these, most of the houses front onto and have direct access from a development or estate road, and the back gardens to the houses are enclosed. Access to the backs of houses and enclosed gardens may be obtained through side passages, covered ways or stores to houses. The dwellings are sited on either side of a street, and separated from each other by front gardens, public footpaths and carriageway. This form of layout can become monotonous unless the houses are skilfully sited and external elevations of houses varied. There is no attempt to secure separation of vehicles and pedestrians.

Culs-de-sac layouts. In the garden cities a conscious attempt was made to produce more attractive layouts by grouping houses around culs-de-sac and grassed areas. These arrangements have been incorporated into many public and private housing schemes during the last forty years. Large numbers of culs-de-sac can however produce a feeling of restlessness and long culs-de-sac without adequate turning areas can be very inconvenient. The provision of culs-de-sac in moderation can be an attractive feature and also assist in securing economical layouts by opening up awkwardly shaped interior plots of land and by improving the house to road ratio, with houses surrounding the head of the cul-de-sac.

Radburn layouts. The first form of Radburn layout was used at Radburn, New Jersey, USA between the wars, in which the houses with their private gardens backed on to culs-de-sac and fronted on to communal landscaped areas, through which footpaths gave direct access to schools, shops and amenities without crossing a road. A variety of forms of Radburn layout are now used but the primary aim of all of them is to separate pedestrians and vehicles and so secure more pleasant and safer residential environments. Indeed the choice of route for pedestrian ways should have priority over roads, with regard to directness of route, gradients and intersections with roads.

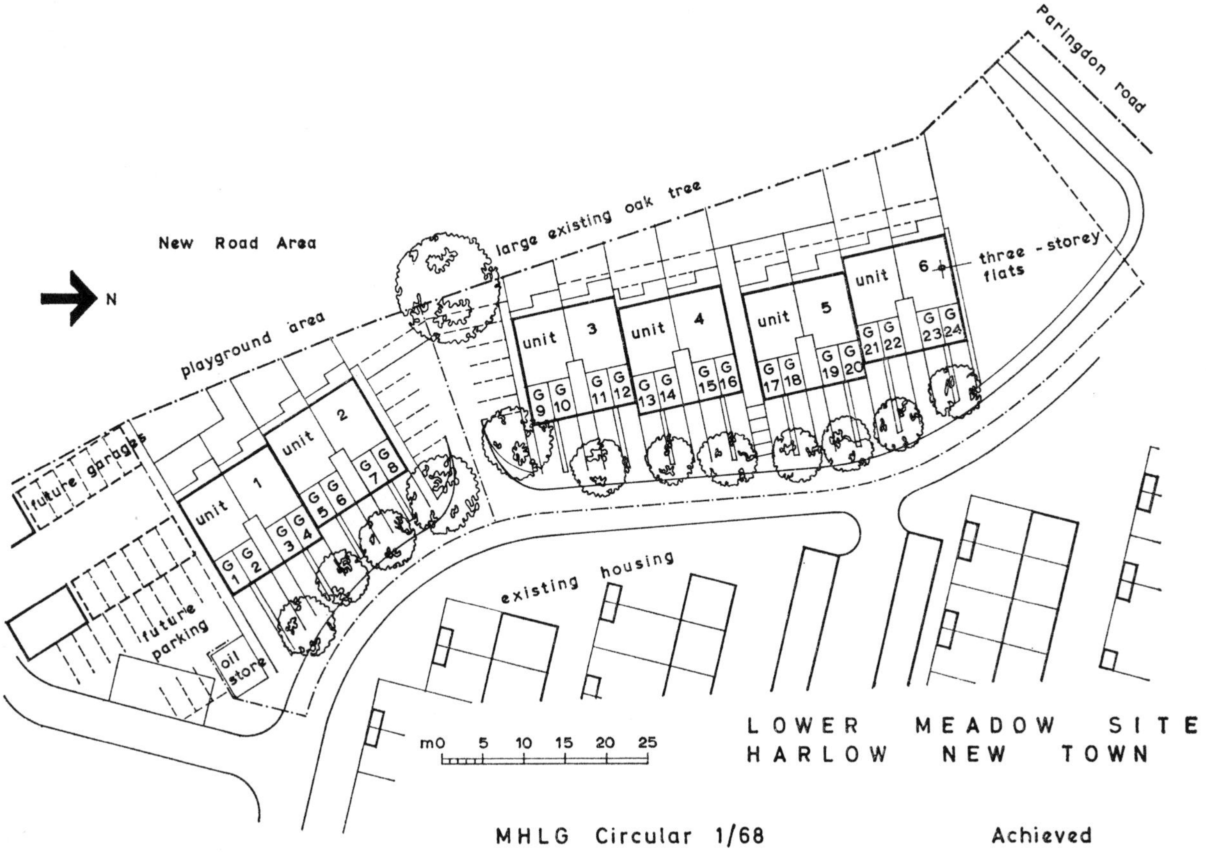

New Road Area

N

playground area

large existing oak tree

future garages

future parking

oil store

unit 1 unit 2 unit 3 unit 4 unit 5 unit 6

G 1 G 2 G 3 G 4 G 5 G 6 G 7 G 8 G 9 G 10 G 11 G 12 G 13 G 14 G 15 G 16 G 17 G 18 G 19 G 20 G 21 G 22 G 23 G 24

three-storey flats

Paringdon road

existing housing

m 0 5 10 15 20 25

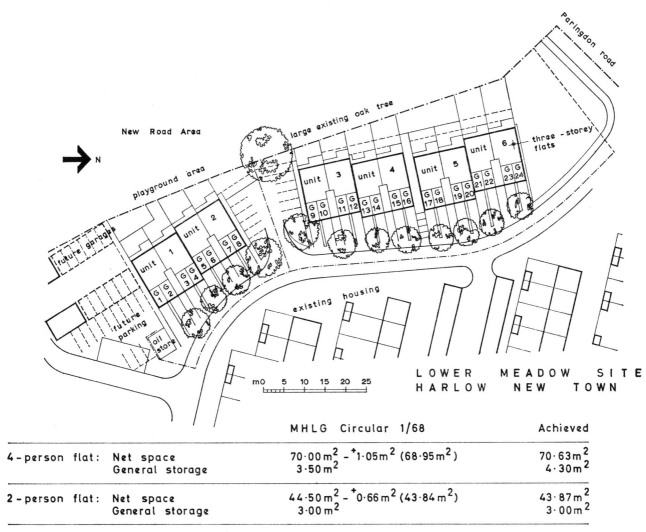

LOWER MEADOW SITE
HARLOW NEW TOWN

		MHLG Circular 1/68	Achieved
4-person flat:	Net space	$70 \cdot 00 m^2 - {}^{+}1 \cdot 05 m^2$ ($68 \cdot 95 m^2$)	$70 \cdot 63 m^2$
	General storage	$3 \cdot 50 m^2$	$4 \cdot 30 m^2$
2-person flat:	Net space	$44 \cdot 50 m^2 - {}^{+}0 \cdot 66 m^2$ ($43 \cdot 84 m^2$)	$43 \cdot 87 m^2$
	General storage	$3 \cdot 00 m^2$	$3 \cdot 00 m^2$

+ Maximum minus tolerance of 1½%.

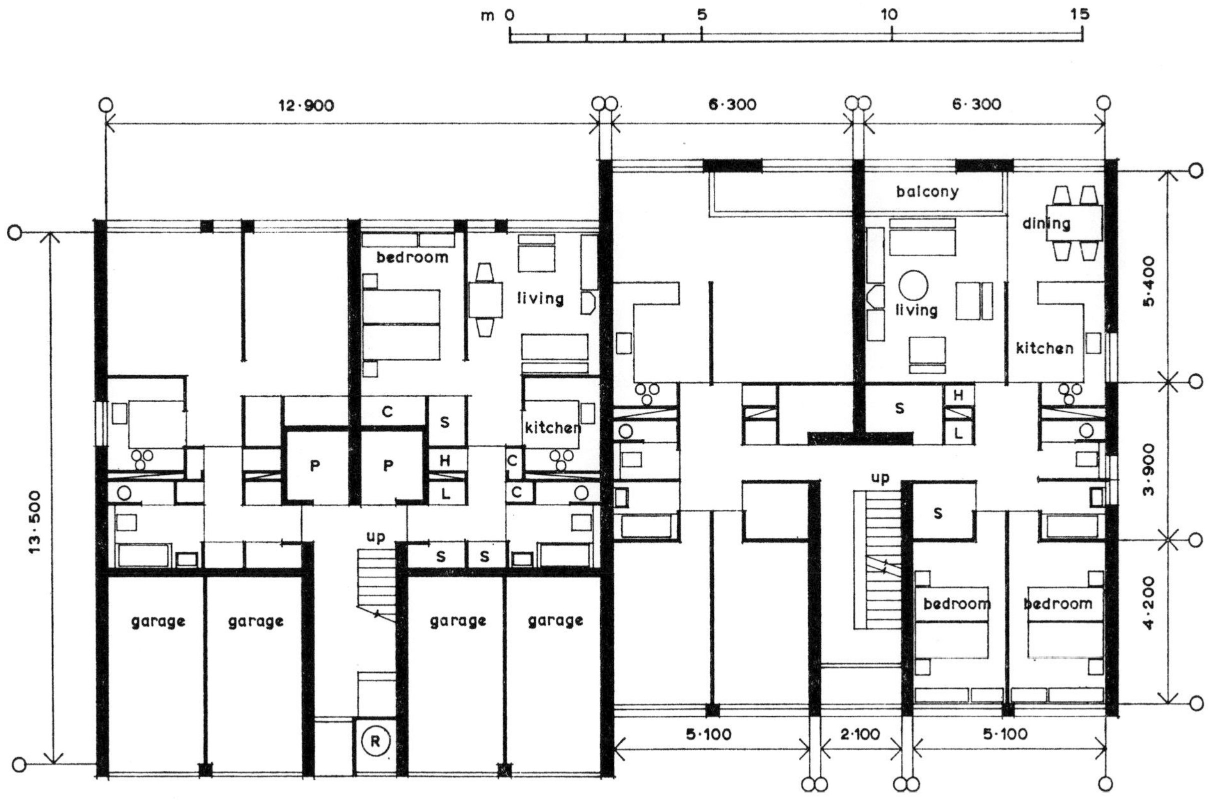

left, ground floor plans; right: first and second floor plans

Figure 5.2 Three-storey flats infill development – Harlow New Town

left, ground floor plans; right, first and second floor plans

Figure 5.3 shows one form of Radburn layout at Cwmbran New Town which clearly demonstrates the dominance of the central footpath system, with the culs-de-sac so aligned that they do not interfere with the flow along the main pedestrian routes. The majority of the houses have an east/west aspect and the private back gardens face west to south for maximum sunlight.

Figure 5.4 illustrates another form of Radburn layout at the expanding town of Andover, Hampshire. It is an interesting scheme consisting of a series of closes extended laterally from short culs-de-sac off perimeter distributor roads. It possesses a compact urban character with vehicles restricted to the perimeter of the residential block. Private back gardens face west or south but quite a high proportion of houses front on to rear boundary fences of other houses.

The Ministry of Housing and Local Government[26] has shown that the grouping of houses in a Radburn scheme is likely to conform to one of three basic types.

(1) Vehicle cul-de-sac, with a turning circle or hammerhead at the end of the carriageway and with individual or grouped garages adjoining it.

(2) Garage court, with the carriageway widened to form a single large enclosure for vehicles and grouped or individual garages.

(3) Pedestrian forecourt, whereby the head of the cul-de-sac or garage court is extended to form a paved pedestrian area from which each house is entered. Garages are grouped away from the houses. Variations on the pedestrian forecourt include the pedestrian link where a pedestrian way, lined with houses, forms a link between two vehicle access points, and the pedestrian passageway where at high densities the forecourt becomes part of a network of footways between houses.

Radburn layouts do, however, increase the complexity of schemes and involve additional external works. The National Federation of Building Trades Employers[27] has suggested that in many cases occupiers express a preference for a more traditional layout and that the extra cost of Radburn schemes is not entirely offset by improved amenity.

Wates[19] postulates that the high cost of building land coupled with the general desire to give coherence to residential neighbourhoods is likely to lead to most new developments being within the gross residential density range of 175 to 225 persons per hectare. To achieve these densities with larger homes, accommodation for two cars per household and parking space for visitors will require skilful planning and must mean that mixed rise developments will become the norm, possibly incorporating an increasing amount of patio-housing.

ECONOMICS OF HOUSING LAYOUTS

General Considerations

As the demand for higher densities increases so the need for deeper cost investigations and tighter cost control becomes that much greater. The design of multistorey housing will condition its cost. Such design features as the number of storeys; form of layout; means of access; shape, size and type of dwelling; form of construction; quantity and quality of fittings and services, and similar matters will have a significant influence on cost. The former RICS Cost Research Panel[28] believed that it was vital that a quantity surveyor should be available to make cost appraisals of such projects at the design stage.

The first stage of cost investigation should consider the most economical use of a particular site. In schemes of comprehensive redevelopment, consideration should be given to the best combinations of high and low blocks in different forms and dispositions, in slab and/or tower form, to meet the site conditions and housing policy and accommodation requirements. In general, high density is achieved most economically by a combination of a few very high blocks and a majority of lower blocks. Such considerations underline the importance of cost investigation into patterns of development. After the basic development and design decisions have been made, there is still ample scope for investigating the use of various materials and forms of construction. Their relative costs can be scheduled in a 'cost study' to enable the architect to select any desired combination in full knowledge

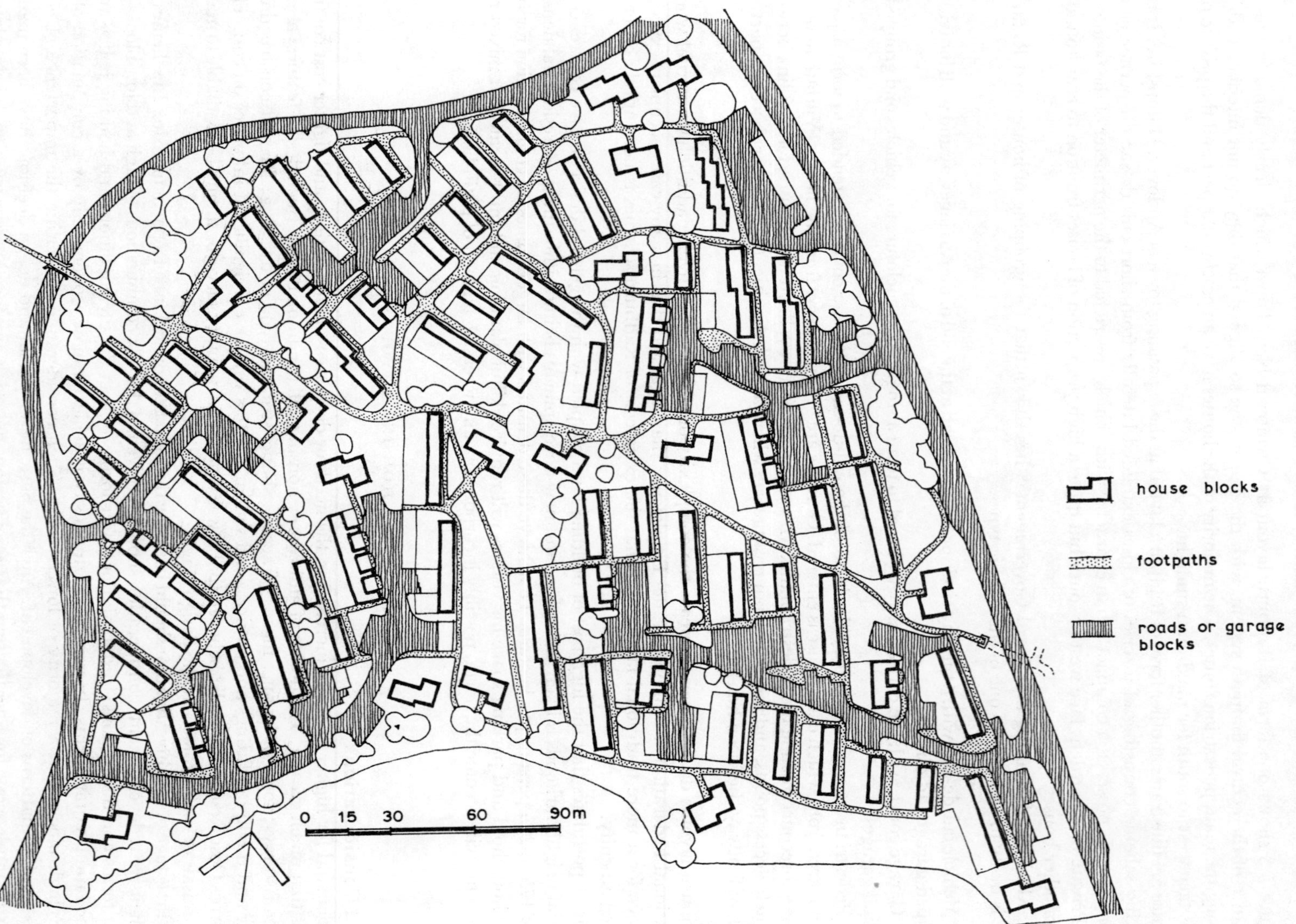

Figure 5.3 Radburn layout – Cwmbran

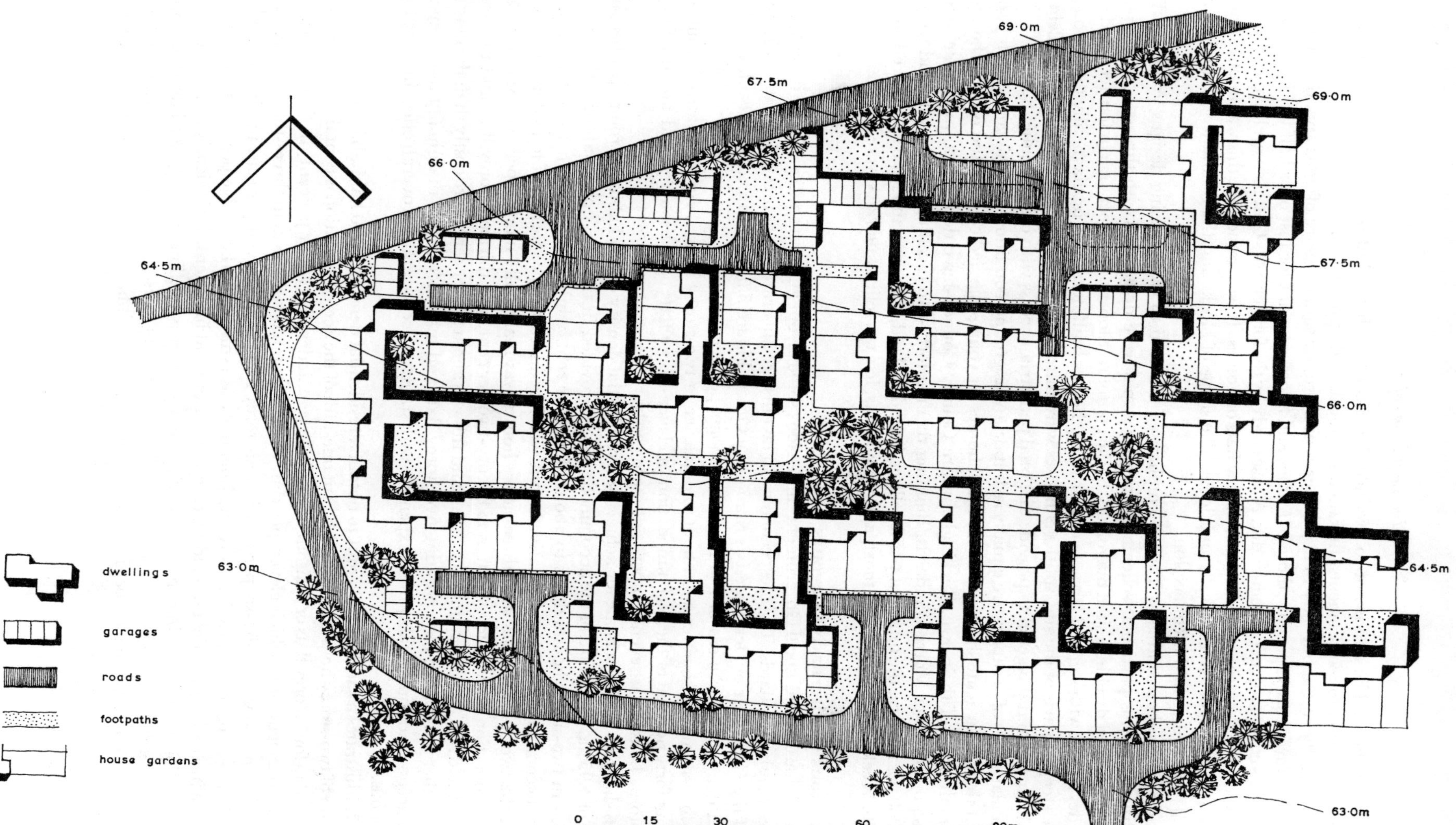

Figure 5.4 Radburn layout – Andover

of their cost implications. The aim is not to produce the cheapest scheme but one which maintains a proper balance of cost in relation to requirements, function, standards and appearance.

High flats generally cost about sixty to eighty per cent more per square metre of floor area than two-storey houses. Superstructure costs, especially those of frames and floors, account for just over half this extra cost, and service and lift costs cover most of the remainder. A case study by the former RICS Cost Research Panel[29] indicated that high flats showed a greater range of costs than low flats or houses. It was found that this was mainly attributable to variations in costs of external walls and floors. The shape of blocks and the resultant wall to floor ratios were of special significance in this connection.

A cost study at Cumbernauld New Town[30] emphasised the need to vary housing designs to meet different ground conditions. In this particular scheme the gentler slopes were developed with three-storey, six-person, narrow-fronted terraced dwellings sited diagonally to the contours. Slopes of between one in fourteen and one in twenty were developed with stepped terraces of two-storey houses built normally, or diagonally to the contours. Whilst on the steepest parts of the site, splitlevel houses were introduced in terraced blocks built parallel to the contours, and stepped and staggered to provide diagonal pedestrian access with easier gradients.

The particular needs of occupiers are equally as important as ground conditions. In this connection Pooley[31] has found that there is an acute shortage of very small houses for young married couples and others who want to purchase a small place of their own. Buckinghamshire County Council have designed a 'minihome' with an area of 48.4 sq m to Parker Morris standards. Building them at forty-two dwellings per hectare and with land costing £25 000 per hectare, the total cost per dwelling, including land, siteworks and legal fees, was estimated at £2950 in 1970. Weekly outgoings in the mortgage repayments, less tax relief, and rates were less than £5.

Cost Implications of Alternative Layouts

The Ministry of Housing and Local Government[8] undertook some layout and cost studies of a site in Birmingham based on three density standards. The most economical layouts are illustrated in figures 5.5, 5.6 and 5.7, by kind permission of the Controller of Her Majesty's Stationery Office. Figure 5.5 shows an economical layout for the site to meet a density requirement of 250 habitable rooms per hectare, with about two-thirds of the accommodation in four-storey maisonettes and about one-quarter in two-storey houses. An alternative layout which provided about forty per cent of the accommodation in eleven-storey flats or maisonettes and another forty per cent in two-storey houses was seventeen per cent more costly.

Figure 5.6 shows a suitable scheme for achieving a residential density of 350 habitable rooms per hectare on the same site of 4.95 ha. One-third of the accommodation is provided in thirteen-storey flats or maisonettes and the proportion in two-storey houses drops to about fifteen per cent, whilst four-storey maisonettes account for one-half of the accommodation. An alternative layout which reduced the proportion of four-storey maisonettes and replaced the two-storey houses with three-storey houses was marginally more expensive. Another scheme increased the proportion of two-storey houses to twenty-two per cent, and thirteen-storey flats and maisonettes then accommodated half the total number of rooms and this was about six per cent more costly than the illustrated layout. Yet another alternative incorporating five-storey flats and maisonettes was marginally more expensive with balcony access, and considerably more expensive with staircase access, and the overall effect in any case was very monotonous.

Figure 5.7. illustrates a satisfactory layout to obtain a residential density of 400 habitable rooms per hectare. The proportion of two-storey houses is negligible, but one-fifth of the accommodation is contained in three-storey houses, one-third in four-storey flats or maisonettes and the remainder (forty-four per cent) in sixteen-storey flats. An alternative arrangement in which eleven per cent of the accommodation was in three-storey houses, ten per cent in four-storey maisonettes, fifty-six per cent in ten-storey maisonettes, seven per cent in nine and thirteen-storey flats and sixteen per cent in twelve-storey maisonettes was estimated to be twelve per cent more expensive.

The effect of increasing densities on building costs at medium and high densities is clearly indicated in Design Bulletin No. 7.[32] For instance, taking an average of three persons per dwelling, the average late 1962 costs per person at a density of 200 persons per hectare were £578, £736 for 300 persons per hectare, £852 for 400 persons per hectare and £920 for 500 persons per hectare. Although prices will have increased significantly since 1962,

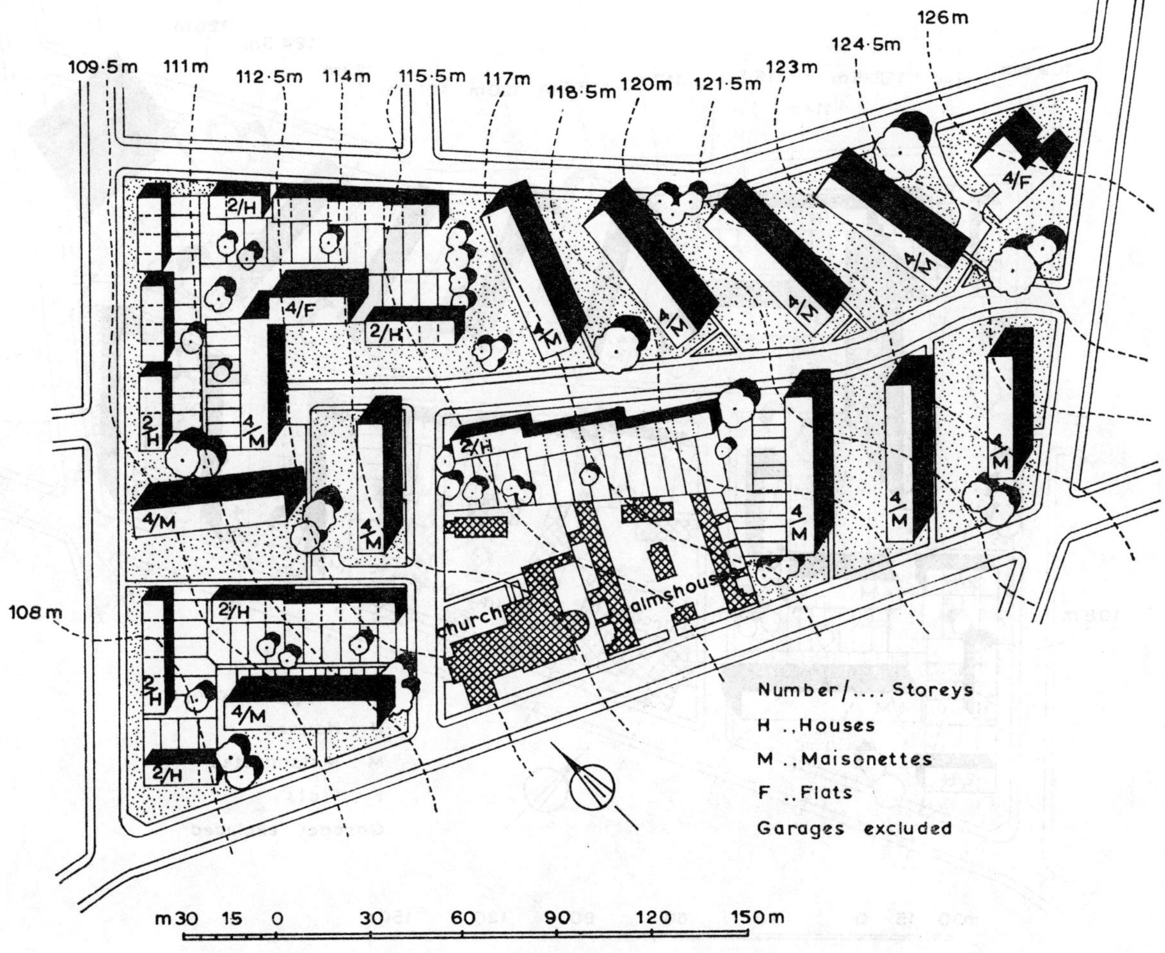

250 habitable rooms per ha: 4·95 ha: 1,234 habitable rooms: 327 dwellings.

Types of dwelling and percentage required	2-Storey houses		4-Storey flats		4-Storey maisonettes		Dwlgs.	Rms.
	Dwlgs.	Rms.	Dwlgs.	Rms.	Dwlgs.	Rms.		
BSR. 1p. 1 rm. 5 %.	-	-	16	16	-	-	16	16
1 BR. 2p. 2 rms. 5 %.	-	-	16	32	-	-	16	32
2 BR. 4p. 3 rms. 20 %.	-	-	8	24	58	174	66	198
3 BR. 4p. 4 rms. 25 %.	-	-	-	-	82	328	82	328
3 BR. 5p. 4 rms. 30 %.	14	56	-	-	84	336	98	392
3 BR. 6p. 5 rms. 8 %.	26	130	-	-	-	-	26	130
4 BR. 7p. 6 rms. 7 %.	23	138	-	-	-	-	23	138
Totals	63	324	40	72	224	838	327	1,234
Percentages	19·3	26·2	12·3	5·8	68·8	67·9	-	-

Figure 5.5 Residential development at 250 habitable rooms per hectare

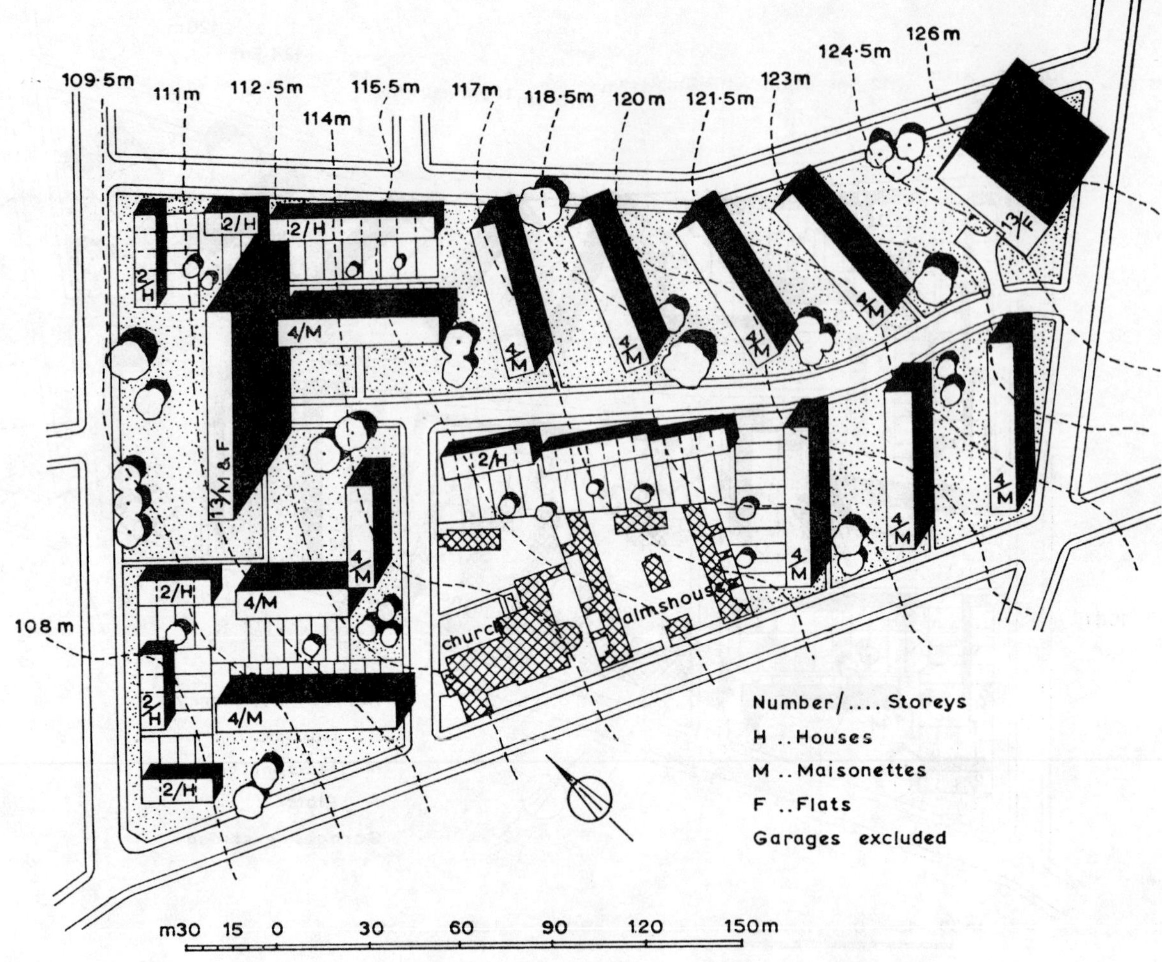

350 habitable rooms per ha: 4·95 ha;
1,716 habitable rooms: 456 dwellings.

Types of dwelling and percentage required	2-Storey houses		4-Storey maisonettes		13-Storey flats		13-Storey maisonettes		Dwlgs.	Rms.
	Dwlgs.	Rms.	Dwlgs.	Rms.	Dwlgs.	Rms.	Dwlgs.	Rms.		
BSR. 1p. 1rm. 5%.	-	-	-	-	24	24	-	-	24	24
1 BR. 2p. 2rms. 5%.	-	-	-	-	24	48	-	-	24	48
2 BR. 4p. 3rms. 20%.	-	-	30	90	59	177	-	-	89	267
3 BR. 4p. 4rms. 25%.	-	-	30	120	-	-	84	336	114	456
3 BR. 5p. 4rms. 30%.	-	-	136	544	-	-	-	-	136	544
3 BR. 6p. 5rms. 8%.	15	75	22	110	-	-	-	-	37	185
4 BR. 7p. 6rms. 7%.	32	192							32	192
Totals	47	267	218	864	107	249	84	336	456	1,716
Percentages	10·3	15·5	47·9	50·3	23·5	14·5	18·4	19·5	-	-

Figure 5.6 Residential development at 350 habitable rooms per hectare

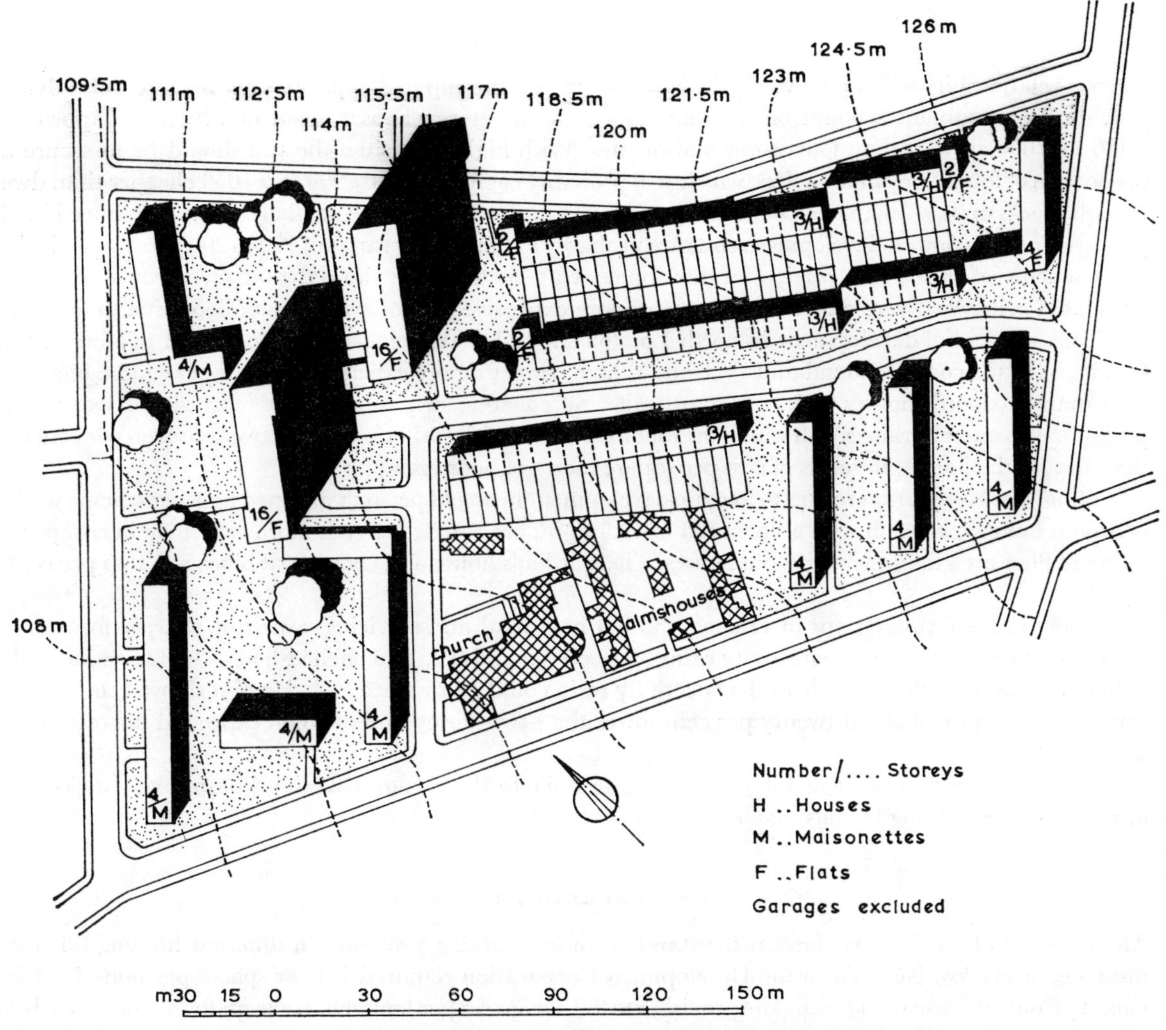

400 habitable rooms per ha: 4·95 ha;
1,966 habitable rooms: 520 dwellings.

Types of dwelling and percentage required	2-Storey flats		3-Storey houses		4-Storey flats		4-Storey maisonettes		16-Storey flats		Dwlgs.	Rms.
	Dwlgs.	Rms.	Dwlgs.	Rms.	Dwlgs.	Rms.	Dwlgs.	Rms.	Dwlgs.	Rms.		
BSR. 1p. 1rm. 5%	-	-	-	-	24	24	-	-	-	-	24	24
1 BR. 2p. 2rms. 5%	6	12	-	-	16	32	-	-	-	-	22	44
2 BR. 4p. 3rms. 20%	-	-	-	-	-	-	-	-	112	336	112	336
3 BR. 4p. 4rms. 25%	-	-	-	-	-	-	-	-	128	512	128	512
3 BR. 5p. 4rms. 30%	-	-	-	-	-	-	152	608	4	16	156	624
3 BR. 6p. 5rms. 8%	-	-	42	210	-	-	-	-	-	-	42	210
4 BR. 7p. 6rms. 7%	-	-	36	216	-	-	-	-	-	-	36	216
Totals	6	12	78	426	40	56	152	608	244	864	520	1,966
Percentages	1·2	0·6	15·0	21·7	7·7	2·8	29·2	30·9	46·9	44·0	-	-

Figure 5.7 Residential development at 400 habitable rooms per hectare

the cost relationships will not be vastly different. With densities up to 250 persons per hectare, it is advisable to provide the maximum amount of accommodation in two-storey houses, each of which are approximately £450 to £600 cheaper than four-storey maisonettes. With higher densities the aim should be to secure a high proportion of four-storey maisonettes, which will probably each be about £900 to £1050 cheaper than dwellings in high blocks of flats. Increased construction costs will at the very least be partially offset by reduced land costs per dwelling. The lowest development costs are probably in the density range of 120 to 200 persons per hectare.

A scheme of mixed rise residential development at Portsdown[33] showed that seventeen-storey blocks in calculated brickwork were only twelve per cent more expensive per square metre of floor area than four-storey blocks. A residential development in north London showed that a nine-storey block with reinforced concrete frame and *in situ* concrete foundations was sixty-three per cent more expensive in costs per square metre of floor area than two-storey houses with brick crosswalls and concrete strip foundations (exclusive of external works, preliminaries and insurances), although the foundation cost of the taller block was twenty-seven per cent cheaper than that of the houses, in terms of cost per square metre of floor area.

A final cost comparison relates to low rise development in an old person's grouped housing scheme where one-bedroom bungalows, each with a floor area of 49.5 sq m, are eleven per cent more expensive in cost per square metre of floor area than two-bedroom houses. The warden's house and communal block are 13.5 per cent more costly.

Jarle[34] in his investigations in Finland showed how road and servicing costs rose sharply from the use of multistorey housing, with an effectivity ratio between 0.5 and 0.7, to detached houses beside residential roads with footpaths on both sides, where the effectivity ratio could be as low as 0.1. He also showed the cost of four-storey dwellings to be about twenty per cent lower than two-storey houses when expressed in costs per square metre of floor area.

The reader might find it useful at this stage to refer to the section covering housing yardsticks and their implications for housing layouts, in chapter 8.

CAR PARKING PROVISION

There seem to be wide variations in the standard of car parking provision in different housing schemes. For instance, at Harlow New Town the Development Corporation required 1.3 car spaces per household but the County Council insisted on raising the provision to 2.0 car spaces per household, except for old person's dwellings where the rate of provision was 0.5 spaces per dwelling unit.[25] By way of contrast, Cumbernauld New Town decided on one car space per household in high rental areas and 1/3 space per household in lower rental areas.[35]

The cheapest form of provision is that of open hardstandings which are often screened with trees for amenity reasons. They give no protection to cars from either the weather or vandals but their cost is unlikely to exceed £60 per car space (1971 costs).

Carports consisting of a roof and possibly one side have become quite popular in recent years. They give only limited protection from the weather and 1971 costs were in the order of £150 each.

Detached garages erected close to dwellings give a maximum of convenience and protection but are costly, probably in the order of £350. The cost will be reduced if the garage is built beside a house or in pairs between houses. Three-storey houses can accommodate garages on the ground floor when costs are likely to be around £375 each (1971 prices).

On local authority housing estates it is customary to provide garages in single-storey blocks to serve groups of houses. Costs are reduced to about £250 per garage and they permit greater flexibility in the letting of dwellings. They are less convenient for occupiers, are more vulnerable to vandalism and result in vehicular noise concentration at specific points on housing estates. Probably the most economical form of lock-up garage layout is that shown in figure 5.8 where constructional costs and apron widths are kept to a minimum. Circular blocks of lock-up garages have been provided at Cumbernauld New Town.

On a high density residential development it may be desirable to provide multistorey garages with two or three storeys, to conserve land and meet increasing car parking needs. The cost is likely to be in the order of

£450 to £500 per car space. Another alternative with high rise blocks is to provide basement car parking areas under some of the blocks, particularly where they are sited on sloping ground. This is a very expensive method of garage provision and could cost as much as £650 per car space (1971 prices).

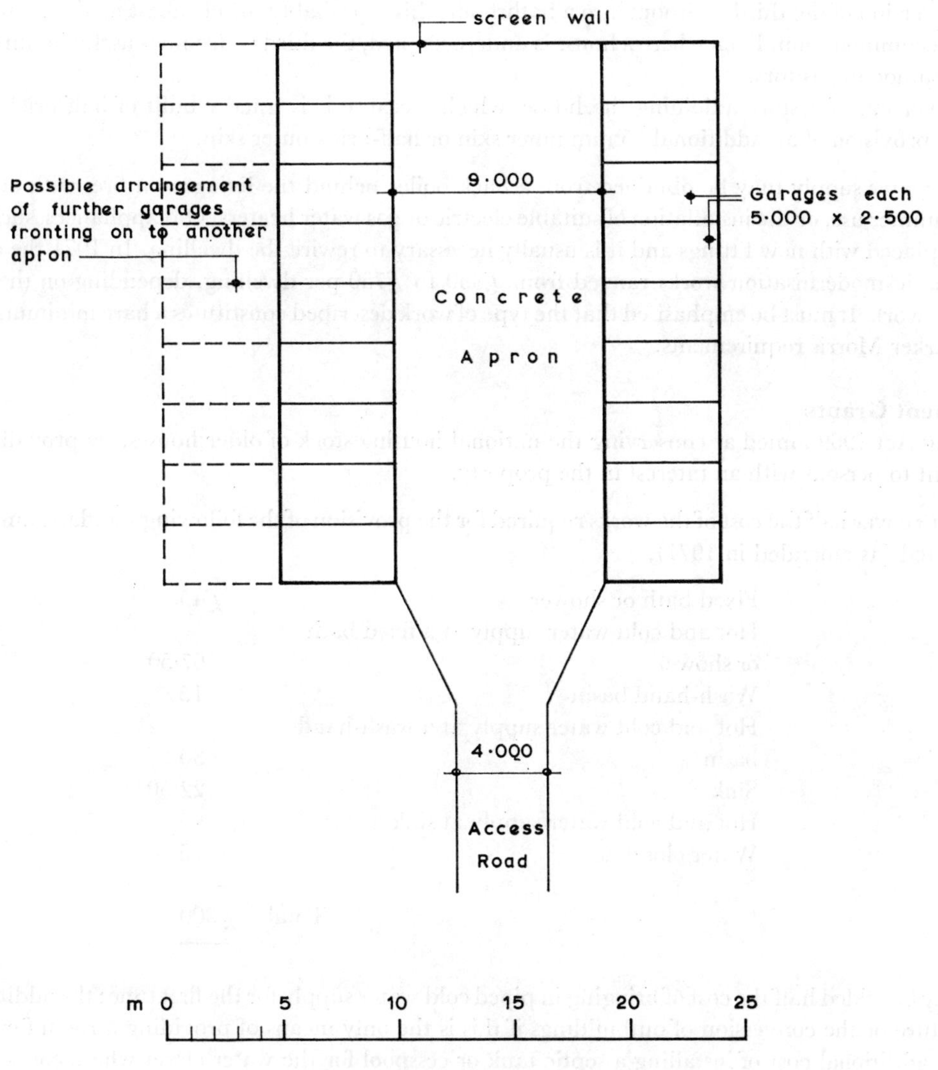

Figure 5.8 Blocks of lock-up garages

REHABILITATION AND MODERNISATION OF OLD DWELLINGS

Modernisation of Older Houses

In 1967 the Minister of Housing and Local Government stated that about five million houses in England and Wales were unsatisfactory to a lesser or greater degree. Approximately four million of these houses lacked one or more of the basic amenities such as internal wc, fixed bath, hot and cold water supply, and wash basin.

Indeed many prewar houses lack proper bathrooms and hot water supply systems, and frequently the only wc in the house is entered from outside. Many of these houses are now being modernised by the provision of modern bathrooms, internal WCs and efficient hot-water supply systems. There are three principal methods of approach in providing minimum basic needs.

(1) An internal rearrangement on the ground floor to provide a bathroom and WC, if sufficient space is available.

(2) Conversion of the third bedroom into a bathroom; this is probably the cheapest method but reduces the amount of accommodation. Even where a house is underoccupied, the third bedroom is useful in times of sickness and to accommodate visitors.

(3) Conversion of a store adjoining the house, which, because it is usually built of half-brick walls, often requires the provision of an additional 75 mm inner skin or half-brick outer skin.

The hot water supply may be obtained from a back boiler behind the livingroom fire, with an immersion heater for summer use, or the installation of suitable electric or gas water heaters. Old appliances, such as shallow sinks, are replaced with new fittings and it is usually necessary to rewire the dwelling. In 1971 the cost of these minimum basic modernisation works ranged from £350 to £750 per dwelling, depending on the extent and nature of the work. It must be emphasised that the type of work described constitutes a bare minimum and would not meet Parker Morris requirements.

Improvement Grants

The Housing Act 1969 aimed at conserving the national housing stock of older houses, by providing for three types of grant to persons with an interest in the property.

Standard grants cover half the cost of the works required for the provision of the following standard amenities up to the limits listed (as amended in 1971),

Fixed bath or shower	£45
Hot and cold water supply at a fixed bath or shower	67·50
Wash-hand basin	15
Hot and cold water supply at a wash-hand basin	30
Sink	22·50
Hot and cold water supply at sink	45
Water closet	75
Total	£300

To these may be added half the cost of bringing in piped cold water supply for the first time; the additional cost of a new structure or the conversion of outbuildings if this is the only means of providing a room for the bath or shower; the additional cost of installing a septic tank or cesspool for the water closet where connection to the main drainage is impossible; and essential additional works up to a total maximum of £675.

Improvement grants, formerly called discretionary grants, remain discretionary and aim at the provision of dwellings by the conversion of houses or other buildings and for improvements not restricted to the standard amenities. Such grants cover half the approved cost of the works which must not be less than £100. The grant is limited to £1200 for dwellings provided by conversion of a building on three or more storeys, or £1000 for other dwellings.

Special grants are also discretionary and cover works required for the provision of standard amenities in houses in multiple occupation. The amount allowable for each amenity is the same as that for standard grants.

The Housing Act 1971 provided that householders in development and intermediate areas could qualify for improvement grants of up to £1500, as against the figure of £1000 contained in the 1969 Act. The areas affected were primarily parts of north-east and north-west England, the whole of Scotland and most of Wales. The reason for this legislation was that the number of grant applications in these areas was much less than in the more prosperous parts of the country. The higher grant limit, coupled with the payment by the Exchequer of seventy-five per cent of the cost of the improvements as against the previous fifty per cent, provided the improvement work was completed in two years, was believed to provide sufficient incentive to speed up the implementation of improvement schemes. The National Federation of Building Trades Employers and Forward Trust Ltd. are willing to provide loans, repayable over a ten-year period, in respect of the quarter share of the cost to be borne by the householder.

Rehabilitation of Twilight Areas

In our towns and cities there are large numbers of terraced houses with small paved rear yards which were built mainly towards the end of the last century and which could, with advantage, be improved to extend their useful lives. Trepass[36] has described how a social survey of the area can reveal the tenant's attitude to such a scheme, his willingness to pay more rent should the scheme be implemented, and the household structure for the area. A survey of this kind undertaken in north Liverpool indicated a considerable attachment to the area by the populace, limited car ownership and a willingness to pay higher rents for a better home.

To improve the environment, there must be a certain amount of selective demolition to create open space, redirect roads to give improved traffic circulation and to eliminate the worst property. An abridged building survey can be used to classify the houses into a limited range of categories according to their condition. Consultation with the rent officer and an examination of the rent registers will indicate rent levels in the district. Where the controlled weekly rent is approximately £1 and the present regulated level is around £2, then the anticipated rent level on improvement is likely to be in the order of £3.50 to £4.00.

From a local authority viewpoint the rehabilitation approach offers a number of attractions. The cost of providing improved houses with lives of thirty years is much less than acquiring sites and providing new units of accommodation at a cost probably in excess of £5000 per unit. Where the area is designated as a general improvement area, further grants of £100 per dwelling can be made available to cover tree planting, provision of open spaces and the general improvement of the area as a whole.

Improvement works may include the removal of rear additions to provide more open space, the conversion of three houses into two in some cases to provide three or four-bedroom houses and the removal of chimney stacks. Rearrangement of streets and/or direction of traffic flows, and the provision of play areas and garages are also important to improve the environment.

The quantity surveyor will exercise cost control of the improvements having regard to the financial resources and Government grants available, preferably by the use of a cost plan. Typical costs of improvements to houses in north Liverpool in 1970 were as follows:

	Three-bedroom house	*Two-bedroom house*
Environmental improvement cost (demolition, road features, making good gables, open space)	£ 100	£ 100
General house improvement	2035	1810
Repairs	375	250
	2510	2160
Less grants (improvement and environmental improvement)	1100	1100
	£1410	£1060

With these costings, weekly rents were likely to increase by about £1.90 for three-bedroom and £1.25 for two-bedroom houses.

Trepass[36] has described how various constraints can collectively prevent a scheme from being viable. These include the phasing of increased rents, improvement grants which are not over-generous, the application of grants to individual properties rather than on an area basis, nonrecognition of need to merge houses on occasions, high interest rates and restrictions on sale. McKie[37] has suggested that a policy of *cellular renewal* which encourages house improvements 'matched to demand so that marginal households are not pushed out of the market, coupled with small-scale development, will make better use of resources than the alternative policies of stagnation, comprehensive improvement or comprehensive redevelopment'. McKie is also critical of 'social reformers' who attempt to raise the housing standards of lower income families beyond their means and of housing policies which aim at Parker Morris standards for all households.

Needleman[38] favours improving the general standard of amenity of older houses following their removal from private ownership. This policy, it is claimed, will reduce the inflationary effect of a large rebuilding programme, produce a quicker solution and cause less social disturbance. He identifies three purely economic factors which are relevant in deciding whether it is cheaper to renovate old properties or to demolish and rebuild them – the rate of interest, the future length of life of the renovated property and the difference between the running costs of the renovated property and those of the property which would be necessary to replace it. Needleman's detailed approach, which indicated that it would be economic to spend up to about £1000 on the improvement of a house with an anticipated life of five years, up to £1600 for ten years and more than £2000 for up to fifteen years, has been strongly disputed by Sigsworth and Wilkinson[39] as being unrealistic and excessively optimistic.

Improvement grants payable to local authorities and housing associations tackling twilight areas are assessed at three-eighths of the combined cost of acquisition and works and are limited to £2500 per unit, or £5000 in Greater London because of high acquisition costs. The grant is paid in the form of annual loan charges over a period of twenty years, which at eight per cent interest would amount to a maximum of £189 per annum in London. The cost of local authority housing is in fact split between the tenant, the local authority and the Government. In Islington the rent officer estimated the value of an improved dwelling to be about ninety per cent of a new one.

Although some long-established tenants prefer improved old houses, many more only regard improvement as a temporary patch-up. Some people see redevelopment as a means of escape from existing neighbours and streets, while those who prefer to remain favour improvement. Improvement does not necessarily have the same appeal to council tenants as to owner-occupiers. The main advantages of redevelopment lie in the provision of parking facilities, possibly pedestrian/vehicular segregation and improved daylighting.

Brown[40] sees the Government's concern as the protection of areas of national historic or architectural importance; ensuring a minimum standard of environment and housing; and accelerating urban renewal projects where rendered necessary by obsolete land use patterns. Given these constraints the Government's task is to devise a subsidy system which will promote the most economical form of development.

In his study of a 377-dwelling site at Islington, Brown[40] found that improvement provided the cheaper overall scheme, although probably more expensive to the local authority, as the housing revenue account deficit reverses the overall economic balance in favour of redevelopment. If intangible elements are included in cost-benefit computations, then both types of scheme are roughly comparable. Average costs per dwelling to the local authority in the Islington scheme follow.

Scheme	*Overall average cost per dwelling*	*Subsidy*	*Net cost to local authority*
Redevelopment	£730	£400	£330
Improvement	£520	£180	£340

The possibility of retaining all the houses in an improvement scheme by omitting amenity space or off-street parking should be ignored, as the potential saving could be offset by reduced rent.

REFERENCES

1. M. PAWLEY. Housing: a continuing story of trials and errors. *Building Design* (1 May 1970)
2. E. and G. MCALLISTER. *Town and Country Planning*. Faber (1941)
3. S. WEBB. The dispossessed. *Architectural Design* (March 1970)
4. MINISTRY OF HOUSING AND LOCAL GOVERNMENT. Homes for today and tomorrow. HMSO (1961)
5. J. R. JAMES. Residential densities and housing layouts. *Town and Country Planning* **35.11** (1967)
6. I. H. SEELEY. *Municipal Engineering Practice*. Macmillan (1967)
7. MINISTRY OF HOUSING AND LOCAL GOVERNMENT. Planning Bulletin No. 2: Residential areas – higher densities. HMSO (1962)
8. MINISTRY OF HOUSING AND LOCAL GOVERNMENT. Flats and houses 1958: Design and economy. HMSO (1958)
9. MINISTRY OF HOUSING AND LOCAL GOVERNMENT. Housing statistics No. 19. HMSO (1970)
10. MINISTRY OF HOUSING AND LOCAL GOVERNMENT. Statistics for town and country planning: series 3. Population and households. No. 1. Projecting growth patterns in regions. HMSO (1970)
11. NATIONWIDE BUILDING SOCIETY. The prospect for housing (1971)
12. W. V. HOLE. The effects of current and future social changes on house design. *Building Research Station Current Paper*, **8/71,** Department of the Environment (1971)
13. MINISTRY OF HOUSING AND LOCAL GOVERNMENT. Design Bulletin No. 2. Grouped flatlets for old people. HMSO (1962)
14. BUILDING RESEARCH STATION. Trends in population, housing and occupancy rates 1861–1961. HMSO (1971)
15. W. V. HOLE. User needs and the design of houses. *Building Research Station Current Paper*, **51/68,** Ministry of Public Building and Works (1968)
16. T. R. W. ROBERTS. Low rise, high density. *Town and Country Planning*, **39.2** (1971)
17. O. HILL. Homes of the London Poor. London (1875)
18. BRICK DEVELOPMENT ASSOCIATION. The way we live now . . . and the way we would like to live (1969)
19. R. W. WATES. Homes for the nineteen-eighties. *The Chartered Surveyor* (April 1969)
20. G. P. WIBBERLEY. Pressures on Britain's land resources. Tenth Heath Memorial Lecture, University of Nottingham, School of Agriculture (1965)
21. TOWN AND COUNTRY PLANNING ASSOCIATION. Memorandum to the Secretary of State for the Environment. *Town and Country Planning*, **38.11** (1970)
22. V. HOLE and P. G. ALLEN. A survey of modern dwellings for old people. *Architects' Journal*, **135** (1962) 1017–1026
23. LORD LLEWELLYN-DAVIES. The problems of design. *The Chartered Surveyor* (April 1969)
24. MINISTRY OF HOUSING AND LOCAL GOVERNMENT. Planning Bulletin 5. Planning for daylight and sunlight. HMSO (1964)
25. E. CORKER. Metric flats project: National Building Agency's low rise infill development. *Building* (26 February 1971)
26. MINISTRY OF HOUSING AND LOCAL GOVERNMENT. Design Bulletin 10. Cars in housing/1. some medium density layouts. HMSO (1967)
27. NATIONAL FEDERATION OF BUILDING TRADES EMPLOYERS. The NFBTE and housing policy: memorandum to Common's Estimates Committee. *Building* (28 March 1969)
28. ROYAL INSTITUTION OF CHARTERED SURVEYORS, COST RESEARCH PANEL. Report on the cost of flats and houses. *The Chartered Surveyor* (July 1958)
29. ROYAL INSTITUTION OF CHARTERED SURVEYORS, COST RESEARCH PANEL. The cost and design of high flats – a case study. *The Chartered Surveyor* (May 1961)
30. J. A. DENTON. The cost planning of Cumbernauld New Town. *The Chartered Surveyor* (July 1963)
31. F. POOLEY. Enter the mini-house. *Town and Country Planning*, **38.11** (1970)
32. MINISTRY OF HOUSING AND LOCAL GOVERNMENT. Design Bulletin No. 7. Housing cost yardstick for schemes at medium and high densities. HMSO (1963)
33. BRICK DEVELOPMENT ASSOCIATION. BDA case study 3: Portsdown 3 – cost planning (Undated)
34. P. O. JARLE. Finland searching for most economical building methods. *Build International*, **2.5** (1969)
35. E. S. TRAILL. Housing layouts: they've changed in 20 years. *Journal of Institution of Municipal Engineers*, **97.8** (1970)
36. B. TREPASS. Twilight to daylight. *Proceedings of Annual Conference of Royal Institution of Chartered Surveyors, University of Warwick* (1970)
37. R. MCKIE. Housing and the Whitehall bulldozer. Institute of Economic Affairs (1971)
38. L. NEEDLEMAN. *The Economics of Housing*. Staples Press (1966)
39. E. M. SIGSWORTH and R. K. WILKINSON. Rebuilding or renovation? *Urban Studies*, **4.2** (1967)
40. P. BROWN. Housing: rehabilitate or rebuild? *Building Design* (1 August 1970)
41. THE TREASURY Economic Progress Report No. 27 (May 1972)

6 APPROXIMATE ESTIMATING

IN THIS CHAPTER WE CONSIDER the function served by approximate estimates, the methods employed and the factors controlling their use. The various methods are compared and applied to practical examples.

PURPOSE AND FORM OF APPROXIMATE ESTIMATING TECHNIQUES

Purpose of, and Approach to, Approximate Estimating

The primary function of approximate or preliminary estimating is to produce a forecast of the probable cost of a future project, before the building has been designed in detail and contract particulars prepared. In this way the building client is made aware of his likely financial commitments before extensive design work is undertaken.

Higgin[1] regarded the intrusion of quantity surveyors with adequate techniques into the estimating field as of considerable significance in the development of a professional role. To extend this role into that of building economist requires the development of understandings and techniques of a kind that will deal, not just with the items which go into the accountancy of a particular building, but with the forces, economic and other, which have determined the nature and relationships among those quantities and costs, and which determine the trends they show. Indeed, economics is the study of all the forces which determine the present functioning and probable future trends of a whole industrial or financial system.

The quantity surveyor performs an extremely important role in cost assessment, giving advice as to the probable cost of a particular design proposal and variations to it, and suggesting how similar objectives could be achieved more economically. It must, however, be emphasised at the outset that no approximate estimate can be any better than the information on which it is based. Indeed, realistic approximate estimating can be achieved only when there is full co-operation and communication between architect, quantity surveyor and building client from the inception of the scheme. It is advisable to dissuade the architect from reporting forecasts of costs to the building client until some drawings, even preliminary sketches, have been prepared and an inspection made of the site. There is a distinct possibility that the building client will endeavour to obtain an independent check on the preliminary cost figures and it is accordingly unwise to supply a high 'cover' figure. On the other hand the submission of too low a figure can lead to recriminations, as the first figure is the one that the building client will always remember. The quantity surveyor should always emphasise that an estimate based on inadequate information cannot be precise, and in such a situation he would be well advised to give a range of prices, as an indication of the lack of precision that is obtainable.

The choice of method employed will be influenced by the information and time available, the experience of the surveyor and the amount and form of the cost data available to him. It is essential to carry out a detailed site survey before a preliminary estimate is prepared as it would be quite unrealistic to assume that the site is level, and free from obstructions, has a water table well below foundation level, and that the soil has an average load-bearing capacity. Similarly, old drawings of existing buildings scheduled for adaptation need checking.

Classification of Approximate Estimating Procedures

James[2] has attempted to classify the various precontract approximate estimating methods.

Single-purpose estimates are aimed at forecasting cost and these can be further subdivided into *preliminary estimates*, which establish the broad financial feasibility of the project, and *later stage estimates* which will produce a figure comparable with that of the lowest tender.

Dual-purpose estimates are aimed at determining total costs and also the various design-cost relationships between possible variants of the projects. These estimates delve into the wider field of cost planning. James[2] has subdivided his second category of estimate into two groups. *Primary comparative-cost estimates*, which indicate the relative costs of different design solutions which will satisfy the requirements of the building client, and *secondary comparative-cost estimates* which apply the financial yardstick to alternatives of construction, finish and service installations applicable to the selected design.

The extent to which these various estimating processes are needed for a given project is very much dependent upon the degree of importance with which the building client views financial considerations and the size of the project. For instance, buildings let to third parties, which must show a profitable return on their capital cost and owner-occupied industrial and commercial projects where capital is limited, all need to be exhaustively and skilfully costed at each stage of development of the design.

The main problems in implementing each of the main categories of approximate estimating are now considered.

Single-purpose preliminary estimates. These are often required before any drawings are prepared and are frequently computed by rather imprecise methods. These methods include unit prices such as the comprehensive price per bed of a hospital or hotel, and an all-in price per cubic metre of a building, whose location, shape, height, site works, services and other important characteristics may not have been determined at the time the estimate is required. These estimating methods are difficult to apply and should be used with the greatest care.

Single-purpose later stage estimates. These should only be prepared by single price-rate methods if the quantity surveyor is very experienced in the use of these methods, or if he has available as a starting point the tender particulars of a recent project which is similar in character and preferably designed by the same architect. In all other circumstances it is better to prepare estimates of this kind by way of priced approximate quantities which is a much more accurate method of estimating. In addition the total estimate can be broken down into convenient parts and justified if required, a process which is hardly feasible with single price-rate methods.

Primary comparative-cost estimates often use single price-rate methods as a basis for the computations. Nevertheless, it is important to make allowances for differing shapes and wall–to–floor ratios of alternative designs. As illustrated in chapter 2, external walling is expensive and, as a general rule, the smaller the ratio of external walling to floor area, the cheaper will be the building. The aim should be to use rectangular buildings with the sides as near equal in length as possible, provided that other aspects such as lighting and ventilation can be dealt with satisfactorily. Similarly, it is usually more expensive to construct accommodation vertically instead of horizontally at the stage when it becomes necessary to provide a structural frame, probably at the two-storey stage in the case of a heavily loaded building, and three to five-storeys with lightly loaded ones.

Secondary comparative-cost estimates are an integral part of cost planning, in comparing the costs of possible variants in construction, finishing and servicing of the selected plan. Comparisons could be made between traditional and nontraditional construction; steel frame, reinforced concrete frame and load-bearing brickwork; various types of pitched and flat roofs; different floor, wall and ceiling finishings; and different forms of heating and lighting.

In all cases the drawing numbers on which the estimate is based should be recorded on the estimate. The date of the estimate and the allowance for price fluctuations between the estimate and tender dates should also be clearly shown. Finally, all supporting data should be filed with the estimate.

UNIT METHOD

On occasions a building client requires a preliminary estimate for a building project based on little more information than the number of persons or units of accommodation that the building is to house. The unit method of approximate estimating seeks to allocate a cost to each accommodation unit of the particular building, be it persons, seats, beds, car spaces or whatever. The total estimated cost of the proposed building is then determined by multiplying the total number of units accommodated in the building by the unit rate. Thus the mathematical process is very simple but the computation of the unit is exceedingly difficult.

The unit rate is normally obtained by a careful analysis of the unit costs of a number of fairly recently completed buildings of the same type, after making allowance for differences of cost that have arisen since the buildings were constructed and any variations in site conditions, design, form of construction, materials, etc. Variations in rates, stemming from differences in design and constructional methods, are difficult to assess and frequently there is insufficient information available to make a realistic assessment. Hence although the method has the great merit of speed of application, it suffers from the major disadvantage of lack of precision and at best can only be a rather blunt tool for establishing general guidelines, more particularly for budgetary estimating on a rolling programme covering a three to five-year period ahead. Because of the lack of precision it is advisable to express costs in ranges, with more precise costs to be determined at a later stage by more reliable estimating methods when much more detailed information is available.

The following 1971 unit rates are given as a general illustration of the application of the method to specific classes of building and are subject to all the limitations previously described.

Hospital ward accommodation	£2160 to £2260 per bed
Church hall	£60 per seat
Primary school	£220 per place
Secondary school	£424 per place
Multistorey car park	£400 to £600 per car space

The main weaknesses of the method lie in its lack of precision, the difficulty of making allowance for a whole range of factors, from shape and size of building to constructional methods, materials, finishings and fittings, and it is not sufficiently accurate for the majority of purposes. It does, however, serve a limited number of uses such as establishing an overall target for a cost plan or calculating a sum for investment purposes, in cases where the building client or his professional advisers have considerable experience of the construction and cost of similar buildings, such as hospitals and schools. Even under these circumstances it is necessary to use this method with the greatest care and skill and with a full appreciation of its limitations.

CUBE METHOD

The cube method of approximate estimating was used quite extensively between the wars but has since been largely superseded by the superficial or floor area method. The cubic content of the building is obtained by the use of rules prescribed by the Royal Institute of British Architects[8] which provide for multiplying the length, width and height (external dimensions) of each part of the building, with the volume expressed in cubic metres. The method of determining the height varies according to the type of roof and whether or not the roof space is occupied. For a normal dwelling with an unoccupied pitched roof, the height dimension is taken from the top of the concrete foundation to a point midway between the apex of the roof and the intersection of wall and roof (half the height of the roof). If the roof space is to be occupied then the height measurement is taken three-quarters of the way up the roof slope, and with mansard roofs it is usual to measure the whole of the cubic contents.

With flat roofs, the height dimension is taken 600 mm above roof level except where the roof is surrounded by a parapet wall which has a height in excess of 600 mm, when the height will be measured to the top of the parapet wall. If the height of the parapet wall is less than 600 mm, the minimum height of 600 mm will still be taken. All projections such as porches, steps, bays, dormers, projecting roof lights, chimney stacks, tank compartments on flat roofs and similar features, shall be measured and added to the cubic content of the main building. On occasions it may be found that a small part of the foundations may be deeper than the remainder, and in this situation it is better to adjust the unit rate rather than to vary the cubic content of the building. Projecting eaves and cornices should be ignored when computing the volume of the building.

Where different parts of a building vary in character or function, such as a workshop with an office block frontage, then the different parts should be separately measured and priced. Basements should also be cubed separately, so that allowance can be made in the unit rate for the increased excavation and construction costs. Features such as piling, lifts, external pavings, approach roads, external services, landscaping and similar works which bear no relation to the cubic unit of measurement, should be dealt with separately by the use of lump sum figures or approximate quantities.

The assessment of the price per cubic metre of a building calls for the exercise of careful judgement coupled with an extensive knowledge of current prices and trends. Unit prices show wide variations between different classes of building and will even vary considerably between buildings of the same type, where such factors as the proportion of walling to floor area and quality of finishings and fittings vary to a significant extent. The greater the proportion of walling in relation to the cubic contents of the building, the greater will be its cost per cubic metre.

The following typical cubic metre rates in 1971 for various building types will give a rough guide

City bank with two floors of offices over	£22 to 30
Church hall	£12 to 16
Hotel	£25 to 35
City office block with shops on ground floor (excluding shop fronts and shop finishings, but including lifts and heating)	£20 to 28
Small shop with one floor of offices over (excluding shop fronts and shop fittings)	£12 to 18

Tremendous variations can occur in the cube rates of buildings of the same type. For instance, city offices can vary from £20 to £35 per cubic metre depending on size, shape, quality of finishings, amount of partitioning, number of fittings and a whole host of other factors, and these must all be taken into account when assessing the unit rate. With single-storey industrial buildings, wide variations in storey height can occur and costs will not vary directly in proportion to height. The roof, foundations and floor remain constant and a comparatively cheap form of cladding may be used for the walls. In this situation the cube method of approximate estimating would be ill-suited and could give quite unrealistic results. A far more satisfactory approach would be to use the superficial method and to adjust the unit rate for increases in height. The cube method is, however, useful for heating and steelwork estimates.

A primary weakness of the cube method is its deceptive simplicity. It is an easy matter to calculate the volume of a building but much more difficult to assess the unit rate on account of the large number of variables which have to be considered. The cubic method fails to make allowance for plan shape, storey heights and number of storeys, which all have an important influence on cost, and cost variations arising from different constructional techniques such as alternative foundation types are difficult to incorporate in a single unit rate. Ideally, the building from which the basic cubic rate is obtained should be of similar shape, size and construction as the one under examination. Other weaknesses are that cubic content does not give any indication to a building client of the amount of usable floor area, and it cannot readily assist the architect in his design of a building as it is difficult to forecast quickly the effect of a change in specification on the cube unit price rate. The following example may help to illustrate the approach.

Example (a)

Figure 6.1 shows a block of six unit factories with an office in each unit. The effective height is 0.600 m (top of foundations to ground level) +3.000 m (ground level to roof) $+\frac{1}{2}/3.000$ m (roof) = 5.100 m.

Volume of six workshops = 60 m × 22 m × 5.1 m
= 6732 cu m

Volume of offices = 60 m × 4 m × 5.1 m
= 1224 cu m

Estimated cost of block would be
6732 cu m @ £9 = £60 588
1224 cu m @ £13 = £15 912
Total £76 500

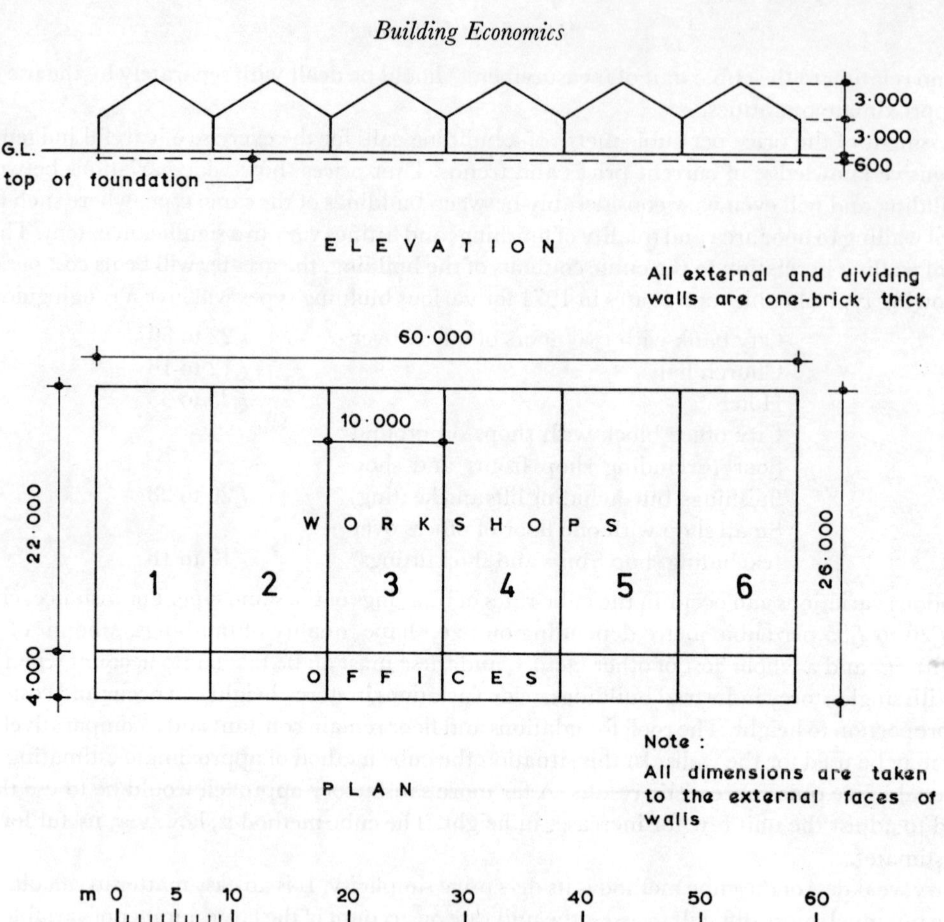

Figure 6.1 Approximate estimating – block of six unit factories

SUPERFICIAL OR FLOOR AREA METHOD

In this method the total floor area of the building on all floors is measured between the internal faces of the enclosing external walls, with no deductions for internal walls, partitions, stairs, landings, lift shafts, passages, etc. A unit rate is then calculated per square metre of floor area and the probable total cost of the building is obtained by multiplying the total floor area by the calculated unit rate. Where the building varies substantially in constructional methods or in quality of finish in different parts of the building, it will probably be advisable to separate the floor areas to enable different units rates to be applied to the separate parts. Consideration must also be given to varying storey heights in assessing unit rates and when extracting rates from cost analyses.

This is a popular method of approximate estimating, as it is comparatively easy to calculate the floor area of a building and the costs are expressed in a way which is fairly readily understood by a building client. Furthermore, most published cost data is expressed in this form. It has advantages over the cube method as the majority of items with a cost impact are related more to floor area than to volume and it is therefore easier to adjust for varying storey heights. Nevertheless, it has a number of inherent weaknesses and, in particular, it cannot directly take account of changes in plan shape or total height of the building. Similarly, difficulties are experienced in building up unit rates from known rates for existing buildings due to the need to make allowances for a number of variables including site conditions, constructional methods, materials, quality of finishings and number and quality of fittings.

As with the cube method, special items such as piling, heating and lift installations are normally covered by lump sums which are added to the overall cost calculated on a floor area basis. The specialist lump sums may be derived from quotations obtained from specialists or be based on information arising from previous contracts. The estimated cost of external works is usually based on priced approximate quantities, an estimating process which is described later in the chapter.

A few typical 1971 unit rates per square metre of floor area follow to indicate the range of prices, but it must be emphasised that wide variations on these rates occur in practice.

Factory workshop	£40 to 50
Semidetached house (estate development)	£35 to 50
Detached house (built singly and including central heating, garage and external works)	£60 to 85
Office block (including lifts and heating)	£60 to 100
Local authority office block (low rise centrally heated)	£85 (average)[9]
Banks	£100 to 150
Shops with flats over (excluding shop fronts)	£40 to 60
Departmental stores	£75 to 90
Hospitals	£90 to 140
Churches	£80 to 120
Schools	£55 to 75
Hotels	£90 to 140

NOTE: Costs expressed in £ per sq ft can be converted to £ per sq m by multiplying by ten and adding ten per cent, for example,

£3.00 per sq ft = (3·000 × 10) + 10% = £33 per sq m.

Example (a)
To illustrate the method of approach, an estimate is now prepared for the block of six unit factories illustrated in figure 6·1 using the floor area method. In this case the measurements are taken to the inside faces of the external walls and no deduction is made for internal walls.

Area of workshops = 60.000 − (2/215) × 22.00 − 215
= 59.570 × 21.785
= 1298 sq m
Area of offices = 59.570 × 3.785
= 226 sq m

Estimated cost of block would be
1298 sq m @ £46 = £59 708
226 sq m @ £65 = £14 690

Total = £74 398

This estimate compares reasonably favourably with the figure of £76 500 obtained by the cube method of approximate estimating.

STOREY-ENCLOSURE METHOD

Objectives
In 1954 James[3] introduced the work of an RICS study group on a new method of single price-rate approximate estimating, termed the *storey-enclosure method*, with the aim of overcoming the drawbacks of the methods so far

described in this chapter. The study group's primary objective was to devise an estimating system which, whilst leaving the type of structure and standard of finishings to be assessed in the price rate, would take the following factors into account in the measurements:

(1) shape of the building (by measuring external wall area);
(2) total floor area (by measuring area of each floor);
(3) vertical positioning of the floor areas in the building (by using greater multiplying factors for higher floors and greater measurement product for suspended floors than nonsuspended);
(4) storey heights of building (proportion of floor and roof areas to external walls);
(5) extra cost of sinking usable floor area below ground level (by using increased multiplying factors for work below ground level).

Nevertheless, the following works would have to be estimated separately:

(1) site works, such as roads, paths, drainage, service mains and other works outside the building (these are best covered by approximate quantities);
(2) extra cost of foundations, which are more expensive than those normally provided for the particular type of building (again this is best covered by approximate quantities);
(3) sanitary plumbing, water services, heating, electrical and gas services and lifts (priced approximate quantities or price from specialist consultant);
(4) features which are not general to the structure as a whole, such as dormers, canopies and boiler flues (separate priced additions);
(5) curved work.

Rules of Measurement

The storey-enclosure method consists basically of measuring the area of the external walls, floor and ceiling which encloses each storey of the building. These measurements are adjusted in accordance with the following set of rules.

(1) To allow for the cost of normal foundations, the ground floor area (measured in square metres between external walls) is multiplied by a weighting factor of two.
(2) To provide for the extra cost of upper floors, an additional weighting factor is applied to the area of each floor above the lowest. Thus the additional weighting factor for the first suspended floor is 0.15, for the second 0.30, for the third 0.45 and so on.
(3) To cover the extra cost of work below ground level a further weighting factor of one is applied to the approximate wall and floor areas that adjoin the earth surface.

Summing up, the procedure is to take twice the area of the ground floor, if above finished ground level, and three times the area if below (measured between external walls). It will take once the area of the roof measured on plan to the external face of the walls (the same area whether a flat or pitched roof), twice the area of upper floors (to cover work above and below, including partitions), plus the appropriate positional factor; it will also take once the area of external walls above ground measured on their external faces (ground level to eaves), and twice the area below ground in basements, etc.

All the areas are multiplied by the appropriate weightings to obtain the storey-enclosure units which are then totalled. To obtain the estimated cost of the building, the total of storey-enclosure units in square metres is multiplied by a single price-rate built up from the costs of previous similar projects. The cost of external works and other special items can be added. Basically, the aim of this method is to obtain a total superficial area in square metres to which a single price-rate can be attached, and the effect of the various rules that have been outlined is to apply a weighted cost factor to each of the main parts or elements of the building.

Comparison with Other Single Price-Rate Methods

The study group considered and analysed ninety tenders for new buildings, by reference to drawings and priced bills of quantities, and compared the results achieved with those operating the cube and floor area methods of approximate estimating. The buildings were classified into a number of general types – houses, flats, schools and industrial buildings, and conversion factors were devised to make allowance for price variations arising from different contract dates, so that all were reduced to a common price time datum. Tender figures, not final account figures, were used in the investigations. The proportional cost of engineering services was assessed separately and was found to differ very widely (the proportion of the cost of services to building work varied from fourteen to thirty-eight per cent).

In these comparisons one of the tests used by the study group was to determine how many of the rates fell into a particular range group, above and below the tender figures. The study group adopted ten per cent plus or minus for houses, flats and schools, where ministerial planning and financial restrictions operate, and twenty per cent plus or minus for industrial buildings. Table 6.1 indicates the results of this test.

TABLE 6.1

COMPARISON OF CUBE, FLOOR AREA AND STOREY-ENCLOSURE APPROXIMATE ESTIMATING METHODS

Type of building	*Total number of cases examined*	*Number of rates within percentage grouping*		
		Cube	*Floor area*	*Storey-enclosure*
Houses	17	8	9	10
Flats	16	9	10	12
Schools	14	9	8	12
Industrial buildings	39	16	24	26

With the storey-enclosure method the prices are nearer to the tender figures and the range of price variation is accordingly reduced. The weighting factor to cope with foundations is a particularly good feature of the system, as sketch plans are often very vague at foundation level. Used with care it could result in a much more realistic method of computing and comparing the cost of building projects than any of the other single price-rate methods.

Unfortunately, it has been little used in practice, mainly because it involves more calculations than either the cube or floor area methods. Furthermore, there are no rates published for this method, hence, a quantity surveyor would have to work floor area and storey-enclosure methods in parallel for a trial period and also work up storey-enclosure rates from past projects. The storey-enclosure method does little to assist either the building client or the architect, does not provide an aid to elemental or comparative cost planning and the effect of changes in specification on the price rate would probably prove difficult to assess. Nevertheless, it does not seem to be generally appreciated that although this method complicated the measurement aspect, it was accompanied by a simplification of the most difficult aspect of pricing.

Table 6.2 provides a comparison of cube, floor area and storey-enclosure price rates.

TABLE 6.2

COMPARISON OF CUBE, FLOOR AREA AND STOREY-ENCLOSURE PRICE RATES

Type of building	*Cube method: cost/sq m*	*Floor area method: cost/sq m*	*Storey-enclosure method: cost/sq m*
Local authority houses	£10 to 12	£30 to 38	£ 8 to 10
Local authority flats	£13 to 18	£40 to 55	£10 to 14
Schools	£18 to 23	£55 to 80	£20 to 24
Industrial buildings	£11 to 18	£35 to 55	£13 to 20

Example (a)

Figure 6.2 illustrates an office block to which the storey-enclosure approach of approximate estimating will now be applied.

	Storey-enclosure units
Floors	
Basement	
Floor area = (34.000 − 760) × (13.000 − 760)	
= 33.240 × 12.240 = 406.86 sq m	
× weighting 3	
Storey-enclosure units	1221
Ground floor	
Floor area as basement = 406.86 sq m	
× weighting 2	
Storey-enclosure units	814
First, second, third, fourth and fifth floors	
Floor area = (34.000 − 500) × (18.000 − 500)	
= 33.500 × 17.500 = 586.25 sq m	
Multiplier for first floor 2.15	
,, ,, second floor 2.30	
,, ,, third floor 2.45	
,, ,, fourth floor 2.60	
,, ,, fifth floor 2.75 ×12.25	
Storey-enclosure units	7182
Roof	
Area = 34.000 × 18.000 = 612.00 sq m	
No multiplier: storey-enclosure units	612
Walls	
Basement	
Wall area = (34.000 × 2) + (13.000 × 2) × 3.000 = 282.00 sq m	
× weighting 2	
Storey-enclosure units	564
Ground floor — *area*; *storey-enclosure units*	
Exposed wall 180 sq m = 180	
Retaining wall 102 sq m × 2 = 204	
Storey-enclosure units	384
First floor to roof	
Wall area = (34.000 × 2) + (18.000 × 2) × (5 × 3.000)	
No multiplier: storey-enclosure units	1560
Total number of storey-enclosure units	12 337

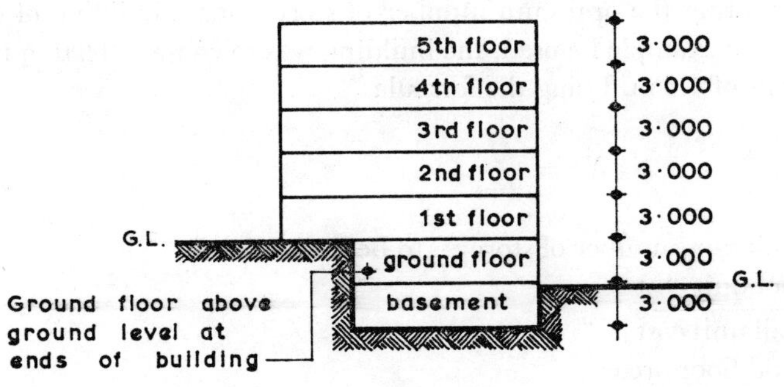

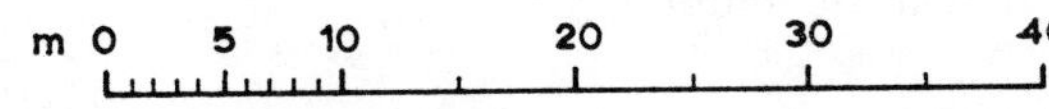

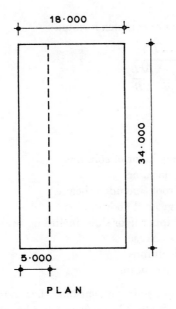

Figure 6.2 Office block – estimating by storey-enclosure method

Estimated cost

12 337 storey-enclosure units @ £22 =	£271 414
Estimated cost of lifts	£18 000
Estimated cost of external works	£25 000
Total estimated cost of work	£314 414
Rounded off to	£315 000

Ferry[4] devised a formula for calculating the optimum number of storeys for a building of given width to accommodate a specific floor area. In the example he used, the building was to contain 9290 sq m of floor area, with a width of 15 m and storey heights of 4.6 m. Using the formula

$$N^2 = \frac{xf}{2ws},$$

where

$N =$ optimum number of storeys (to be determined)
$x = \dfrac{\text{roof unit cost}}{\text{wall unit cost}}$
$f =$ total floor area
$w =$ width of block
$s =$ storey height

the optimum number of storeys was seven.

Bathurst[5] subsequently produced a much more complex formula

$$N^2 = \frac{f(R + F - U)}{2wsC + 2sS + \dfrac{fskL}{2B}}$$

where

$f =$ total area
$w =$ width
$s =$ storey height
$k =$ strengthening factor of columns
$C =$ unit cost of cladding
$R =$ unit cost of roof (including beams)
$F =$ unit cost of ground floor slab
$U =$ unit cost of upper floor slab (including beams)
$S =$ unit cost of staircase
$L =$ unit cost of columns
$B =$ floor area per column

Bathurst believed it is essential to take upper floors, stairs and various other factors into account. When the latter formula was applied to the example previously described the optimum number of storeys was shown to be between three and four and this is likely to be a more realistic solution.

APPROXIMATE QUANTITIES

Approximate quantities priced at rates produced at the time the quantities are computed provide the most reliable method of approximate estimating, possibly using a single price-rate method as a check. It does however involve more work than any of the methods previously described in this chapter and there are occasions when lack of information precludes its use. The method is sometimes described as *rough quantities* and the pricing document resembles an abbreviated bill of quantities. It provides, however, an excellent basis for cost checking during the detailed design of a project.

Composite price rate items are obtained by combining or grouping bill items. James[2] has described the process as one of coalescing of items in single omnibus description measurements. For example, a brickwork item measured in square metres will normally include all incidental labours and finishings to both wall faces. Doors and windows are usually enumerated as extra over the walling and associated finishings, thus avoiding the need to make adjustments to the walling. Furthermore a door item will be a comprehensive one, including the frame or lining, architraves, glazing, ironmongery and decoration.

It is good practice to build up a series of prices for composite items from a number of priced bills and to examine critically the range of variation between these prices and to establish the underlying reasons for them, and the relationships between the net cost of the main components, such as brick walls of various thicknesses and the gross cost of composite or all-in items, such as brick walls including finishings and all incidental labours. Sundry labours are often covered by the addition of a percentage to the composite items. The priced preliminaries in a bill of quantities must be examined before any priced rates are extracted, because of the different approaches adopted by contractors, with a view to securing comparability of rates as between different contractors.

James[2] has drawn attention to the desirability of developing an instinct for forecasting the total cost of small jobs from an examination of the drawings prior to receipt of tenders. Afterwards these estimates can be usefully compared with the tenders and the reasons for any disparities then assessed. This process will help a surveyor to acquire a feel for building prices and to build up expert intuition.

Example (a)

Figure 6.3 contains a plan and section of a small factory check office for which the estimated cost is to be obtained using the approximate quantities approach. The normal practice is followed of using paper with dimension columns and provision for squaring on the lefthand side of the sheet, and quantity, rate and pricing columns on the righthand side. Although prices are built-up for each composite item, this would rarely be necessary in practice as a set of prices would already be established.

Estimate for Factory Check Office

Drawing No. 6.3 — Date . . . .

Ref. no.	*Dimensions*	*Extension*	*Description and price build-up*			*Quantity*	*Rate (£)*	*(£)*
				Floor Slab	*Base*			
			add	2.750	2.100			
			walls 2/215	430	430			
			proj. conc. 2/100	200	200			
				3.380	2.730			
1.	3.38 2.73	9.23	Excavate oversite, remove surplus spoil, lay 150 mm bed of hardcore, polythene membrane and 150 mm bed of concrete (1:2:4) floated to smooth finish and splayed top edge to projecting part of slab.			9 m²	2.69	24.21
			Price build-up					
			1 m² excavate oversite and disposal		£0.40			
			1 m² hardcore		0.45			
			1 m² concrete bed		1.40			
			1 m² polythene membrane 100 mm wide		0.08			
			1 m² floated finish to concrete		0.12			
					£2.45			
			Sundry labours – 5%		0.24			
			Cost per m²		£2.69		c.f.	£24.21

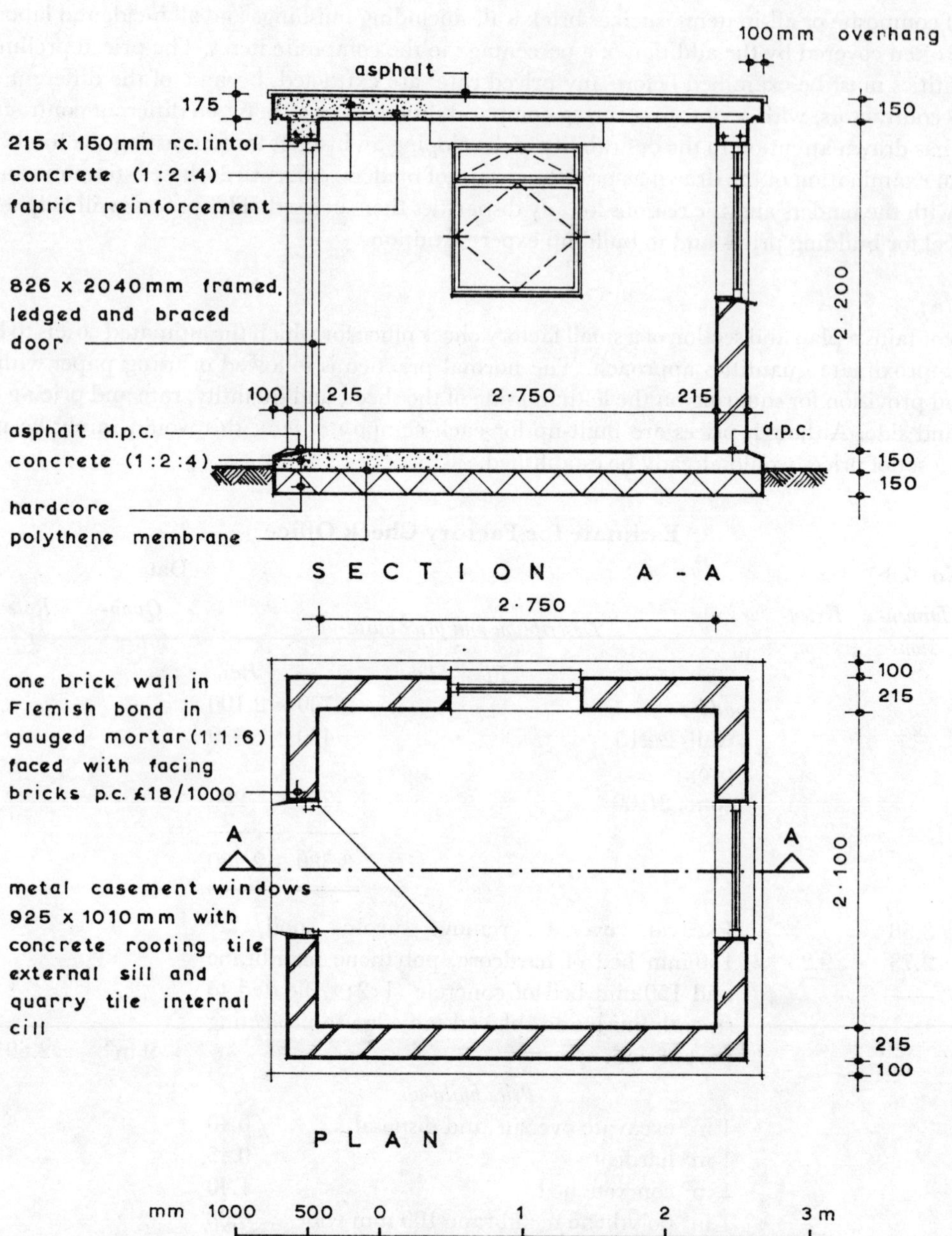

Figure 6.3 Factory check office – approximate quantities

Estimate for Factory Check Office—(*contd.*)

Drawing No. 6.3 Date . . . .

Ref. no.	*Dimen-sions*	*Exten-sion*	*Description and price build-up*		*Quan-tity*	*Rate* (£)	(£)
				Walls			
			Walls	2.750			b.f. £24.21
				2.100			
				2/4.850			
				9.700			
			add corners 4/215	860			
				10.560			
2.	10.56 2.20	23.23	One brick wall in Flemish bond in gauged mortar (1:1:6), faced externally with facing bricks, pc. £18/1000 and flush pointed and two coats of emulsion paint internally.		23 m²	7.45	171.35
			Price build-up				
			1 m² of one-brick wall	£4.40			
			1 m² extra over common brickwork for facings	1.90			
			1 m² flush pointing	0.36			
			1 m² emulsion paint	0.35			
			0.5 m² dpc.	0.26			
				7.27			
			Sundry labours – 2½%	0.18			
				£7.45			
			Metal windows				
3.	2	2	Extra over last for metal window size 925 × 1010 mm, including glass, painting, etc.		2	5.22	10.44
			Price build-up				
			One no. window and fixing	£3.50			
			1 m² glass	1.50			
			2 m² painting	2.00			
			1 m concrete roofing tile sill	0.65			
			1 m quarry tile sill	0.55			
			2.9 m emulsion reveal	0.80			
			1.9 m facework to reveal	0.26			
			1.2 m concrete lintel	3.10			
				£ c.f. £12.36			c.f. £206.00

5

Estimate for Factory Check Office—(*contd.*)

Drawing No. 6.3 Date . . . .

Ref. no.	*Dimensions*	*Extension*	*Description and price build-up*		*Quantity*	*Rate (£)*	*(£)*
				b.f. £12.36			b.f. £206.00
			Sundry labours – 2½%	0.31			
				12.67			
			less 1 m² of brick wall with finishings	7.45			
			Extra cost of window over wall	£5.22			
			Door				
4.	1	1	Extra over brick wall for framed, ledged and braced door, including frame, painting, etc.		1	7.20	7.20
			Price build-up				
			One no. door	£6.20			
			3.4 m² painting	2.00			
			One pair cross garnets	0.15			
			One no. lock	1.70			
			5 m frame	4.50			
			10 m painting frame (both sides)	1.50			
			5 m emulsion reveal	1.50			
			4 m facework to reveal	0.60			
			1.1 m concrete lintel	2.90			
				21.05			
			Sundry labours, cramps, etc. – 5%	1.05			
				22.10			
			less 2 m² of brick wall with finishings	14.90			
			Extra cost of door over wall	£7.20			
			Roof				
5.	3.38 2.73	9.23	Reinforced concrete roof, average 162 mm thick, reinforced with fabric reinforcement and covered with two coats of asphalt.		9 m²	6.54	58.86
			Price build-up				
			1 m² RC roof slab average 162 mm thick	£0.85			
			1 m² fabric reinforcement	0.70			
			1 m² wrought formwork to soffit	2.40			
			1 m² two coats of asphalt	1.90			
			1 m² emulsion paint	0.38			
				£ c.f. £6.23			c.f. £272.06

Estimate for Factory Check Office—(*contd.*)

Drawing No. 6.3 Date . . . .

Ref. no.	*Dimen-sions*	*Exten-sion*	*Description and price build-up*		*Quan-tity*	*Rate* (£)	(£)
				b.f. £6.23			b.f. £272.06
			Sundry items – 5%	0.31			
			Cost per m²	£6.54			
6.			Electrical work				30.00
7.			Telephone				15.00
8.			Preliminaries				25.00
9.			Contingencies				20.00
			Estimated cost of building				£362.06

(NOTE: No external works are included in the estimate) Rounded off to £360

(£62.5/sq m).

The priced rates for a job of this nature will vary tremendously with the particular circumstances. Considerations include whether or not it can be included as part of a larger contract, the state of the market, the time of the year when the work is to be undertaken, and particularly the availability of small builders in the locality if it is to form a separate contract. Very small jobs are usually more expensive in terms of cost per unit of floor area because of the limited amount of work to be carried out in each trade and the larger proportional amount of overheads. It should also be borne in mind that it may be necessary to adjust the estimate in order to make allowance for possible increased costs occurring between the date of preparing the estimate and the date of letting the contract.

Other composite or all-in items follow, in further amplification of the approximate quantities approach to approximate estimating with 1971 guide figures (excluding preliminaries which often amount to about six per cent). Readers requiring further cost data on approximate quantities are referred to *Spon's Architects' and Builders' Price Book.*[6]

	Unit	*£*
Strip foundations. Excavating trench 1 m deep in heavy soil; levelling; compacting; planking and strutting; backfilling; disposal of surplus material from site; concrete foundations 300 mm thick; hollow brickwork in cement mortar to 150 mm above ground level; bitumen hessian-based horizontal dpc; and facing bricks externally. (£2.75 for each additional 300 mm in depth).	m²	11.80
Hollow ground-floor construction. Excavation; disposal of surplus; hardcore 100 mm thick; concrete bed 150 mm thick; half-brick sleeper walls, honeycombed, at 2 m centres; horizontal dpc; 100 × 50 mm plates; 100 × 50 mm joists at 400 mm centres; and 25 mm tongued and grooved softwood boarded flooring.	m²	6.25
Upper floor. 175 × 50 mm joists at 400 mm centres with ends creosoted and built into brickwork; 50 × 25 mm herringbone strutting; trimming to openings; 25 mm tongued and grooved softwood boarded flooring; plasterboard; one coat of 5 mm gypsum plaster; and two coats of emulsion paint.	m²	4.70

	Unit	£
Pitched roof (measured on flat plan area). 75 × 40 mm plates; TRADA trussed rafters at 2 m centres; 100 × 32 mm rafters and ceiling joists at 500 mm centres; 150 × 40 mm purlins; 125 × 50 mm binders; 150 × 25 mm ridge; 40 × 19 mm battens; felt; and concrete interlocking tiles nailed every second course.	m^2	5.30
Stairs. 900 mm wide to BS 585; treads and risers, winders, balustrade one side; plasterboard and one coat of gypsum plaster to soffit and painting; rising 2600 mm. (£6.00 for each additional 300 mm in height.)	No.	60.00
Lavatory basin. White gvc, waste fitting, cantilever brackets and pair of taps (pc £10 complete); trap; and copper waste pipe. (Comparable costs of other sanitary appliances would be £30 for a sink; £65 for a bath; £25 for a ground floor WC and £60 for a WC on an upper floor, including a cast iron soil pipe.)	No.	26.00
Electrical installations. These are most conveniently priced on a cost per point basis, for example lighting points at £6.60 each and double 13-amp switched socket outlets wired in a ringmain circuit at £7.00 each.		
Drainage. 100 mm gvc pipes and fittings to BS 65; excavating trenches average 1 m deep in heavy soil; grading bottoms; trench timbering; backfilling; removal of surplus spoil; and 150 mm concrete beds and benchings. (£0.60 for each additional 300 mm depth of trench not exceeding 1.50 m deep and £1.20 for each 300 mm between 1.50 m and 3.00 m deep.)	m	3.00
Manhole 665 × 450 × 900 mm deep internally in one-brick walls in engineering bricks on 150 mm thick concrete base; with 100 mm half-section channel and branches; concrete benchings; 600 × 450 mm cover and frame; and all necessary excavation, backfill, disposal of surplus and planking and strutting. (£9.00 for each additional 300 mm of depth up to 1.50 m deep internally.)	No.	38.00

ELEMENTAL COST ANALYSES

Another approximate estimating method uses elemental cost analyses for previous similar jobs as a basis for the estimate. The cost is computed on a superficial or floor-area basis but the overall superficial unit cost is broken down into elements and subelements. At this lower level of division it is possible to make cost adjustments for variations in design in the new project as compared with the previous job. It will also be necessary to update the costs to take account of increased costs which have occurred since the tender date of the job for which the cost analysis is available. Elemental cost planning, cost analyses and building cost indices will be considered in some detail in later chapters. A cost analysis relating to a factory is shown below to illustrate the general method of approach, although it will be appreciated that in practice the differences between the old and new projects will generally be much greater and the problems involved preparing the estimate much more complex. This example has intentionally been kept fairly simple to avoid becoming involved in excessive detail in the costing of each element. This approach involves a close examination of the design and cost aspects of major parts or elements of the building and the information thus produced can form a preliminary cost plan. Changes in storey heights, floor loadings, column spacings or the number of storeys can have varying effects on costs as indicated in the Wilderness study.[7]

Example (a)

An estimate is required for a new two-storey factory with a floor area of 1500 sq m. The cost analysis which follows relates to a factory of 1600 sq m of similar shape and specification to the proposed factory, for which tenders were received in July 1970. Adjustments need to be made for the following factors.

(1) The weighted building cost indices relating to this class of building were 178 in April 1970 and 196 in July 1971 – an increase of

$$\frac{196}{178} \times 100 = \text{ten per cent.}$$

(2) The ground floor of the new factory has a storey height of 4.5 m (300 mm lower).
(3) The wall/floor ratio = 0.78.
(4) Facing bricks are £5/1000 cheaper.
(5) Increase in amount of partitioning by ten per cent.
(6) Twenty-five per cent increase in number of sanitary appliances.
(7) Five per cent increase in waste pipes stemming from greater number of sanitary appliances.
(8) Three goods lifts (two in existing factory).
(9) Sprinkler installation required at estimated cost of £3000.
(10) Twenty per cent reduction in amount of siteworks (smaller site).

Cost Analysis of Existing Factory
Tender date: April 1970.
Contract: Lowest of six selected tenders – standard form of building contract with quantities – fourteen months contract period.
Floor area: Ground floor: 800 m², first floor: 800 m²; total floor area: 1600 m².
Wall/floor ratio = 0.750.
Storey heights: Ground floor: 4.80 m, first floor: 3.60 m.

Summary of element costs

Element	*Existing factory*		*New factory*	
	Cost per m² of gross floor area	*Group element total*	*Cost per m² of gross floor area*	*Group element total*
1. *Substructure*	£6.22	£6.22	£6.84	£6.84
2. *Superstructure*				
2A. Frame	10.30		10.99	
2B. Upper floors	2.87		3.16	
2C. Roof	1.29		1.42	
2D. Stairs	0.23		0.24	
2E. External walls	4.26		4.43	
2F. Windows and external doors	0.36		0.41	
2G. Internal walls and partitions	1.54		1.80	
2H. Internal doors	0.52	21.37	0.57	23.02
3. *Internal finishes*				
3A. Wall finishes	1.35		1.56	
3B. Floor finishes	4.65		5.12	
3C. Ceiling finishes	1.74	7.74	1.91	8.59
4. *Fittings and furnishings*	—		—	

Summary of element costs—(contd.)

Element	Existing factory: Cost per m² of gross floor area	Existing factory: Group element total	New factory: Cost per m² of gross floor area	New factory: Group element total
5. *Services*				
5A. Sanitary appliances	0.83		1.14	
5B. Services equipment				
5C. Disposal installations	1.90		2.19	
5D/G. Water, heating and ventilating installations	27.84		30.62	
5H. Electrical installations	11.05		12.16	
5I. Gas installations				
5J. Lift and conveyor installations	3.45		5.68	
5K. Protective installations	—		2.00	
5L. Communication installations	0.04		0.04	
5M. Special installations	—	£45.11	—	£53.83
(Builder's work, profit and attendance on services included with services elements)				
Sub-total		80.44		92.28
6. *External works*				
6A. Site work	5.67		4.99	
6B. Drainage	2.42		2.66	
6C. External services	—		—	
6D. Minor building works	—		—	
		8.09		7.65
Preliminaries		4.50		5.00
Total (*less* contingencies)		£93.03		£104.93
			to nearest pound	£105/m²

The elemental costs relating to the new factory have been inserted in the summary alongside the figures for the existing factory. The techniques employed are now described.

All elemental costs are increased by ten per cent to update them, and then adjusted in other ways if required.

1. *Substructure:* ten per cent period price increase: £6.22 + 0.62 = £6.84.

2A. *Frame:* £10.30 + 10% = £11.33 − 3.33% (1/30) (lower storey height) = £10.99.

2D. *Stairs:* £0.23 + 10% = £0.25 − 3.33% = £0.24.

2E. *External walls:* £4.26 + 10% (period costs) + 4% (wall/floor ratio) − 3.33% (lower storey height) − 6.66% (cheaper facing bricks) = £4.26 + 4% net = £4.43.

2F. *Windows and external doors:* £0.36 + 14% (period costs and wall/floor ratio) = £0.41.

2G. *Internal walls and partitions:* £1.54 + 20% (period costs and increased quantity) − 3.33% (lower storey height) = £1.54 + 16.67% net = £1.80.

3A. *Wall finishes:* £1.35 + 10% (period costs) + 7% (increased partitions and wall/floor ratio) − 3.33% (lower storey height) = £1.35 + 13.67% net = £1.56.

5A. *Sanitary appliances:* £0.83 + 10% (period costs) = £0.91 + 25% (increased quantity) = £1.14.

5C. *Disposal installations:* £1.90 + 10% (period costs) = £2.09 + 5% (increased quantity) = £2.19.

5J. *Lift and conveyor installations:* £3.45 + 10% (period costs) = £3.79 + 50% (increased quantity) = £5.68.

5K. *Protective installations (sprinkler system):* $\frac{£3000}{1500} = £2.00$.

6A. *Site work:* £5.67 + 10% (period costs) = £6.24. − 20% (smaller site) = £4.99.

The estimated cost of the new factory:

£105 × 1500 =	£157 500
add contingencies (about 1.6%)	2500
Total estimated cost	£160 000

COMPARATIVE ESTIMATES

Another method of approximate estimating is to take the known cost of a similar-type building as a basis and then to make cost adjustments for variations in constructional methods and materials. For this purpose it is advisable to build up costs usually related to a square metre of finished work for a whole range of alternatives, to enable speedy adjustments to be made when preparing approximate estimates. These comparative costs will also be useful for costing alternative proposals as the detailed designs are developed. It is important to consider possible side effects of alternative choices as various forms of cladding may offer differing degrees of thermal insulation, and the adoption of a cheaper solution may result in greater heat losses and increased heating costs once the building is occupied. The weight of the cladding may also affect the design of the foundations. The speed of erection of any specific part of the building may have important cost implications and may also affect progress in construction of other sectors of the building. Table 6.3 contains a number of comparative prices to illustrate the form that a comparative cost schedule would take.

INTERPOLATION METHOD

A variant of the comparative method is the *interpolation* method whereby, at the brief and investigation stages in the design of a project, an estimate of probable cost is produced by taking the cost per square metre of floor area of a number of similar type buildings from cost analyses and cost records and interpolating a unit rate for the proposed building. This method looks deceptively easy, but no two buildings are the same and it is difficult to make adjustments to the unit rate to take account of the many variables that are bound to occur between the buildings for which known costs are available and the project under consideration. In practice it often may be necessary to use a method which is a combination of both the interpolation and comparative approaches.

TABLE 6.3

SCHEDULE OF COMPARATIVE COSTS OF DIFFERENT CONSTRUCTIONAL METHODS

Typical 1971 London prices for medium-sized job.

	Cost per m²
Substructure	
Solid ground floor, including excavation but excluding finish	£ 3.80
Hollow ground floor, ditto.	5.00
Upper floors	
Softwood joists and herringbone strutting (depending on sizes of joists)	2.00 to 3.30
Hollow tiles, 150 mm thick	4.70
Reinforced concrete, 150 mm thick	6.00
Flat roof finishes	
Three-layer bitumen felt roofing finished with granite chippings	1.30
19 mm mastic asphalt on felt underlay	1.80
0.91 mm aluminium	6.00
0.56 mm copper	9.20
1.7 mm lead	10.50
Pitched roofs (measured per m² of plan area) (all to 40° pitch)	
Timber roof construction (TDA or traditional)	3.00
Concrete interlocking tiles to 75 mm lap	1.90
Concrete plain tiles to 63 mm lap	3.15
Machinemade sandfaced plain clay tiles to 63 mm lap	3.70
Handmade sandfaced plain clay tiles to 63 mm lap	5.00
Welsh slates (400 × 250 mm) to 75 mm lap	4.80
Westmorland green slates, random sizes, to 75 mm lap	11.20
External walls	
Hollow brick wall	5.30
Hollow wall with outer half-brick skin and inner skin of 75 mm hollow clay blocks	4.40
Hollow wall with outer half-brick skin and inner skin of 75 mm thermalite blocks	4.40
Hollow wall with outer half-brick skin and inner skin of 100 mm clinker concrete blocks	4.80
One-and-a-half brick wall	7.00
150 mm reinforced concrete wall	8.90
Tile hanging including battens	3.00
25 mm western cedar boarding and battens	3.25
50 mm precast concrete facing slabs	9.40
75 mm Portland stone facing slabs	19.20
Galvanised steel curtain walling	19.80
Anodised aluminium curtain walling	22.80
6 mm polished plate glass	4.70
Double glazing of two skins of 6 mm polished plate	14.50
Internal walls and partitions	
75 mm hollow clay block	1.90
75 mm clinker concrete block	1.95
Half-brick wall	2.50
Stud partition	0.95
55 mm Paramount dry partition	3.15
50 mm demountable steel partition	8.40
Doors (762 × 1981 mm)	*(each)*
35 mm flush door, cellular core, hardboard faced, with 100 mm steel butts and painted three coats	7.00
Ditto, plywood faced	7.85
Ditto, cedar faced and wax polished with brass butts	10.40
50 mm softwood purpose made four-panel door with 100 mm steel butts and painted	14.10

TABLE 6.3 (*contd.*)

	Cost of each
45 mm flush door, solid core, walnut veneered plywood faces, with brass butts and wax polished	£ 17.30
55 mm flush door, one hour fire check, plywood faced, with 100 mm steel butts and painted	17.70
106 × 25 mm softwood lining with grounds and painting	6.80
Ditto, in mahogany and wax polished	8.60
106 × 75 mm softwood rebated and rounded frame, including fixing and painting	10.70
Ditto, in mahogany and wax polished	15.30
Wall finishes	*Cost per m²*
Render brickwork in cement, lime and sand and set in gypsum	£ 0.75
10 mm gypsum lath fixed to studs (not included) and skim coat of gypsum	1.00
13 mm insulation board and battens plugged to wall	1.10
13 mm softwood wall lining	2.25
6 mm white glazed wall tiles on cement-sand backing	3.45
Floor finishes	
19 mm cement-sand screeded bed	0.40
3 mm thermoplastic tiles (grade B)	0.90
25 mm granolithic paving	0.90
2 mm vinyl tiles	1.25
16 mm red pitchmastic	1.60
5 mm cork tile flooring and polishing	2.25
22 mm red quarry tile paving	2.85
5 mm rubber tile flooring	3.50
25 mm softwood tongued-and-grooved flooring including fillets (£0.55)	2.10
25 mm Iroko block flooring and polishing	4.00
25 mm maple strip flooring including fillets and polishing	5.10
16 mm terrazzo paving in squares	5.30
25 mm teak block flooring and polishing	5.70
Ceiling finishes	
10 mm gypsum lath fixed to joists (not included) and skim coat of gypsum	1.00
Suspended ceiling of 19 mm insulating board tiles, flameproofed, and fixed in steel tees	2.50
Ditto, with 19 mm acoustic tiles	3.00
Decorations	
Prepare, and two coats distemper on plastered surfaces	0.25
Ditto, but emulsion paint	0.30
Prepare, prime and three coats oil paint on metalwork	0.85
Knot, prime, stop and three coats oil paint on woodwork	0.90
Prepare, stain and wax polish on hardwood	1.00
Pavings	
50 mm gravel paving	0.35
50 mm tarmacadam in two coats	0.95
50 mm precast concrete paving slabs	1.35
50 mm York stone paving slabs	2.25
75 mm cobble paving	4.40
Drains	*Cost per m*
100 mm drains including excavation average 1 m deep:	
Pitch fibre including 50 mm bed of sand	£ 2.30
Second quality GVC including 150 mm concrete bed and benching	3.10
BS quality GVC including 150 mm concrete bed and benching	3.20
Spun iron pipe to BS 1211	4.70

REFERENCES

1. G. HIGGIN. The future of quantity surveying: some reflections of an outsider. *The Chartered Surveyor* (October 1964)
2. W. JAMES. The art of approximate estimating. *The Chartered Surveyor* (October 1955)
3. W. JAMES. A new approach to single price-rate approximate estimating. *The Chartered Surveyor* (May 1954)
4. D. FERRY. The building and its envelope. *The Chartered Surveyor* (March 1966)
5. P. E. BATHURST. The building and its envelope. Correspondence in *The Chartered Surveyor* (June 1966)
6. DAVIS, BELFIELD and EVEREST (Editors). *Spon's Architects' and Builders' Price Book*. Spon (1971)
7. THE WILDERNESS COST OF BUILDING STUDY GROUP. An investigation into building cost relationships of the following design variables: storey heights, floor loadings, column spacings and number of storeys. Royal Institution of Chartered Surveyors (1964)
8. ROYAL INSTITUTE OF BRITISH ARCHITECTS. RIBA rules for cubing buildings for approximate estimates, D/1156/54 (1954)
9. DEPARTMENT OF THE ENVIRONMENT. Local authority offices: areas and costs (1971)

7 COST PLANNING THEORIES AND TECHNIQUES

THERE IS NO universal method of cost planning which can be readily applied to every type of building project. Buildings have widely varying characteristics, perform a diversity of functions, serve the needs of a variety of building clients, and their erection is subject to a number of different administrative and contractual arrangements. Hence, it is not surprising that a wide range of cost planning techniques has been devised to meet the needs of a variety of situations.

The quantity surveyor frequently acts as specialist adviser to the architect on all matters concerned with building costs and it is vital that he should be involved at the earliest possible stage in the design of the building. He can offer considerable assistance to the architect in advising on the financial effect of design proposals and so help in ensuring that the money available is put to the best possible use and that the tender figure is close to the initial estimate.

The three main themes of a conference on integrated building design at Loughborough in 1971[23] were that cost control, including costs in use, should be based on giving the client value for money; that design should be a team effort involving new attitudes and relationships; and that improved design is a learning process based on feedback.

PLAN OF WORK

The Royal Institute of British Architects[1] has formulated a suggested pattern of procedure for architects in the preparation and implementation of building schemes, based on a hypothetical £300 000 job. This plan of work represents a sound and practical analysis of the operations and has been applied successfully on many contracts. The plan is outlined in table 7.1, together with possible activities by the quantity surveyor to provide an effective contribution in cost control.

At the inception stage of a building contract, the building client considers his building requirements and appoints an architect. The next stage in the plan of work is the feasibility stage at which an effective cost control mechanism needs to be established, and, in particular, a realistic first estimate is produced. The cost limit will be very much influenced by the floor space required, the standard of accommodation and the function of the building.[3] The cost limit is often assessed from a comparative study of known costs of similar type buildings, sometimes referred to as the interpolation method. In other cases it may be established by the building client by means of a developer's budget calculation of the type described in chapter 13 and occasionally described as the financial method.

At scheme-design stage the brief is completed and the design team develops the full design of the project. The cost plan is now formulated, which consists of a statement showing how the design team proposes to distribute the available money over the various elements of major parts of the building. Typical cost plans are illustrated later in this chapter and in chapter 9. The cost plan is used continuously throughout the detail design stage of the project as a means of checking that the detail design is kept within the cost framework. The quantity surveyor translates the design team's decisions on each element into cost targets, to which an allowance for design and price risk, (often about five per cent) is added to produce cost limits.

At the detail-design stage final decisions are made on all matters relating to design, specification, construction and cost. Every part of the building must be comprehensively designed, its cost checked and the design adjusted if

TABLE 7.1

PLAN OF WORK FOR BUILDING PROJECT

	Stage	Purpose	Suggested cost activities by quantity surveyor
Brief	1. Inception and feasibility (these two processes may be separated)	To prepare general outline of requirements and to provide building client with an appraisal of the project and a feasibility report.	General contribution towards preparation of *feasibility report.* Give cost range with indication of quality or advise on building client's cost limit. Unit, cube or floor-area estimating methods may be used.
Sketch plans	2. Outline proposals	To determine general approach to layout, design and construction and to obtain approval of building client to outline proposals.	Confirm cost limit or give *firm estimate* based upon user's requirements and outline designs and proposals. Prepare *outline cost plan* in consultation with design team, either from comparison of requirements with analytical costs of previous projects or from approximate quantities based on assumed specification.
	3. Scheme design	To complete the brief and determine specific proposals, including planning arrangement, appearance, constructional method, outline specification and cost, and to obtain all approvals.	Prepare *draft,* and later, *final cost plan* on basis of scheme design and statements of quality standards and functional requirements received from architects and engineers, for building client's final approval.
Working drawings	4. Detail design (later part of this stage is sometimes referred to as production information)	To obtain final decision on all matters relating to design, specification, construction and cost, and to prepare production information.	Carry out cost studies and *cost checks,* and obtain quotations from specialist subcontractors. Inform architects and engineers of results and give cost advice.
	5. Bills of quantities	To prepare and complete all information and arrangements for obtaining tenders.	*Cost checks* during bill preparation.
	6. Tender action	Action as recommended in *Selective Tendering.*[2]	*Check cost plan* against priced bill of quantities.
Site operations	7. Project planning	To plan efficient execution of contract.	Prepare *cost analysis* of accepted tender.
	8. Operations on site	To secure effective implementation of contract and translation of contract particulars into building work.	Maintain close *check* on all financial aspects of contract. Provide monthly statements, value variations and report to design team.
	9. Completion and feedback	Completion of contract, settlement of final account and feedback of information from job to assist future designs.	Prepare final account, *final cost analysis* and deal with settlement of contractual claims.

necessary. Since detail designs and specifications are available at this stage, the most suitable and accurate estimating technique is the approximate quantities method described in chapter 6. The quantity surveyor prepares approximate grouped or composite quantities from the detailed designs for each element in turn and prices these quantities on the basis of priced bill rates for previous similar jobs. The estimate is then compared with the cost target for the element in the cost plan and this constitutes a cost check. In this way the design team is able to exercise effective cost control. Where the estimated cost exceeds the cost target then either the element must be redesigned or other cost targets reduced to make more money available for the element in question, but leaving the overall cost limit unaltered. The quantity surveyor must keep the rest of the design team informed of any actions of this kind.

Where the estimated cost of the design element is within the cost target in the cost plan, the design should be confirmed in writing as suitable for the preparation of production drawings, sometimes described as the production information stage. If the estimated cost is substantially below the cost target, then surplus funds may be released for other elements. Each element is investigated in turn to arrive at a logical and balanced distribution of costs throughout the major parts of the building. Finally, after checking the costs of each design element, a final cost check of all elements is made. Where a number of design elements would have to be adjusted to keep costs within the total cost limit and this would result in a building of undesirably low quality, the design team should request additional funds from the building client.

Programmes or timetables of cost advice to be given during the development stage of a project have also been prepared by Waters[4] and the former Ministry of Education.[5] There is a certain similarity of approach between these two programmes, and a comparative schedule indicating the main stages of each programme is given in table 7.2.

TABLE 7.2

OUTLINE TIMETABLE OF COST ADVICE DURING THE DEVELOPMENT OF A PROJECT

	Ministry of Education programme[5]	*Programme suggested by A. B. Waters*[4]
Approximate estimating	*Client's brief* Architect receives particulars from building client. QS – guide figures.	*Client's requirements* Architect consults with building client. QS – guide figures.
	Investigation Preliminary study of forms of construction and preparation of sketch plan. QS – approximate estimate or first cost plan.	*Brief* Requirements agreed and general design established with some thoughts on construction. QS – approximate estimates generally on comparative basis.
Cost planning	Sketch plan finalised. QS – first cost plan checked.	*Sketch design* Major planning resolved, sketch design finalised. QS – first cost plan.
	Design Detailed study on draft drawings of all parts of building. QS – cost checks.	*Constructional design* Detailed study of construction and design, sub-contractors, etc. QS – cost checks.
	Working drawings Production of final drawings. QS – final cost check.	*Working drawings* Preparation of production drawings. QS – final cost check.

Nott[6] has emphasised that the essence of cost planning is to enable the architect to control the cost of a project (within the target) while he is *still* designing. The earlier this process is introduced, the greater the measure of control that can be exercised over ultimate cost, quality and design. Cost planning should be a continuous process, progressive checks being made from time to time in relatively more detail on perhaps smaller sections of the project as the design is finalised. Another merit of cost planning is that it introduces a procedure into the design stage where previously nothing systematic had existed.

Nisbet[7] has also described the way in which cost plays its part throughout the design process. In the first instance, it influences the size of the project and its general form, then later indicates the type of structure and subsequently affects the choice of services and finishings. Cost is, therefore, a continuing influence but it has two distinct phases. During the *brief* and *investigation* stages, the building client and architect have the joint responsibility of deciding just how much the project should cost, or alternatively of deciding what size and quality of building can be provided for a given sum. During the *design* and *working drawings* phases, the architect has the responsibility of designing a building in such a way that the tender will not exceed the client's budget or cost limit. The costing process therefore has two distinct parts – firstly, the determination of total cost, and secondly the costing of the design within the total sum. As indicated in table 7.2, the first stage (that is to decide how much to spend) is the approximate estimating stage, whereas the second stage (that is to decide how the budget is to be spent) is the cost planning stage. Hence these two terms are not synonymous, as they have different meanings and apply to separate aspects or phases of the costing process. They are, however, mutually supporting.

It would be useful at this stage to examine the various activities that take place during the five design phases listed in table 7.2.

(1) Client's Brief

At this stage the architect and the building client are busily endeavouring to establish the client's requirements and, in particular, to distinguish between desires or whims and essential needs. The architect frequently finds it very difficult to determine just how much the building client is prepared to spend and to reconcile the two major factors of cost and quality, which are so closely related.

Often the client is pressing for an assessment of cost before any drawings have been produced. The quantity surveyor can render valuable service at this stage by supplying cost information based on the actual cost of previous buildings of a similar type. He will make allowance for such factors as differences of location, site conditions, market conditions and quality of job, and so arrive at a provisional estimate on a comparative or interpolation basis.

(2) Investigation

The building client's requirements have now been definitely established, the site has been surveyed and the architect begins to consider the various alternative ways in which the building can be designed and constructed. With industrial projects the architect will probably decide on a single-storey building and will then determine which heights and span give the most useful and economical arrangement. With blocks of offices and flats the number of storeys will have to be decided, having regard to such matters as soil conditions, type of foundation to be used and constructional form and costs.

Some drawings will be produced at this stage and the quantity surveyor will be in a position to give general guidance on costs and, in particular, to evaluate the financial effect of different solutions to any specific design problem.

(3) Sketch Design

During this stage the major planning problems will be resolved and the outline designs will emerge. The sketch designs will include sections and elevations, and services and finishings will be considered in addition to the form of the structural framework. For instance, if it is a framed building, it will be necessary to consider the relative merits of steel or reinforced concrete. Consultants will be brought in at this stage in order that their requirements may also be investigated.

The quantity surveyor checks on his approximate estimate figure and, with the aid of extensive cost information, prepares an initial cost plan with provisional target cost figures set down for each element or major part of the building.

(4) Constructional Design

Sketch plans are now finalised and some working details are prepared. It is most desirable that these should be approved by the building client to avoid the possibility of future alterations. Outline schemes will be prepared by consultants and designing subcontractors and provisional estimates supplied in some cases.

The quantity surveyor will be called upon to give comparative costs of different forms of construction, materials and service layouts and will adjust the distribution of costs in the cost plan if required. It is to be hoped that these comparative cost studies will include probable running and maintenance costs wherever they are likely to have a significant effect on the outcome. Future costs will be examined in some depth in chapter 11. Continuous cost checks by the quantity surveyor will ensure that the development of the design remains compatible with the cost plan, and this process is sometimes described as cost reconciliation.

(5) **Working Drawings**
The final working drawings (production drawings) will now be prepared from which bills of quantities can be produced. Consultants, subcontractors and suppliers will be required to supply full information at this stage, including realistic quotations. The quantity surveyor continues his cost checks and ultimately produces the final cost plan. He will also be available to give advice to the architect on any financial or contractual matters associated with the project, including the terms and conditions of the main contract and subcontracts and on the selection of tenderers.

COST CONTROL PROCEDURE

The actual procedure involved in controlling the cost of a project depends to some extent upon the stage at which the cost limit is determined by the building client. There are three main arrangements and these are now outlined.

(1) The accommodation requirements are prescribed by the building client at the outset and the cost limit is established early in the design stage, after the sketch proposals and approximate estimate of cost are approved by the building client. Cost will be an important factor and the estimate must be realistic.

The architect and building client will begin by discussing the client's requirements, and the quantity surveyor may be asked to give general information on the known costs of similar buildings. Once the basic form of the building has been decided, the quantity surveyor should prepare an approximate estimate, in particular taking full account of shape, number of storeys and structural form.

Outline plans will usually be drawn to a scale of 1:200. The estimate may be based on cube, floor-area, storey-enclosure or approximate quantities methods depending on the amount of information available, with assumptions having to be made concerning finishings, fittings, services and similar matters. These estimating methods are described and illustrated in chapter 6. The sketch plans will then be prepared, usually to a scale of 1:100. The quantity surveyor should be available with comparative costs of alternative forms of construction, service arrangements, etc. By the end of the sketch plan stage, a more accurate estimate will be prepared usually based on approximate quantities. The plans and estimate are then submitted to the building client for approval.

During the detailed design stage the cost of the project can be influenced appreciably by the choice of materials and constructional methods, for example, aluminium, steel or wood casements; timber, concrete or metal cladding panels; inclusion or otherwise of a parapet wall; and many other alternatives. Continual reference to a cost plan is essential as the details of each part of the project are finalised with possible adjustments to design or distribution of costs as the process of reconciliation proceeds. A final overall cost check can be made when the design stage is complete. It will be appreciated that the amount of cost investigation work is dependent to a large extent upon the size and nature of the project. A small and relatively straightforward job requires only a minimum of cost investigation.

(2) The accommodation requirements and cost limits are both determined by the building client. This arrangement is most likely to operate where the client has considerable experience of the past costs of similar buildings. A common example is that of educational buildings but it could also apply to commercial, industrial and residential buildings. It is important to check the accommodation requirements against the cost limit to be sure that the project is feasible, and the usual approach is to make a comparison with the known costs of similar buildings. The remainder of the procedure will be identical to that described in (1).

(3) Where available funds are limited and the building client wishes to carry out as much building work as possible, he may prescribe a cost limit and ask the architect for details of the size and quality of building that can be provided for this sum of money. In this case it is necessary to compute the amount of floor area that can be provided by reference to the known costs and floor areas of similar buildings. The subsequent procedure will be as described in (1).

Berryman[22] has defined cost control as an umbrella term embracing all contributory stages such as cost analysis, cost planning, cost comparisons, cost checking, cost reconciliation at the tendering stage, and cost monitoring at the postcontract stage in the manner indicated below.

(1) Initial cost budget and cost plan prior to detailed design, often based on functional unit costs, elemental cost analysis, or some similar method.

(2) Preliminary cost studies, usually based on measured approximate quantities, to compare alternative materials and systems in terms of capital, operating, maintenance and depreciation costs.

(3) Cost checking detailed designs as they are produced, again usually on the basis of measured approximate estimates, to ensure that any underdesign or overdesign relative to the budget is corrected early enough to avoid delays and abortive effort.

(4) Preparation of tender documents (for example, plans and specification; schedules of rates; bills of quantities) for pricing by tendering contractors.

(5) Checking and reporting on competitive tenders or agreeing negotiated tenders.

(6) Preparing and agreeing precontract variations (if required) to balance tenders with budgets.

(7) Analysing accepted tenders into functional or elemental costs (or both), comparing tenders with estimates, and modifying basic estimating data as necessary for future reference.

(8) Calculating final costs (including variations, remeasurements, PC accounts, dayworks, increased or decreased costs and authorised overtime) in two stages – rough estimates reported quickly for budget control; and accurate valuations agreed with the contractor for final payment.

(9) Checking and negotiating contractors' claims.

INFORMATION REQUIRED BY ARCHITECT AND BUILDING CLIENT

Need for Team Approach

When it is proposed to carry out private development in the central areas of large towns and cities, it is often found that the cost of the land is far in excess of the cost of the building work. Therefore it is imperative to obtain the most suitable site available at a fair market price and to use the site for the most advantageous permitted use. In some cases the cost of the building is relatively small by comparison with the cost of the land and in the last two decades land costs have generally risen at a much faster rate than building costs. Building land in expanding towns, without services, increased from about £250 per hectare in the early nineteen-fifties to around £7000 per hectare in the late nineteen-sixties, although admittedly part of this increase is due to the revised basis of compensation for the acquisition of land. Land for a small shop in the centre of a provincial town could cost as much as £20 000 to £50 000, whereas the cost of the building may not exceed £10 000; a ratio of between two and five to one. It is vital that the quantity surveyor should appreciate this type of relationship and its general significance.

This set of relationships also serves to indicate the need for a team effort when formulating development proposals, particularly those connected with the implementation of costly redevelopment schemes in the central areas of large cities. The team should include a valuer, architect and quantity surveyor. The valuer can advise on the best type of development and its ultimate value, and can give useful information on the most suitable type of construction and quality of finishings. The architect advises on building designs and the quantity surveyor on building costs. Together they can supply all the information necessary to prepare a complete budget of probable expenditure and revenue and are able to make recommendations on the advisability, or otherwise, of proceeding with the development before the land is purchased by the building client. The client must be satisfied that the projected development is in all respects sound and that it will produce a worthwhile return, before he makes an offer for the land.

The team of professional advisers has to consider a number of related matters, such as town planning requirements, services to be provided, operative rights of way and of light, road widening proposals and provision of access and car parking space. If the budget calculations prove to be satisfactory, a target cost will be established for the building work and this will constitute the total sum in the cost plan which subsequently emerges. The cost plan will break down this total sum over the main component parts or elements of the building.

Building Client's Main Needs

Building clients often want the best possible quality but are not prepared to pay for it. This frequently results in the architect's major problem being not one of design, but of cost. The quantity surveyor can be of tremendous assistance to the architect in the earliest days of a project, armed with his comprehensive records of historical cost information giving actual costs and essential details of buildings as constructed.

The main requirements of a client in connection with a building project can be conveniently listed as follows.

(1) The building must satisfy his needs, otherwise the architect has failed in his design function.

(2) The building should be available for occupation on the specified completion date if humanly possible.

(3) The final cost of the building should be very close indeed to the original estimate given to the building client.

(4) The building should be maintainable at reasonable cost.

The quantity surveyor is primarily concerned with the last two requirements. In particular he is anxious to ensure that the initial estimate and final account figures are closely related. The building client will always remember the initial estimate, as this will be the sum on which he has based all his calculations. The quantity surveyor needs the fullest constructional information if he is to provide a realistic estimate, and he often has to press the architect very hard indeed to obtain adequate information. In like manner the architect frequently experiences great difficulty in obtaining sufficient particulars from the building client in the early stages of a project. It is essential that the estimate prepared by the quantity surveyor should be as accurate as his skill and knowledge of the cost of previous jobs will permit.

Settling the Brief

Hogan[8] has aptly described how the initial brief for a building project must of necessity be a broad and flexible statement of objectives in fairly abstract terms, defining such matters as the site, building type, space requirements, general comfort standards, desired approximate total budget, timescale for design and construction, and the estimated useful life of the building. It is desirable for the brief to be drafted on the basis of a questionnaire prepared by all the design team. The next stage is to establish the external factors which define the physical limitations of the building, such as availability of public services; site survey; plot ratio, floor space index or residential densities; site coverage; building lines; maximum building height; adjoining buildings; rights of light and of way; road widening proposals; permissible points of vehicular access; extent of control of external elevations by local planning authority; and established environmental character of the area. These aspects will be considered in greater detail in chapter 12. The architect has then to consider the internal planning of the building related to both horizontal and vertical circulation networks. This will lead to a number of possible solutions which will need to be tested for general viability against the external site factors described previously.

Lucas[9] has described how the shape of some buildings is determined by the nature of the process carried on within them. For example, breweries are traditionally constructed on several floors so that the product will gravitate from floor to floor between successive stages of manufacture. Many factories, such as meatworks, car assembly plant, paper works and steel mills, all need to be designed so that each part of the manufacturing process follows in logical sequence. Similarly offices need planning to allow maximum ease of communication for their occupants. Lucas[9] considers the case of a company who wishes to build a new head office, covering the various policy decisions that have to be made, including an estimate of projected growth in staff over say a ten-year period and the resultant accommodation needs and their financial consequences. The success of speculative ventures will be dependent largely upon the developer's ability to identify the needs of potential users.

Falconer[10] has described how the techniques of warehousing and distribution are changing as mechanical and electronic equipment take over work currently performed by hand and human calculation. In drawing up the brief the architect must be aware of the problems of loading and unloading vehicles by mechanical means; economical storage and handling of goods within the warehouse; and making up diverse orders related to journey planning and mechanised stock control. Costs of the handling equipment must be balanced against the cost of the building to ensure that the most economical overall scheme is being adopted commensurate with inevitable future improvements in handling techniques. Basic requirements for warehouses include a light, insulated weatherproof single-storey shell with as large a stanchion grid as possible, demountable partitions and high specification floor, with provision for 100 per cent expansion. It is worth noting that an increase in storey height from 4.50 to 9.00 m will result in increased costs of only about five per cent.

A survey of architects' offices[11] reported that the critical relationship between architect and client/user was often unsatisfactory, and this highlights the difficulties encountered in this important phase of the design process. The Building Research Station[12] carried out a case study of a civic design commission undertaken by a private architect for a large local authority covering the redevelopment of an urban site. It was found that the architect's preference for exercising his design function led to a rigid concentration on design work, which in turn resulted in a lack of awareness of the management opportunities open to the architect, and of the consequences of neglecting them. The study postulated that the architect failed to realise how much pertinent information lay hidden in the client's briefing documents and the minutes of initial briefing meetings. Better communications between specialist sections of the client organisation and the designer could have resulted in better judged decisions in important matters such as allocation of resources, procedure and time schedule.

Planning of Services

Hogan[8] has described the sort of problems that arise in reconciling service arrangements with architects' preliminary building designs. The services consultant commences with minimum statutory comfort levels as prescribed by codes of practice, Acts of Parliament and by-laws at one end of the scale, and ideal conditions at the other, to arrive at a broad environmental performance specification for discussion with the architect. This specification will include air temperatures, humidity, ventilation (natural or mechanical), lighting (natural and artificial), aspect and the thermal capacity of the structure. These could be expressed as maxima and minima to create a range of options.

The services consultant tests his performance specification against the architect's alternative diagrammatic layouts and, with assistance from the quantity surveyor, comparative cost factors will emerge. The primary objective at this stage is to find an ideal cost balance between building form and services installation, within the overall budget and the site parameters.

Subjective factors can make decision-making difficult. For example, the design team may submit that a higher level of factory lighting will increase productivity and reduce staff turnover; or that a hotel needs to be air conditioned to prevent excessive annoyance to occupants from city centre traffic noise; or that reduced glazing in an office building will produce improved internal comfort, whilst the letting agent believes that this will adversely affect its appearance and ease of letting.

The physical space needs of the services must be considered. A relatively cheap structural proposal of loadbearing crosswalls and a flat slab creates serious problems if services are to be concealed. The location of lift motor rooms (top, bottom or side), boiler plant room (roof or basement), substation with its special access requirements, escape routes from basement garages and plant rooms, and layout of basement ventilation ducts, all have cost implications which may give rise to further options needing early evaluation. Finally, the service installation ought to be reasonably flexible to meet future changed requirements such as increased demand for power outlets or a change from space heating to air-conditioning.

Conflicting Claims

It is important that the building client shall allow adequate time for all the desirable feasibility studies to be undertaken and for many alternatives to be investigated during the design of a project, even although he is anxious to see the job designed and constructed as speedily as possible. A prime objective should be to produce a building which offers the best solution to the problem in all its aspects and this requires thorough and careful planning.

At any stage one particular factor may exert a dominant influence on design. The requirements of the local authority, the desire to achieve a particular aesthetic effect, and convenience of construction are factors which may tend to exert undue influence at the expense of the best building in terms of cost to the building client, contribution to the overall environment and convenience of the user. The aim must be to balance these and other factors satisfactorily and this can only be achieved through effective team working, competence and understanding.

ROLE OF THE QUANTITY SURVEYOR DURING THE DESIGN STAGE

Nisbet[7] has described how building costs are now scrutinised more closely and with greater skill and accuracy as buildings have become larger, more complex and more expensive, and building clients have become more exacting in their requirements. These and other factors have compelled the architect to design with greater care and in more detail and, since the ultimate cost of a building is determined during the design stage, the architect has increasingly directed his attention to cost control procedures and invited assistance from the quantity surveyor. If costs are to be effectively controlled, the quantity surveyor must be closely associated with the design process, as he is the recipient of a large volume of cost data extracted from priced bills of quantities. He has available costs of complete buildings, of different structural forms, varying service layouts and a wide range of materials and components. In addition he is aware of the cost effects of variations in the shape, height and other characteristics of buildings and it is in everyone's interests that this information should be fed back to the design process.

Hence it has become increasingly apparent that the architect and quantity surveyor should be working together as a team during the design stages of a project. Where both professions are housed under the same roof as is often the case with local authorities and government departments, co-ordination of activities is that much simpler. Where practising professional firms are employed, a conscious effort is necessary to secure ample contact and co-operation throughout the design stage. Pott[13] asserts that more effective teamwork is necessary in order to raise individual and collective efficiency and to give better service to the building client. This entails the introduction of the quantity surveyor in a constructive role at the beginning of the design process and his active participation throughout the detailed design stage. The architect must surely benefit from the quantity surveyor's knowledge and skill in ensuring that the design is founded on a sound economic base from the outset, by the ability to make major decisions in full knowledge of their economic consequences, and by formulating the design against a cost background so that a balanced and consistent design is secured. Stevens[14] has also emphasised how the architect has a moral obligation to design buildings efficiently and economically. Whilst not wishing to inhibit the enterprise and creative powers of the individual designer, it is nevertheless incumbent upon an architect to take positive steps to secure overall economy in design, and the quantity surveyor can make a valuable contribution towards it.

Before the architect can settle a number of fundamental design issues he will frequently need assistance from a quantity surveyor on the various cost implications. The following examples will serve to illustrate the type of decisions that have to be made.

(1) What are the cost relationships of, say, three blocks 15 m high, two blocks 22 m high and one block 45 m high, where all three schemes will give the same floor area?

(2) At what height of building will it be necessary to introduce a reinforced concrete or structural steel frame?

(3) To what extent should precast concrete units be used as against *in situ* construction?

(4) Would it be best to use a solid reinforced concrete slab floor, hollow pots or some form of patent flooring?

The quantity surveyor needs to keep up to date on the latest constructional processes and techniques, materials and components. In conjunction with architects he will conduct cost studies into various forms of construction, internal finishings, cladding, infill panels, etc. Probably one of the biggest difficulties facing an architect is the choice of infill panels from the wide range of available materials which range from glass and other manufactured materials of every colour and texture to concrete panels incorporating a large variety of aggregates. The architect has to consider a number of factors including appearance, acoustic properties, thermal insulation, fire resistance, durability, strength and cost. The quantity surveyor can help by advising the architect as to whether he is obtaining good value for money.

COST PLANNING TECHNIQUES

The customary method of submitting tenders based on bills of quantities which have been prepared from fairly detailed drawings means that the cost of a project is not clearly established until after the design has been finalised. This is obviously an extremely bad practice but the main deficiency can be overcome by the use of cost planning which enables the cost to be established before final design decisions are made and the cost effects of each decision can be clearly seen before the decision is implemented.

The architect and quantity surveyor should be continually examining the cost aspects throughout the design process. Typical questions are: 'Is a particular feature, material or component really giving value for money, or is there a better way of meeting the particular need?' or 'Is a certain item of expenditure really necessary?' Cost planning establishes the needs, sets out the various solutions and the cost implications of these solutions, and finally produces the probable cost of the project. At the same time a sensible relationship must be maintained between cost on the one hand, and quality, utility and appearance on the other.

In recent years various methods of cost planning have been evolved but there is no universal system which can be satisfactorily applied to every type of job. Southwell[21] has described how cost control, operating at various stages of the design process, will require different techniques according to whether the only information is about function or whether both function and the building morphology are known, or whether, finally, function, morphology and structural information is available. The method using cost plans broken down into elements has proved to be very suitable for school projects but is not necessarily the ideal system for use with industrial buildings. There are, broadly, two basic methods of cost planning currently in use, although in practice variations of these methods have been introduced.

One method has been described as *elemental cost planning*, *elemental target cost planning* or the *Ministry of Education system*. This method was introduced by the Ministry of Education (now the Department of Education and Science) and has been used with some variations by many local education authorities for the cost control of school projects. The system is fully described in Building Bulletin No. 4.[5]

Briefly, in the elemental system sketch plans are prepared and the total cost of the work is obtained by some approximate method, such as cost per place or per square metre of floor area. The building is then broken down into various elements of construction or functional parts such as walls, floors, roof, etc., and each element is allocated a cost based on cost analyses of previously erected buildings of similar type. The sum of the cost targets set against each element must not exceed the total estimated cost. Cost checks are made throughout the design stage and lastly a final cost check is made of the whole scheme. Thus the system incorporates a progressive costing technique with the establishment of cost targets and the use of constant checks to ensure that the design is kept within the cost targets.

Another method is generally described as the *comparative* or RICS *Cost Research Panel system*. It was first introduced by the Cost Research Panel of the Royal Institution of Chartered Surveyors (now Quantity Surveyors' Research and Information Committee) and used on the Park Hill housing scheme at Sheffield. This method also stems from sketch plans but does not use a fixed budget like the elemental system. Instead a cost study is made showing the various ways in which the design may be performed and the cost of each alternative approach. The cost study will indicate whether the project can be carried out within the cost limit laid down by the building client and the cost of each of the major parts of the building. The cost study is usually based on approximate quantities and constitutes an analysed estimate.

The cost study provides a ready guide to design decisions and it enables the architect to select a combination of alternatives which will satisfy the financial, functional and aesthetic considerations. The selection thus made becomes the working plan and operates as a basis for the specification and working drawings. The quantity surveyor will need to carry out cost checks periodically throughout the design stage as with the elemental system, to ensure that the architect's proposals are being kept within the total cost limit agreed with the building client. The essential difference between these two methods of cost planning is that with the elemental system the design is evolved over a period of time within the agreed cost limit, whereas in the comparative system the design is fairly clearly established at the sketch plan stage, after the choice of various alternatives has been made, and is not generally materially altered after this stage. The elemental system has been described as 'designing to a cost' and the comparative system as 'costing to a design'.

Each of the two cost planning methods will now be examined in some depth accompanied by examples.

Elemental Cost Planning

Approximate estimate stage. The cost limit for the building will be determined either by the building client or by the joint action of the architect and quantity surveyor for the project. With educational buildings, cost limits are often expressed in pounds per place, and approximate estimates are largely based on the known costs of similar or comparable projects.

It is often considered good policy to break down the estimate over the building elements, as in this way the architect will be aware of how much he can spend on a particular element before he settles his design and specification. The elemental costs are generally expressed as the cost per square metre of floor area and constitute, in effect, a preliminary cost plan. This procedure avoids the possibility of an architect incorporating an expensive feature in his design which the target cost cannot possibly accommodate.

When the sketch plans are complete, various quantity factors are checked against those assumed in the estimate and any necessary adjustments to costs will then be made where differences occur. The quantity factors include the ratio of enclosing walls to floor area and the ratio of roof to floor area, and they have an important influence on cost.

Cost planning stage. In the cost plan the sum allocated to each element is usually expressed in pounds per square metre of floor area. The number of elements used on school projects varies considerably from the thirty originally envisaged by the former Ministry of Education to the sixteen subsequently adopted by Hertfordshire County Council. Nott[6] has described how the choice of elements can be varied by subdivision or grouping as required, but points out that the greater the amount of subdivision the more a surveyor has to pick his way through analyses or cost plans to determine the total cost effect of a design decision with possibly increased chance of error. He favoured the inclusion of both wall finishes and windows and doors with the walling element.

Nisbet[7] believes that the elements to which costs are assigned must be related to the way in which an architect builds up his detailed design. For example, an architect may consider the enclosing walls as an indivisible element which he must design as an entity, while in other cases it may be desirable to have two separate elements, such as one for windows and another for solid cladding. It is possible, for different types of buildings to have different sets of elements.

In like manner it is reasonable to postulate that the element of upper floors includes the total sandwich of construction which divides one space from the space above or below it, which should form the unit for comparison by cost or performance. For instance, the sound transmission qualities are influenced by floor and ceiling finishings and so these need to be included in the upper floors element. The same criteria apply to all the main structural elements and this indicates that finishes ought not to be shown as separate elements.

The cost allotted to an element may be based on a cost analysis of a similar building or on approximate quantities but in either case the architect will not be bound to the same material. The intention is to provide an allowance which is adequate to cover the functional requirements of the element related to a definite standard of quality. In this way the architect will not be tied to any preconceived ideas and will be free to make a decision at a later and more suitable stage in the design process.

The cost of elements is then converted into unit costs for the work involved, such as walling as £*x* per sq m, doors £*y* each and windows £*z* each, with the dimensions taken off sketch plans. Each element is examined in detail and the design and costs built up and checked against the original estimate. Some amendment to the distribution of costs between elements is bound to occur as the detailed design evolves. Close collaboration between architect and quantity surveyor in the transfer of sums between elements is essential. The introduction of target costs for elements has accelerated the use of new constructional techniques, materials and components and, in some cases, has resulted in commonly used materials being used advantageously in new ways.

Nisbet[7] has described how the cost checking process is vital to cost planning as it provides the means by which the cost design is controlled. As each element is designed on draft drawings, these are passed to the quantity surveyor who prepares an estimate based on approximate quantities. If the estimate should exceed the cost target, adjustments are made to the design until the cost target is reached. If the estimate is less, then a better quality can be provided, or the saving can be used to improve other elements. When all the elements have been satisfactorily checked, the working drawings can be prepared without fear of substantial revisions, and both the architect and building client can be reasonably confident that the tender will not exceed the approximate estimate or cost limit.

Cost plans must be regarded as flexible and it must never be the aim of the quantity surveyor to instruct the architect as to where and how money can be spent. The intention of cost planning is that the quantity surveyor shall assist the architect in designing the building, by giving him full information on the cost implications of his design decisions.

Table 7.3 contains a summary of a final cost plan relating to a secondary school, using 1970 prices, from which the breakdown of items into essential elements and their further grouping into structure, finishings and services can be seen. The first part of the table shows the initial cost plan with the total cost and cost per square metre of floor area inserted against each element. Square metre costs of groups of elements are also incorporated and totalled to give an estimated total cost per square metre of £67.01, which includes allowances for preliminaries, contingencies, design risk and price fluctuations. Working across the table, the totals of the final omissions and additions are entered against each of the elements and on the extreme right-hand side are the final cost plan figures which are determined when all the working drawings are complete. Most local education authorities use a much smaller number of elements than are shown in table 7.3, but the latter largely conforms to the BCIS arrangement for cost analyses.

Table 7.4 contains a detailed cost plan for the work below the lowest floor level of the secondary school, of which the final cost plan summary was detailed in table 7.3, to illustrate a way in which the information can be obtained and recorded. The detailed cost plan consists of schedules of approximate quantities making up each element and incorporating the quantity of each basic item, its unit rate and the total cost involved. These costs are totalled to give the cost of the appropriate element. The unit rates will be built up from the average billed rates for similar jobs suitably adjusted to take account of period increases, site differences, etc.

Cost analyses record the costs of jobs for which priced bills of quantities are available, and the costs are grouped in a similar way to those shown in the cost plan in table 7.3. Detailed cost analyses also incorporate quantity factors, such as wall to floor ratios and specification notes to increase their value as a basis for cost planning new projects. The preparation of cost analyses will be examined in chapter 8.

Comparative Cost Planning

Grafton[15] has asserted that comparative cost planning assumes that initial feasibility studies and cost advice have determined the general layout and arrangement of the building in the light of its total estimated or prescribed cost limit, and sets out to examine what could be described as a market of alternatives open to the designer in respect of each part of the building, and which are both feasible and acceptable to him. This study of alternative design solutions takes account of all the consequential effects of decisions on various parts of the building, relating to one particular part. The information concerning alternatives is set out in a manner which enables the architect to make rational decisions in the light of their individual order of cost and their cumulative effect on total cost, before he starts developing his design. Having settled his design decisions he then develops them, and only if he changes from them should the cost plan need adjustment.

The comparative method does not seek to enforce rigid cost limits for the design of particular elements, but rather to maintain flexibility of choice of a combination of possible design solutions, that will serve the purpose to be achieved. It is more concerned with the comparison of alternative possibilities within a total sum, rather than attempting to control the design piecemeal in relation to targets for limited sections of the work. Its object is not necessarily to show how cheaply a building can be produced but to show the spread of costs over various parts of the building and what economies are feasible. This enables the architect, within his cost terms of reference, to use

TABLE 7.3

FINAL COST PLAN SUMMARY FOR SECONDARY SCHOOL

	Initial cost plan (area – 3360 m²) Cost (£)	Cost in £ per m²		*Final checks* Omit	Add	*Final cost plan (area – 3374 m²)* Cost (£)	Cost in £ per m²	
Structure								
1. Work below lowest floor level	17 800	5.29		—	30	17 830	5.30	
2. Frame	14 000	4.17		—	70	14 070	4.17	
3. Upper floors	6700	2.00		—	180	6880	2.04	
4. Roof	13 500	4.02		100	—	13 400	3.97	
5. Roof lights	700	0.21		—	—	700	0.21	
6. Staircases	2400	0.71		60	—	2340	0.69	
7. External walls	5600	1.67		200	—	5400	1.60	
8. Windows	11 500	3.42		150	—	11 350	3.36	
9. External doors	220	0.07		—	—	220	0.07	
10. Internal load-bearing walls	3000	0.89		20	—	2980	0.89	
11. Partitions	6500	1.93		—	100	6600	1.96	
12. Internal doors	1600	0.48		—	80	1680	0.50	
13. Ironmongery	4100	1.22	26.08	—	50	4150	1.23	25.99
Finishings								
14. Wall finishes	5500	1.63		150	—	5350	1.59	
15. Floor finishes	8600	2.56		—	100	8700	2.58	
16. Ceiling finishes	4800	1.43		—	40	4840	1.44	
17. Decoration	3300	0.98		—	50	3350	1.00	
18. Fittings	14 000	4.16	10.76	—	200	14 200	4.21	10.82
Services								
19. Sanitary appliances	3200	0.95		50	—	3150	0.93	
20. Waste, soil and overflow pipes	2500	0.74		—	80	2580	0.77	
21. Cold water services	2800	0.83		—	100	2900	0.86	
22. Hot water services	4600	1.37		—	80	4680	1.39	
23. Heating services	19 000	5.65		—	400	19 400	5.75	
24. Ventilation and air-conditioning	—	—		—	—	—	—	
25. Gas services	1900	0.56		—	50	1950	0.58	
26. Electrical installation	15 000	4.47		—	700	15 700	4.66	
27. Drainage	6000	1.78	16.35	—	150	6150	1.82	16.76
	178 820		53.19			180 550		53.57
28. External works	16 000	4.76	4.76	300	—	15 700	4.65	4.65
	194 820		57.95			196 250		58.22
29. Preliminaries	20 000 }	7.74	7.74	—	—	20 000 }	7.72	7.72
30. Contingencies	6000 }					6000 }		
	220 820		65.69	1030	2460	222 250		65.94
31. Design risk (1½%)	3310	0.99						
32. Fluctuations (½%)	1100	0.33	1.32					
Total net cost	£225 230		£67.01					

the money to the best advantage in interpreting his design. This should lead to economy in design and will assist in the comparison of elemental cost apportionment as between one building and another.

The comparative method of cost planning differs from the elemental system in that although the building may be broken down into similar elements for the consideration of cost implications, a theoretical cost allocation

TABLE 7.4

DETAILED COST PLAN: WORK BELOW LOWEST FLOOR LEVEL TO SECONDARY SCHOOL

Quantity	*Unit rate*	*Cost (£)*
Excavate over site for removal of vegetable soil 225 mm thick		
3500 sq m	£0.15/sq m	525
Excavation and disposal		
600 cu m	£1.10/cu m	660
Excavation for foundation trench, including planking and strutting, compacting, removal of surplus and backfilling		700
Concrete (1:12) in blinding bed		
250 cu m	£6.50/cu m	1625
Concrete (1:3:6) in foundations		
350 cu m	£8/cu m	2800
Concrete edge beam, 275 × 275 mm including formwork to both sides and two coats of bitumen paint, 150 mm high one side, trowelled smooth on top.		
800 lin m	£2/lin m	1600
125 mm concrete slab on and including 150 mm of hardcore and ashes.		
3500 sq m	£2/sq m	7000
Ducts. Foundation ducts size 900 × 600 mm with 150 mm concrete bottom and 100 mm sides, including 50 mm precast concrete covers.		
200 lin m	£12/lin m	2400
Ditto, 600 × 300 mm		
52 lin m	£10/lin m	520
Total cost of element		£17 830

to a particular element based on previous experience, is not accepted as a valid factor for controlling the design of the element. Tender pricing is less consistent for individual items even when related to similar buildings, than is the cost arrived at for the building as a whole. Instead, the cost implications of feasible alternative solutions for the elements are considered in the light of their own cost and their effect on the cost of other elements, which their adoption would involve.

Thus in the one case there is a rather arbitrary pattern of cost distribution throughout the building set up to control the design. In the other, a market of alternative solutions is established from which the architect can decide upon a combination which provides, in terms of cost, an optimum design solution for the complete building. In both cases specific cost exercises may have assisted in determining the basic design before the more detailed design is considered. Similarly, once a cost plan is set up, cost checks help in both cases to keep the development of working drawings related to it. The cost plan is only part of the whole process. Basically the elemental system is probably better suited for use with educational buildings where comparisons can be made with similar buildings which have much in common, and this would not for instance apply to industrial buildings.

The comparative system endeavours to show the architect the cost consequences of what he is doing and what he can do. It shows the effect of choice of design for one component of the building on others. It is important to use some method, such as a check list, to obtain all necessary information from the architect at the earliest stage. In deciding the order and scope of the items to be included in the cost study, the quantity surveyor should consider these important points: In what order does the architect require cost information to fit in with his development of design, production of drawings, etc.? What alternatives are both practicable and worthy of consideration?

TABLE 7.5

PART OF COST STUDY USING COMPARATIVE METHOD

	Initial solution		*Alternatives*	*Consequential adjustment of other sections*			*selected solution*						*Remarks*
								Adjustment from other sections			*Net cost*		
	£	*£ per sq m*	£		*Add*	*Omit*	*Section Ref.*	*Ref.*	*Add*	*Omit*	£	*£ per sq m*	
1. *Substructure* (below basement floor level). Foundations to walls and columns; hardcore under basement floor; all excavation, backfill and disposal, including that for the basement below ground level.	13 600	5.24					1				13 600	5.24 (5.00)	Figures in brackets represent percentages
2. *Basement walls and floor* (*a*) 250 mm reinforced concrete walls reducing to 140 mm and 110 mm brick facings at 75 mm below ground level; floor of 75 mm concrete blinding and 150 mm reinforced concrete slab with waterproofer.	8000	3.10					2(a)				8000	3.10 (2.95)	
(*b*) 328 mm brick walls (class B engineering bricks), tanking-asphalt and 110 mm brick lining; floor as (*a*) but with asphalt damp-course under slab and around column bases.			9800	*Section 1* Extra excavation offset by saving in wall foundation.	100								
(*c*) As (*b*) but 328 mm wall in London stocks			9200	*Section 1* Ditto.	160								
c.f.	£21 600	£8.34									c.f. £21 600	£8.34 (7.95)	

The divisions of the building for cost study purposes will normally reflect the functional requirements of the building and generally follow the broad pattern of structure, cladding, finishings, fittings and services. Within each section, however, the subdivisions will differ according to the type of building, the constructional techniques under consideration and the alternatives which it is desired to investigate.

Table 7.5 illustrates part of a cost plan or more accurately a cost study using the comparative system and the form of tabulation introduced by Grafton.[16] This table shows all the cost information relating to a particular element, with alternatives and the effect on cost of the choices made. It provides the cost of both initial and final solutions, with columns added to show the cost effect on other elements of the various alternatives. Costs are also expressed as cost per square metre of floor area for ease of comparison. Its main advantage over percentages is that the cost of an element can be adjusted without affecting the unit expression of others. Nevertheless, the percentage expression can be useful for comparison purposes and can with advantage be added when all the costs have been computed.

The rates inserted in table 7.5 include an allowance for preliminaries and insurances. This is often considered desirable to provide a common basis for assessment of rates, which might be taken from one of several bills for similar work, where preliminaries are dealt with in differing ways. Yet there may on occasions be a case for extracting and dealing separately with some aspect of preliminaries which is peculiar to the site and significant in cost. An approximate assessment of cost worked up from preliminary drawings cannot be guaranteed to include each and every item that occurs in practice. Hence it is customary to add a sum at the end of the cost plan to include such items and the usual contingency sum (possibly around ten per cent).

When the final choice is made, the columns on the righthand side of the cost study are completed and the initial cost of each selected solution has to be reassessed, where necessary, in the light of the consequential adjustments resulting from selected solutions in other sections. The final stage is the preparation of the revised cost plan, which contains the selected solutions from the preliminary cost plan, with the figures subsequently revised in the light of the later and more precise information obtained as the working drawings are developed and quotations obtained from specialists. Cost checks should be made against the elements in the cost plan as and when more up-to-date information shows changes in earlier actual or assumed data, which could have a significant effect on cost. This permits discrepancies, which might affect choice in other directions, to be considered as early as possible. If the contract is to operate on a firm price basis, then an adjustment of the revised estimated figures is needed to take account of fluctuations in the cost of labour and materials which may take effect between the date of preparation of the first cost plan and the end of the contract period. In this way an architect is able to deal with changes in the light of a reasonably sure knowledge of their effect on total cost and can act accordingly. All cost data used in the preparation of cost plans should be carefully preserved, scheduled and adjusted, as necessary, for use in preparing future estimates and cost plans.

BUILDING INDUSTRY CODE

A subcommittee of the Technical Co-ordination Working Party at the Department of Education and Science,[24] on which all the educational building consortia were represented, established a framework for a building industry code designed to be applicable to all forms of building and building documentation. The main aims were to achieve greater uniformity in presentation and classification of detail, enabling wider interchange of information and greater use of computer systems. The primary facets of the code related to the type of building and the use of the spaces within it. The primary elements comprise the structure; external envelope; internal subdivisions; services and drainage; fixtures, fittings and equipment; and site. As the code is now used extensively by educational building consortia the full list of elements which differ from the BCIS grouping are listed, with the kind permission of the subcommittee.

(A) General.
(B) Site preparation.
(C) Foundations.
(D) Frame.
(E) Roofs.
(F) External walls
(G) Internal walls.
(H) Upper floors.
(I) Staircases.
(J) Fixtures, fittings and equipment.
(K) Furniture and furnishings.

(L) Hot and cold water and sanitary installations.
(M) Heating installations.
(N) Mechanical ventilation and air-conditioning.
(O) Electric lighting and power installation.
(P) Communications installation.
(Q) Mechanical services (lifts, hoists, escalators, etc.).
(R) Special services (gas, air, etc.).
(S) External services and connections.
(T) Drainage.
(U) Site furniture and surface coverings.

Within each BIC element will be a number of features, each of which is a 'distinguishable unit of building, being an aggregate of parts or components which together have significance in the total building process'. Typical examples of features are doors, windows and strip foundations, and each one of them is suitably coded.

COST PLANNING OF MECHANICAL AND ELECTRICAL SERVICES

Hogan[8] has criticised the way in which architects tend to approach the initial design of a project on the basis of the physical space requirements and how the services engineer may not be called in until the working drawings are complete, possibly without serious thought being given to the space needs of the service installations, without a full analysis of the expected environmental comfort levels, without full consideration of the thermal efficiency of the structure, and possibly with an inadequate sum of money allocated to the services budget. This situation can give rise to major problems. If the desired comfort levels are to be maintained and the building budget is fixed, the building will be cheapened, the standard of services reduced or the building client asked for more money.

Fletcher[17] has aptly described how it has been a matter of concern amongst all who participate in the design and construction of buildings, that the engineering services element of building cost has not proved so susceptible to control as the remainder. The general lack of use of cost control techniques emanates mainly from the early appointment of consulting services engineers and the prevalence of engineering contracts based on drawings and specifications. Furthermore, with engineering services, initial capital cost is rarely in itself a sufficient criterion.

Azzaro[18] believes that mechanical services cost records should show the overall cost of heating installations in such a manner that different types of installation can be compared. Furthermore, these costs need adjusting to take account of differences in size, shape and insulation of buildings. To operate a satisfactory cost control mechanism the following techniques are required.

(1) Methods of estimating the overall costs of heating installations from minimum data.
(2) Simple methods of adjusting estimates to evaluate the cost of different building shapes and sizes.
(3) Ability to compare costs of different installations.
(4) Ways of establishing the target cost in a form capable of being cost checked as the design develops.
(5) Knowledge of detailed costs and access to a viable cost-checking procedure.

The total cost of a heating installation is made up of the quantity of heat required multiplied by the cost per unit. The quantity of heat is measured in joules and cost is conveniently related to kilojoules (kJ). It is necessary to relate this to the cost per square metre of gross floor area and to be able to add other elements to give the overall cost of the building per square metre. The conversion factor employed is the number of joules per square metre of floor. For a given type of installation, a building which requires 600 J/h per square metre of floor will cost nearly double to heat than one which requires 300 J/h. It should be borne in mind that buildings in excess of 15 m deep (front to back, containing two or more storeys) with no roof ventilation, need artificial ventilation, and the cost of the installation together with the heat loss resulting from the increased air change may be greater than the additional costs accruing from the restricted building width. The larger the building, the greater the proportion of heat loss by ventilation, and the less the difference in cost stemming from alterations in shape, if the storey height and floor area remain constant.

Azzaro[18] has suggested that the total cost of a heating installation should be broken down into subelements: heat source – boilers, fuel storage; distribution – pumps, pipework, calorifiers; and emission – radiators, convectors.

This form of analysis is fully detailed in a MPBW research and development paper[25] and has two main advantages: it fits reasonably well into common design procedures for heating installations, enabling cost targets to be set and cost checking to take place during design; it groups together those parts of the system whose costs are likely to rise or fall together.

There is some evidence to show that as the size of a heating scheme increases so the cost of the heat source falls, but that of distribution and emission rises. In theory, distribution costs should fall as the cost of pipes does not increase proportionally to size, but this may be more than offset by the greater complexity of the control mechanism on the larger installation. The cost per emission unit is influenced by the type of unit, architectural appearance and average size of unit. The main cost of builder's work lies in the boiler room and chimney stack and it is often difficult to separate this cost from the main building work.

A study group of quantity surveyors[17] recommended that cost analyses of engineering services should relate the costs to the square metre of floor area and also to a specific unit for each service. The units recommended were draw-off points for cold and hot water services, kJ (1000 BTU) for heating distribution and emission systems, kJ/h for boilers, cu m/min for ventilation and air-conditioning, lighting points for lighting installations and points for power installations. It is further suggested that quantity factors should also be introduced to assist in comparing the costs of engineering services in different buildings. For instance, with a heating installation a quantity factor of 0.75 indicates that either only seventy-five per cent of the building is heated or that the remaining twenty-five per cent is heated (or air conditioned) by other means and is costed separately. With cold water services, a quantity factor of 0.10 indicates a good supply of outlets with a density of one in every 10 sq m (gross). Element unit rates equivalent to those employed in building work would be the unit rate per kJ for heating and per point for cold water services. Berryman[19] has suggested enumerating all equipment and measuring pipework, ductwork and electrical circuits in linear metres when preparing approximate estimates for engineering services. As indicated in chapter 8, later sources of information on the cost of heating, including the Building Cost Information Service, recommend the use of the kW as the appropriate unit.

Berryman[22] believes that there is a vital need for extensive cost feedback in the design of engineering services to secure efficient cost control. This feedback must incorporate a wide variety of cost control techniques, including summarising costs into functional unit rates; gross floor area rates; treated floor area or volume rates; rates per point, rates per kW, m^3/s, lux; unit rates for pipework, ductwork and equipment; and cost ratios of subsidiary items such as fittings, supports and insulation to principal items such as pipework and ductwork.

THE APPLICATION OF COMPUTERS TO COST CONTROL WORK

Bennett[20] has described how the design process starts with a brief, which in essence consists of a list of the spaces needed by the building client and a statement of the performance that he requires from each space. Most cost plans and analyses are based on elements, whereas the architect and client are concerned with the enclosed spaces. To be really effective the cost control mechanism should be based on space use and performance and this can probably only be achieved with the use of computers and mathematical models, whereby a computer representation of the reality can be manipulated by testing assumptions on the model which reacts in a manner closely resembling the real situation. A computer representation of the building is needed which will respond to the flow of design decisions in a way which closely reflects the actual cost implications.

Computers can be used to assist with cost planning work by storing and analysing vast amounts of cost data. As the size, range and complexity of building jobs increase it becomes more difficult to deal effectively with the cost aspects by manual processes, and there is a real need to harness the immense calculating and storing capacity of the computer in this work. It is anticipated that the main lead will come from the larger public offices who already possess the larger and more powerful computers with their amazingly fast speeds of operation.

COST CONTROL DURING EXECUTION OF JOB

The quantity surveyor's cost control function does not terminate at the tender stage but continues throughout the execution of the job. At the time the contractor commences work on the site, the quantity surveyor must have carefully scrutinised the priced bills, schedules of basic rates, insurances and other relevant documents. He should at an early stage agree ground levels with the contractor and suitable arrangements for dealing with daywork vouchers and claims for increased costs. An accurate record of drawings should be maintained with revisions to drawings noted and costed and variation orders costed and filed, and at the same time the architect should be supplied with relevant cost information. The opportunity should be taken on the occasion of site visits for measurements and interim valuations to note any matters such as labour strength, plant in use, weather conditions and causes of delay, which may subsequently have a bearing on the subject matter of claims. Throughout the contract period the quantity surveyor should maintain effective cost control arrangements to keep a constant check on costs and to supply cost advice to the architect in ample time for any necessary action to be taken without adverse effects on the job.

The cost plan also has its uses during the postcontract or posttender period. When a tender is accepted the priced bill can be analysed in a similar manner to that of the cost plan. A comparison of the priced bill and final cost plan is most valuable in that it shows up the differences between the cost plan and the tender, and so assists in preparing future cost plans. When work on site is commenced the cost analysis can be used for controlling variations. The analysed tender provides a framework of costs which can help to provide a running forecast of total costs as the job proceeds.

REFERENCES

1. ROYAL INSTITUTE OF BRITISH ARCHITECTS. Handbook of architectural practice and management (1965)
2. NATIONAL JOINT COUNCIL OF ARCHITECTS, QUANTITY SURVEYORS AND BUILDERS. A code of procedure for selective tendering (1969)
3. MINISTRY OF PUBLIC BUILDING AND WORKS. *R and D Building Management Handbook 4: Cost Control in Building Design.* HMSO (1968)
4. A. B. WATERS. Cost planning: what information does the architect need from the quantity surveyor? *The Chartered Surveyor* (November 1959)
5. DEPARTMENT OF EDUCATION AND SCIENCE. Building bulletin 4: cost study. HMSO (1972)
6. C. M. NOTT. The development of cost planning during design stages. *The Chartered Surveyor* (February 1960)
7. J. NISBET. The role of the quantity surveyor during the design stage. *The Chartered Surveyor* (July 1959)
8. B. HOGAN. Services – the problems of reconciliation. *Journal of the Institution of Heating and Ventilating Engineers* (February 1971)
9. V. G. LUCAS. Pre-design decision process. *Building Technology and Management,* **8.10** (1970)
10. P. FALCONER. Building for production, storage and distribution. *Architecture East Midlands,* **34** (1971)
11. ROYAL INSTITUTE OF BRITISH ARCHITECTS. The architect and his office: survey of organisation, staffing, quality of service and productivity (1962)
12. J. N. N. O'REILLY. Briefing and design – a case study. *Building Research Station Current Paper,* **34/69** (1969)
13. A. POTT. Architect and quantity surveyor teamwork. Cost planning: Postgraduate cost planning course. Royal Institution of Chartered Surveyors (1961)
14. J. O. STEVENS. Architect and quantity surveyor teamwork. Cost planning: Postgraduate cost planning course. Royal Institution of Chartered Surveyors (1961)
15. P. W. GRAFTON. Cost planning. *The Chartered Surveyor* (May 1966)
16. P. W. GRAFTON. An example of cost planning. *The Chartered Surveyor* (April 1960)
17. RESEARCH AND INFORMATION GROUP OF THE QUANTITY SURVEYORS' COMMITTEE. Cost planning mechanical and electrical engineering services in connection with building projects. Royal Institution of Chartered Surveyors (1968)
18. D. W. AZZARO. Cost planning mechanical services. Nottingham regional conference No. 5, Trent Polytechnic. Royal Institution of Chartered Surveyors (1968)
19. A. W. BERRYMAN. Counting the costs of engineering services. *The Quantity Surveyor,* **27.4** (1971)
20. J. BENNETT. Cost control in the public sector. Nottingham regional conference No. 6, Trent Polytechnic. Royal Institution of Chartered Surveyors (1969)
21. J. SOUTHWELL. Building cost forecasting. Quantity Surveyors' Research and Information Committee. Royal Institution of Chartered Surveyors (1971)
22. A. BERRYMAN. Controlling the costs of engineering services in buildings. *The Chartered Surveyor* (August 1971)
23. UNIVERSITY OF LOUGHBOROUGH/HEATING AND VENTILATING RESEARCH ASSOCIATION. *Proceedings of a Conference on Integrated Design of Building, University of Loughborough* (July 1971)
24. DEPARTMENT OF EDUCATION AND SCIENCE. Technical co-ordination working party. *Building Industry Code* (1969)
25. MINISTRY OF PUBLIC BUILDING AND WORKS. R and D paper: Heating installations; the analysis of costs; cost control during design (1967)

8 COST ANALYSES, INDICES AND DATA

THIS CHAPTER IS concerned with the compilation of cost analyses and other data for assessing costs and preparing cost plans of future building projects. It also examines the methods of compiling and applying cost indices as a means of updating past costs of buildings. The cost limits and yardsticks employed in the public sector are also investigated.

COST ANALYSES

Nature and Purpose

Swett[1] has defined cost analysis as 'the systematic breakdown of costs, according to the sources from which they arise'. He postulates that cost analysis should provide information for any immediate problem and can be performed in a number of ways. It can, for example, permit detailed comparisons to be made between different projects and isolate the causes of differences. These may arise from a variety of causes, such as differences in basic design or details of design; differences in regional pricing or differences in contracting conditions. Indeed, they could arise from a whole range of factors which tend to make every scheme unique in one respect or another. Cost analysis must be the basis of cost control.[25]

Probably the forerunner of most systems of cost analysis operating today was the elemental system introduced by the then Ministry of Education.[2] In this system the analysis was by elements or functional parts of a building, with the aim of providing data for the establishment of cost targets for designing buildings. Sweett[1] has emphasised that it is, however, one of a wide range of methods which could be used for collecting, collating and classifying the data contained in priced bills of quantities. Sweett asserts that the variety of cost problems which arise in building work makes necessary the profusion of methods of analysis, and that all problems cannot be solved by a single method. He favours modifying the analytical process to suit the needs of a particular problem.

The major difficulty in analysing a bill by elements has been the definition and demarcation of elements. In any given building the component parts do not simply or logically subdivide into the various elements or functional parts. Various parts perform more than one function; for instance, a crosswall performs two functions – load-bearing and as an internal partition. Other parts seem to fall between two elements; a concrete lintel could conceivably be included in a 'windows' element or in 'external walls'. Separation on a functional basis may on occasions give rise to difficulties; thus windows provide natural daylighting but so also do rooflights, yet one would not expect to see them in the same element. Furthermore, many elements are interdependent and a change in one may materially affect others. These problems have now been largely resolved by the Royal Institution of Chartered Surveyors' introduction of a Standard Form of Cost Analysis[3] which provides standardisation of elements and will be examined in some depth later in this chapter.

With cost control, a restricted form of cost analysis, usually by elements, is required in the first place to set targets, and thereafter, a wider field of information is needed to check the cost of detailed aspects shown on working drawings against their targets. Much analysed data is obtained from priced bills and, as will be described later in the chapter, may need considerable adjustment prior to its use for cost planning. Furthermore, the pricing of a bill is the outcome of the estimating method used by the successful contractor. Pricing is not only a personal process but it also reflects the varied approaches and policies of different contracting organisations. The estimating and pricing may not always truly reflect the actual costs on site. For these reasons Sweett[1] has wisely suggested that it would be more accurate to describe 'cost analyses' derived from priced bills as 'price analyses'.

Cost studies prepared for architects mainly show major sources of costs and the alternatives available to the architect. Hence the method of analysing the data will often vary from project to project. For example, if the project incorporates crosswalls, then their costs should be separated from those of the infill panels; similarly, costs of load-bearing external walls should be kept separate from minor internal partitions. In some cases, finishings will have special significance as, for example, where a building is to contain an impressive entrance hall with expensive floor, wall and ceiling finishes. The cost of the finishes in this part of the building should be separated from finishes elsewhere, as it could constitute a major item of cost.

For cost planning to be effective, banks of cost or price data are needed because the average quantity surveying office cannot possess sufficient cost information to provide an adequate base for cost plans covering a wide range of building types. It was for this reason that the Building Cost Information Service of the Royal Institution of Chartered Surveyors, described latei in this chapter, was established. One of its main functions is to publish and circulate cost analyses to subscribing members. It is interesting to note that the Institute of Quantity Surveyors[4] describes the processes as *design cost planning* and *design cost analysis* to distinguish between cost to the client and cost to the contractor.

A cost analysis shows how costs of a building are distributed over elements and groups of elements. A meaningful conclusion cannot, however, always be drawn from analyses unless full regard is paid to the quality of the work. When making cost comparisons of buildings, it is advisable to separate those parts of the work affected by site conditions such as foundations and drainage, to permit the comparisons to proceed on a similar basis. The possibility of expressing elemental costs as percentages was considered by the former Ministry of Education (now Department of Education and Science) but was rejected as it tended to conceal the actual cost per element. For example, an element costing £5/sq m in each of two jobs, one with a total cost of £60/sq m and the other £70, would give percentages of 8.3 and 7.15 respectively. Thus the element would appear cheaper in the second job, although in fact it is costing the same in both projects. Or again an element, such as electrical services, could account for roughly the same percentage but in buildings of widely different total costs.

Another purpose of a cost analysis is to show where reductions could most beneficially be made, should the tender unfortunately prove to be too high. The greater the number of cost analyses prepared and circulated, the more extensive will be the body of available cost information and the greater the opportunity for cost comparisons leading to more effective cost control.

A bill of quantities normally provides a cost breakdown of a job on the basis of work sections in accordance with the Standard Method of Measurement of Building Works. The compilation of element costs thus becomes a process of abstracting in reverse – abstract sheets are given elemental headings and then the prices of items, or more usually groups of items, are transferred to the elemental sheets. The elemental costs are totalled and checked against the total cost of the job. The floor area is measured (from the inside face of the external walls) and the cost of elements divided by the area. It will expedite the work if cost analyses can be compiled by the same persons who prepared the original bills. An analysis for a primary school takes about two days to prepare. The use of elemental bills would drastically reduce the time required to produce a cost analysis, but when used in practice the disadvantages seemed to outweigh the advantages. A compromise has been sought in the sectionalised trade bill.[25]

Cost Analysis of Educational Buildings

Vast experience in the preparation of cost analyses has been obtained in connection with educational buildings. This stems from the lead given by the former Ministry of Education in 1950 and has been assisted by the large and continuous volume of school building work and the similarity of many of the buildings. The amplified analyses contain a wealth of information covering not only the cost of elements related to a square metre of floor area, but also quantity factors such as wall/floor ratios and element unit quantities (for example area of each type of internal partition in square metres and number of lighting points), and specification notes describing the main constructional methods and materials used. This form of analysis is invaluable in making cost comparisons between buildings with varying quantitative and qualitative factors. Case studies involving the practical application of cost comparison techniques will follow later in this chapter and also in chapter 9.

Table 8.1 illustrates a typical amplified cost analysis of a primary school.

TABLE 8.1

AMPLIFIED COST ANALYSIS OF A PRIMARY SCHOOL

Element	*Cost per sq m*	*Specification*
1. *Preliminaries, insurance and contingencies*	£2.20	
2. *Foundations and site slab* GF area single storey 753 sq m Bearing pressure 250 kN/m^2 Nature of soil — Sand Site levels — Slope approx. 1 in 20 Water table — Nil	3.00	Strip surface soil 200 mm deep; excavate to reduced level, or fill and consolidate; building paper and 125 mm thick reinforced concrete floor slab, with BSS. 130 mesh; additional reinforcement under stanchion positions; retaining walls in one-brick walls in blue bricks; precast concrete cladding plinth unit at all perimeter edges of site slab.
3. *External walls, external doors and windows* (including glazing, ironmongery and decorations) External wall including door panels, window breast and fascia Ratio: $\frac{\text{Panels}}{\text{Floor area of school buildings}} = 0.821$ Ratio: $\frac{\text{Windows (Net)}}{\text{Floor area of school buildings}} = 0.471$ Ratio: $\frac{\text{Doors (Net)}}{\text{Floor area of school buildings}} = 0.047$	11.40	1. Concrete cladding slabs 65 mm thick, (mainly 425 × 3000 mm and 425 × 2000 mm) with 75 mm gypsum plaster panels as inner lining (33 sq m) 2. Softwood weatherboarding on timber framed panels with 10 mm asbestolux inner linings (210 sq m) 3. Fibrous plaster stanchion casings internally, wood casings externally. Softwood window frames with aluminium horizontal sliding casements and glass louvre ventilators; fascia panels and panels below windows in 16 G. vitreous enamelled steel sheet bonded to insulation board; glazed with 3 mm and 4 mm clear sheet and 6 mm polished plate. Glazed or solid hardwood doors with softwood frames and teak sills, 6 mm polished georgian wired plate; woodwork painted gloss; walls painted flat, semigloss and full-gloss oil.
4. *Steel frame single-storey* *Storey heights* / GF *floor area* / *Imposed load* / *Span* 3.000 / 600 sq m / 1.2 kN/m^2 / 2 m to 8 m and 12 m 5.000 / 153 sq m	7.00	1000 mm modular steel frame stove enamelled consisting of 110 mm sq cold formed box stanchions for single-storey, and 110 mm welded box stanchions on multi-storey; latticed cold-formed floor and roof beams, 100 × 50 mm channel eaves and gable ties, dowel plate fixing to site slab, plates bolted at corner stanchion. Fixed wind braces in cold formed steel; steel angle droppers to support cladding, 50 × 50 × 6 mm angle cladding rail over window heads.
5. *Roof construction* (including finish) Area — 672 sq m Eaves — 250 lin m	5.20	Prefabricated timber panels (mainly 2 m × 2 m and 3 m × 2 m) formed of 19 mm boarding on 100 × 50 mm and 125 × 38 mm joists; asbestos cement decking over boiler house area (16 sq m), three-layer roofing felt with white chippings finish; timber fascia and asbestos soffit to eaves and gutter sole.
6. *Roof lights* (including glazing and gearing) No. 17 single light fixed No. 19 single light opening	1.40	Steel single pitch roof lights, proportion ventilated, and controlled with rod gearing, timber upstands.
	c.f. £30.20	

TABLE 8.1 – *cont.*

Element	*Cost per sq m*	*Specification*
	b.f. £30.20	
7. *Internal partitions and doors* (including ironmongery and decorations) 75 mm bellrock 91 sq m 150 mm bellrock 125 sq m Glazed screen 38 sq m Fibrous plaster double partition 45 sq m Parana pine lining 70 sq m Single doors 30 Double doors 1	5.60	75 mm and 150 mm honeycomb gypsum plaster panels, self-finished both sides; fibrous plaster panels in wind-bracing conditions; fibrous plaster stanchion casings; hardwood glazed internal screens; stud partitions covered one side with Parana pine, other side mainly with asbestolux. 40 mm hardwood veneered flush doors with softwood cores; softwood frames; hardwood architraves.
8. WC *cubicles and screens* No. of cubicles 16	0.60	Galvanised metal-faced chipboard cubicles and screens.
9. *Floor finishes and skirtings* (including screeds)	4.40	20 sq m 15 mm pitchmastic flooring. 160 sq m 150 × 150 × 12 mm clay floor tiles. 66 sq m 150 × 150 × 15 mm quarry floor tiles. 220 sq m thermoplastic tiles; 290 sq m hardwood block flooring; 10 sq m granolithic paving, (maximum thickness screed and finish 44 mm); 75 mm hardwood or tile skirting; hardwood steps on softwood carriage (no. 1 short flight of steps at change of level).
10. *Ceilings* (including decorations) Combined plain and 150 sq m acoustic fibrous plaster Combined plain and 550 sq m acoustic plasterboard	3.40	Perforated and plain plasterboard panels, suspended from roof or floor deck with aluminium hangers, and aluminium cover strips, plain and slotted fibrous plaster bearer and filler beams, from lower flanges of steel beams; small fibrous plaster cornice throughout; two coats emulsion paint throughout except: kitchen – two coats oil paint; changing room – two coats Corktex B paint.
11. *Fittings and furniture*	5.20	Hat and coat hooks on hardwood rails; cloak fittings and seating; fitted wall benching in hardwood, with teak-faced blockboard tops, sinks inset; fitted store cupboards, bookshelves, display shelves in hardwood, kitchen preparation benches; slatted and solid storage shelving in softwood; pin-up panels in 10 mm medium hardboard with 50 × 25 mm hardwood edging; servery units, venetian blinds and curtain tracks.
12. *Plumbing* (*internal and external and sanitary appliances*) Type of supply: Low pressure gravity feed and high pressure Storage capacity: 7300 litres Location of fittings: Dispersed in eleven units No. of sanitary appliances: 47	4.00	75 mm diameter light alloy rainwater pipes and fittings (40 lin m); galvanised steel gutter outlets. CW storage cisterns – galvanised mild steel; wastes – copper; supply piping – copper; soil pipes – cast iron; mirrors and towel rails, toilet roll holders, etc. – white glazed fireclay and stainless steel sinks.
	c.f. £53.40	

TABLE 8.1–(*contd.*)

Element		*Cost per sq m*	*Specification*
		b.f. £53.40	
13. *Electrical installation*		2.80	Supply – 415/240 volts, two phase 4 wire, fifty cycles from electricity board; low tension distribution system. Wiring – PVC with all insulated accessories, simple line glassware fittings; asbestos-type cables to stage wiring where necessary; sub-main connections between main switchboard and distribution boards will be MICC cable, distribution boards of sheet metal construction; main switchboard will incorporate ironclad light duty switch gear contractor-made on site; wiring to heating chamber and unit heaters to weatherfoil specification.
Lighting points	147		
General purpose power points	24		
Stage outlet	1		
Electric kitchen	1		
Local storage water heaters/immersion heaters	11		
Oil tank heater	1		
$3\frac{1}{4}$ kW kiln	1		
Radio system outlets	5		
Class change bell system	2		
Wiring to weatherfoil heating system	8		
Clocks	3		
Water pump with float switch	1		
14. *Heating and hot water installation*		8.20	Heating: thermostatically controlled warm air system. Central oil tank, oilfired boilers, and calorifiers (some hot water provided by electric water heaters in electrical elements). 640 mm diameter steel flue.
Temperature criteria	16°C in classrooms when external temperature is 0°C		
Air changes	Three per hour		
U value of walls	av. 1.60 W/m² deg C		
U value of roof	1.30 W/m² deg C		
15. *Drainage (net cost)*		2.00	Pipes. soil: pitchfibre 75, 100 and 150 mm with vitrified clay fittings; stormwater: as above; manholes: brick.
16. *Playground and paved areas*		1.60	*Playground.* Strip surface soil 200 mm deep. 5 mm cold asphalt paving. 150 × 50 mm pre-cast concrete edging.
Paved surface of playground	700 sq m		
Perimeter strip and thresholds	60 sq m		
17. *Special decoration*		0.20	Allowance of £160
	net cost per square metre	£68.20	

STANDARD FORM OF COST ANALYSIS

The standard form of cost analysis was published by the Building Cost Information Service of the RICS in late 1969,[3] with the full support of the main users of cost data, for controlling building costs during the design stage. Previously, cost analyses had been prepared on a number of different forms and on the basis of a variety of instructions and element lists which detracted from their value for cost comparison purposes. Quantity surveyors are now able to compile their records of building costs knowing that whatever their source, the analyses have been prepared on the same principles and that the historical data derived from them will be comparable.

The information obtained from these analyses will assist and improve cost planning advice and will enable the design team to control more effectively the cost of a building during the early design stages, to ensure that the building client receives value for money in terms of aesthetics, space use and constructional form against initial capital and subsequent running costs. The primary purpose of cost analysis must be to provide data which allows comparison to be made between the cost of achieving various building functions in one project with that of achieving equivalent functions in other projects. To this end the costs are analysed by elements which group together items fulfilling a specific function. The standard form of cost analysis is in various stages of detail related to the design process; broad costs are needed during the initial period and progressively more detail is required as the design is developed. Information in this form is essential to quantity surveyors' cost control techniques. The Building Cost Information Service collects and disseminates cost information within the quantity surveying profession and the latest form of cost analysis is expected to encourage this by facilitating the establishment of a large library of cost or price data, expressed in metric terms and to two decimal places of a pound per square metre of gross internal floor area.

In the standard form the element list has been divided into six groups, five of which cover the building and the sixth external works. The substructure is a single element group, whilst the superstructure comprises frame, upper floors, roof, stairs, external walls, windows and external doors, internal walls and partitions, and internal doors; internal finishes embrace wall, floor and ceiling finishes. Fittings and furnishings appear as a collective single element. The largest single group covers services, which is subdivided into fifteen elements or components, some of which have limited application.

The principles and definitions incorporated in the standard form have been formulated by a working party of the RICS. They should help to remove some of the problems and ambiguities which can arise, particularly when analysing a complex building.

The analysis is in two stages – a brief analysis of four pages and an amplified analysis in much greater detail. The brief analysis constitutes a summary to the amplified analysis. The amplified analysis provides for the inclusion of quantity factors and other basic details, total elemental costs, element costs per square metre of gross floor area, element unit quantity, element rate and specification particulars in respect of each element, and all-in unit rates for subelements. In the brief form of analysis (summary of element costs) there is provision for two sets of elemental figures – one with the preliminaries shown separately and the other with the preliminaries apportioned amongst elements. In the amplified analysis preliminaries are shown separately as a percentage of the remainder of the contract sum (contract sum less preliminaries). The standard form of cost analysis probably entails more work in its compilation than the previous BCIS analysis form, when analysing bills of quantities produced in work sections. However, computers are being used to an increasing extent for bill preparation and they can be readily programmed to provide alternative sortations of information for use in amplified cost analyses, provided the quantities are prepared and coded with this in mind.

The Royal Institute of British Architects expressed regret that the list of elements in the BCIS cost analysis was at variance with the Cl/SfB classification system. The Institute expressed the view that architects coding their drawings and specifications in accordance with the latter system would experience difficulty in making the fullest use of their quantity surveyor's services on design cost problems. Polding[5] asserts that the reservations expressed by the RIBA at the sequential coding used in the analysis would be more valid if there were a readily identifiable alternative, and points out that each element is unique and, in fact, the analysis could be used without any form of reference system. In practice it seems unlikely that architects will be prevented from receiving cost advice in whatever form they want it.

The more important contents of the standard form of cost analysis and the method of approach to be adopted in each case is now outlined.

General Principles

(1) The elemental costs are related to the gross internal floor area and also to a parameter more closely identifiable with the element's function (element's unit quantity).

(2) Supporting information on contract, design/shape and market factors are defined so that the costs analysed can be fully understood.

(3) Professional fees are not included in the analysis and contingency sums are shown separately.
(4) In amplified analyses, design criteria are inserted against each element.

Coding

The BCIS reference code classifies buildings by form of construction, number of storeys and gross internal floor area in square metres. There are four constructional types: A – steel-framed; B – reinforced concrete framed; C – brick; D – light framed steel or reinforced concrete. Hence a two-storey steel-framed building with a gross internal floor area of 6000 sq m would be coded as A–2–6000.

Supporting Information

Cost analyses contain the following supporting information.

(1) A brief description of the building with reference to any special or unusual features affecting overall cost.

(2) Site conditions including access, proximity to other buildings, construction difficulties associated with topographical, geological or climatic conditions, and existing site conditions.

(3) Market conditions, including level of tendering, availability of labour and materials, keenness and competition.

(4) Contract particulars such as type of contract, basis for pricing, method of tendering, firm price/fluctuations, number of contractors invited, number of tenders received, contract periods, list of tenders and whether contractors are local or national.

(5) Design/shape information including general description of accommodation with gross floor area of each type, approximate percentages of building having a different number of storeys, and storey heights.

Amplified Analysis

The principal expressions used in the amplified analysis and the method of computing the various factors and costs are now examined in some detail.

Design criteria and specification. Design criteria covering the requirements, purpose and function of the element are listed under each element. The specification notes reflect the architect's solution to the conditions expressed by the design criteria, indicate the quality of the building, provide a check list of the items to be included with each element, and describe the form of construction and quality of material sufficiently to explain the costs in the analysis.

Element unit quantity and rate. The cost of an element is also expressed in suitable units which relate solely to the quantity of the element itself. For example, with floor finishes the element unit quantity is the total area of the floor finishes in square metres, whilst with a heat source it is kW.

All areas must be the net area of the element, for instance external walls will exclude window and door openings. Cubes for air-conditioning and similar systems are measured as the net floor area of the part of the building concerned multiplied by the height from floor finish to underside of ceiling finish.

The element unit rate is the total cost of the element divided by the element unit quantity. Hence the element unit rate for floor finishes is the total cost of the floor finishes divided by their net area in square metres. Where various forms of construction or finish exist within one element, provision is made for their separate listing together with their individual all-in rates.

Floor areas. The gross floor area is made up of all enclosed spaces fulfilling the functional requirements of the building measured to the internal structural face of the enclosing walls. It includes the area occupied by partitions, columns, chimney breasts, internal structural or party walls, stairwells, lift wells, and the like, as well as lifts, plant, and tank rooms above roof level.

Net floor area is subdivided between *usable*, *circulation*, *ancillary* and *internal divisions*. Usable area is that fulfilling the main functional requirements of the building, such as office or shop space. Circulation relates to entrance halls, corridors, staircases, lift wells, etc. Ancillary covers lavatories, cloakrooms, kitchens, cleaners' rooms, lift, plant, tank and similar rooms. Internal divisions relate to partitions, columns, chimney breasts and internal structural or party walls. The sum of the areas in the last four categories will be equal to the gross floor area.

In the case of residential buildings reference is made to net habitable floor area which covers the floor area within the enclosing walls, including partitions, chimney breasts and the like, but excluding balconies, public access areas, communal laundries, drying rooms, and lift, plant and tank rooms.

Roof and wall areas. Roofs are measured by the plan area (flat) across the eaves overhang or to the inner face of parapet walls, including rooflights.

The external wall area is measured on the outer face of external walls, including window and door openings.

Ratios. The wall to floor ratio is obtained by dividing the external wall area by the gross floor area to three decimal places. Element ratios are calculated by dividing the net area of the element by the gross floor area to three decimal places.

Related matters. The storey height is measured from floor finish to floor finish, except in the case of single-storey buildings and the top floor of multistorey buildings, where it is taken from floor finish to underside of ceiling finish. The internal cube is measured as the gross internal floor area of each floor measured by its storey height.

A functional unit is expressed as net usable floor area (offices, factories, public houses, etc.) or as the number of units of accommodation (seats in churches, school places, persons per dwelling, etc.).

Elemental divisions/design criteria. It might be helpful to mention that decorations are included in the finishings element and that windows and external doors include lintels, sills/thresholds, cavity damp-proof courses and work to reveals of openings, in addition to the window/door, frame, ironmongery and glazing.

Design criteria for substructures include soil loading, nature of soil, bearing strata depth, site levels, water table depth and average pile loading. Design loads are included for upper floors and roofs. The appropriate ratios are inserted against external walls, windows and external doors, and internal walls and partitions. Entries for services elements include the number of draw-off points/power outlets.

Examples of Brief and Amplified Analyses

Table 8.2 comprises a completed brief cost analysis for an office block and is made up of basic information about the project, a summary of element costs and specification and design notes. The majority of the BCIS cost analyses are of this type. Table 8.3 contains an amplified cost analysis, this time covering two blocks of unit factories. The basic information and summary of element costs is compiled in the same way as in a brief analysis. Then follow detailed design and specification particulars relating to each element coupled with elemental cost information, including element unit quantities and rates, and all-in unit rates for subelements where appropriate. Both of these sample analyses are produced with the kind permission of the RICS Building Cost Information Service, and other organisations listed in the Acknowledgements section of the book.

COST YARDSTICKS

Housing Cost Yardsticks

The housing programme is by far the largest user of the country's building resources. Furthermore, there are a number of factors which tend to cause housing expenditure to increase independent of any increase in building prices. These include the raising of housing standards, the erection of more houses in highly priced areas, at higher densities and in redevelopment schemes, increased provision for cars and greater use of the more costly pedestrian segregated layouts. In the absence of effective cost control, the annual total housing expenditure could reach excessively high proportions resulting in a substantial reduction in the housing programme or large rent increases. Hence it is vital that some method should be applied to ensure that the best possible value is obtained in dwellings provided and that housing needs are met.

For these reasons housing yardsticks were first introduced in 1963 and were subject to a major review in 1967. In the Ministry circular 36/37[6] the objectives which the yardstick sought to achieve were stated as 'to keep down

TABLE 8.2

BRIEF COST ANALYSIS FOR OFFICE BLOCK

Code: B – 2 – 610

Job title: Office block	Client: Owner occupier
Location: South-west England	Tender date: mid-1969

Information on total project

Project and contract information

Project details and site conditions
The erection of a two-storey office block extension to a factory. There were no topographical or climatic difficulties. The foundations were the subject of a separate contract.

Contract
Standard Form of Building Contract, private edition with quantities. Firm price selected tenders, seven tenders issued, six received, all local contractors; lowest accepted, second lowest approx. four and a half per cent above lowest. Contract period stipulated by client and offered by builder – six months.

Market conditions
Work at this time was at a premium and tenders were slightly lower than estimated. There are no shortages of labour in this area.

Contract particulars

Contract particulars					
Type of contract: RIBA (with quantities)		Cost fluctuation	YES ☐ NO ☒		
		LABOUR	☐		
Basis of tender: ~~Open~~/Selected competition	☒	MATERIALS	☐		
		Adjustments based on formula	YES ☐ NO ☒		
Bill of quantities ☒	Negotiated ☐	Provisional sums	£	120	
Bill of approx. quantities ☐	Serial ☐	Prime cost (PC) sums	£	12 815	
Schedule of rates ☐	Continuation ☐	Preliminaries	£	nil	
		Contingencies	£	940	
Contract period stipulated by client: six months.		Contract sum	£	28 616	

Contract period offered by builder: six months.
Number of tenders issued: seven
Number of tenders received: six

Competitive tender list

£	N/L
28 616	L
29 942	L
30 467	L
31 396	L
32 380	L
32 453	L

Analysis of single building

Design/shape information

Accommodation and design features
Two-storey L-shaped office block. Ground floor accommodation consists of two drawing offices, three offices, print room, entrance and staircase. First floor accommodation consists of eight offices off a central corridor, meeting room, library, toilets and staircase.

Areas

Basement floors	– m²
Ground floor	293 m²
Upper floors	317 m²
Gross floor area	610 m²
Usable area	507 m²
Circulation area	74 m²
Ancillary area	13 m⁹
Internal division	16 m²
Gross floor area	610 m²
Floor spaces not enclosed	– m²
Roof area	353 m²

Functional unit: 507 m² (usable floor area)

$$\frac{\text{External wall area}}{\text{Gross floor area}} = \frac{540}{610} = 0.885$$

Internal cube = 1771 m³

Storey heights

Average below ground floor	– m
at ground floor	3.253 m
above ground floor	2.745 m

Design/shape:
Percentage of gross floor area

(a)	Below ground floor	– %
(b)	Single-storey construction	8 %
(c)	Two-storey construction	92 %
(d)	-storey construction	– %
(e)	-storey construction	– %

Brief cost information

Contract sum	£ 28 616			
Provisional sums	£ 120			
Prime cost sums	£ 12 815			
Preliminaries	£ Nil	being –%	of remainder of contract sum.	
Contingencies	£ 940	being 3.39%		
Contract sum less contingencies	£27 676			

Functional unit cost excluding external works: Tender £48.90; Base date £48.31

TABLE 8.2 (*continued*)

Summary of element costs

Gross internal floor area: 610 m²

Tender date: mid 1969

Element 1	Preliminaries shown separately: Total cost of element £	Preliminaries shown separately: Cost per m² gross floor area £	Element unit quantity	Element unit rate £	Preliminaries apportioned amongst elements: Total cost of element £	Preliminaries apportioned amongst elements: Cost per m² gross floor area £	Cost per m² gross floor area at 1st quarter 1969 £
1. *Substructure*	176*	0.29*			176*	0.29*	0.29*
2. *Superstructure*							
2A. Frame	4319	7.08			4319	7.08	
2B. Upper floors	1154	1.89			1154	1.89	
2C. Roof	2074	3.40			2074	3.40	
2D. Stairs	60	0.10			60	0.10	
2E. External walls	1615	2.65			1615	2.65	
2F. Windows and external doors	2672	4.38			2672	4.38	
2G. Internal walls and partitions	1720	2.82			1720	2.82	
2H. Internal doors	1414	2.32			1414	2.32	
Group element total	15 028	24.64			15 028	24.64	24.49
3. *Internal finishes*							
3A. Wall finishes	1127	1.85			1127	1.85	
3B. Floor finishes	1160	1.90			1160	1.90	
3C. Ceiling finishes	1172	1.92			1172	1.92	
Group element total	3459	5.67			3459	5.67	5.60
4. *Fittings and furnishings*	520	0.85			520	0.85	0.83
5. *Services*							
5A. Sanitary appliances	189	0.31			189	0.31	
5B. Services equipment	–	–			–	–	
5C. Disposal installations	132	0.22			132	0.22	
5D. Water installations	–	–			–	–	
5E. Heat source } 5F. Space heating and air treatment } 5G. Ventilating system }	2606	4.27			2606	4.27	
5H. Electrical installations	2500	4.10			2500	4.10	
5I. Gas installations	–	–			–	–	
5J. Lift and conveyor installations	–	–			–	–	
5K. Protective installations	2	0.00			2	0.00	
5L. Communication installations	–	–			–	–	
5M Special installations	–	–			–	–	
5N. Builder's work in connection with services	181	0.30			181	0.30	
5O. Builder's profit and attendance on services	–	–			–	–	
Group element total	5610	9.20			5610	9.20	8.99
Sub-total excluding external works, preliminaries and contingencies	24 793*	40.65*			24 793*	40.65*	40.20*
6. *External works*							
6A. Site work	2160	3.54			2160	3.54	
6B. Drainage	173	0.28			173	0.28	
6C. External services	–	–			–	–	
6D. Minor building works	550	0.90			550	0.90	
Group element total	2883	4.72			2883	4.72	4.69
Preliminaries	–	–			–	–	–
TOTALS (*less* contingencies)	£27 676*	45.37*			£27 676*	45.37*	44.89*

* These figures are exclusive of the foundations, which were the subject of a separate contract.

TABLE 8.2 (*continued*)

	Specification and design notes
1. *Substructure*	All work up to and including ground floor slab was executed as a separate contract prior to the commencement of these works. 3 mm bituminous membrane dampproof course to ground floor and Astos dampproof course to walls.
2. *Superstructure*	
2A. Frame	Frame, erected by nominated subcontractor, of precast concrete construction with an *in situ* ground floor slab.
2B. Upper floors	A hollow pot and beam floor by nominated subcontractor.
2C. Roof	Woodwool slabs, on Metsac joists; 600 mm fascia; three-layer felt roofing (355 sq m); 40 mm and 65 mm screeded beds laid on insulation; ornamental tiles (4 sq m); 22 g copper between layers of felt; aluminium roof trim; cast-iron rainwater heads and pvc roof outlets; seven double-glazed rooflights.
2D. Stairs	See 2A.
2E. External walls	255 mm hollow walls of brick and block, faced one side; one-brick walls.
2F. Windows and external doors	Forty-four metal windows executed by nominated subcontractor; special window by nominated subcontractor (£660); aluminium louvre windows; glazed entrance screen and window to foyer; precast concrete lintels; csg and georgian polished glass. 50 mm hardwood pair of doors and three coats varnish.
2G. Internal walls and partitions	Half- and one-brick walls; 75, 100 and 150 mm block partitions; glazed screens; 25 mm expanded polystyrene insulation.
2H. Internal doors	One pair softwood doors; ten hardwood doors; twenty-five flush doors, softwood and hardwood frames; concrete lintels.
3. *Internal finishes*	
3A. Wall finishes	Softwood and hardwood skirtings; three coats plaster on walls, etc., and plasterboard and thistle on partitions; 100 × 100 mm wall tiles (8 sq m) and three coats emulsion on walls; Sandtex finish on 66 sq m of walls externally.
3B. Floor finishes	Cement and sand screeds; 150 × 150 mm floor tiles (8 sq m); pvc floor tiles (415 sq m).
3C. Ceiling finishes	12.7 mm plasterboard (20 sq m); 12.7 mm foil-backed plaster lath (300 sq m); 100 × 100 mm acoustic tiles (35 sq m) and two coats emulsion paint. A PC sum of £725 was included for a metal suspended ceiling and acoustic tile ceiling.
4. *Fittings and furnishings*	Pay window (£50); shelving (£30); hardwood pelmets (£56); and various sundries (balustrades, etc.); louvre blinds (£160 for three).
5. *Services*	
5A. Sanitary appliances	Three lavatory basins; two WC suites and one 1050 mm long urinal; 22/28 mm copper overflow pipes and capillary fittings.
5C. Disposal installations	35/42 mm plastic wastes and fittings; pvc soil and vent pipes.
5E. Heat source 5F. Space heating and air treatment	Nominated subcontractor executed mechanical installation and hot and cold water installation (PC £2600); existing heat source.
5H. Electrical installations	Nominated subcontractor executed electrical installation (PC £2500).
5N. Builder's work in connection with services	Not priced.
5O. Builder's profit and attendance on services	Not priced.
6. *External works*	
6A. Site works	Reduced level excavation; retaining wall (£85); paving (£350).
6B. Drainage	50 lin m 100 mm cast-iron drain pipes, five soakaways.
6C. External services	External duct (£1750).
6D. Minor building work	Work to old buildings totalled £558.
Preliminaries	Nil.

TABLE 8.3

AMPLIFIED COST ANALYSIS FOR TWO BLOCKS OF UNIT FACTORIES

Code: A – 1 – 5202

Job title: Twelve factory units
Location: Midlands

Client: Public authority
Tender date: late 1970

Information on total project

Project and contract information

Project details and site conditions
Erection of twelve advance factory units with attached offices, in two blocks, together with associated external works and drainage.

Access from a hard-surfaced road running along the site boundary. Rough grassed, gently undulating virgin site with no adverse or suspect subsoil.

Contract
Standard Form of Building Contract, Local Authority edition with quantities; firm price selected tenders, six issued, six received, three local and three national contractors, second lowest accepted – lowest tender failing to obtain performance bond – after addition of increased contingencies and a sum for increased costs. Contract period stipulated by client: twelve months, and offered by contractor: ten-and-a-half months.

Market conditions
Six tenders ranging from £174 000 to £202 000; labour and materials fairly readily available; keen and competitive tendering from the lowest two contractors.

Contract particulars
Type of contract: Standard Form, Local Authority edition
Basis of tender: ~~Open~~/Selected competition ☒
Bill of quantities ☒ Negotiated ☐
Bill of approx. quantities ☐ Serial ☐
Schedule of rates ☐ Continuation ☐
Contract period stipulated by client: twelve months.
Contract period offered by builder: ten-and-a-half months.
Number of tenders issued: six
Number of tenders received: six

Cost fluctuation YES ☐ NO ☒
LABOUR ☐
MATERIALS ☐
Adjustments based on formula YES ☐ NO ☒

Provisional sums	£	3880
Prime cost (PC) sums	£	47 130
Preliminaries	£	15 659
Contingencies	£	8324
Contract sum	£	181 117*

Competitive tender list

£	N/L
201 828	L
193 783	N
189 500	N
187 521	L
178 117*	N
174 007	L

* £2500 added for increased contingencies, £500 added for extra costs caused by delay in starting due to second lowest tender being accepted.

Analysis of single building

Design/shape information

Accommodation and design features
Twelve factory units with offices attached, in two single-storey blocks of six units each. The blocks are approximately 92 m × 36 m × 4.3 m high for the factory and 2.3 m high for the offices.

Areas

Basement floors	– m²
Ground floor	5202 m²
Upper floors	– m²
Gross floor area	5202 m²
Usable area	4796 m²
Circulation area	128 m²
Ancillary area	158 m²
Internal division	120 m²
Gross floor area	5202 m²
Floor spaces not enclosed	– m²
Roof area	5367 m²

Functional unit:

$$\frac{\text{External wall area}}{\text{Gross floor area}} = \frac{2786}{5202} = 0.536$$

Internal cube = 80 160 m³

Storey heights;
Average below ground floor – m
at ground floor offices 2.29 m
factory 4.34 m

Design/shape:
Percentage of gross floor area:

(a) Below ground floor	–	%
(b) Single-storey construction	100	%
(c) Two-storey construction	–	%
(d) -storey construction	–	%
(e) -storey construction	–	%

Brief Cost Information

Contract sum	£	181 117	
Provisional sums	£	3880	
Prime cost sums	£	47 130	
Preliminaries	£	15 659	being 9.97% of remainder of contract sum.
Contingencies	£	8324	being 5.30% of remainder of contract sum.

Contract sum less contingencies £172 793.

Functional unit cost excluding external works	
tender	£26.79
base date	£22.98

TABLE 8.3 (*continued*)

Summary of element costs

Gross internal floor area: 5202 m^2

Tender date: late 1970

Element	*Preliminaries shown separately* Total cost of element	Cost per m^2 gross floor area	Element unit quantity	Element unit rate	*Preliminaries apportioned amongst elements* Total cost of element	Cost per m^2 gross floor area	Cost per m^2 gross floor area at 1st quarter 1969
	£	£		£	£	£	£
1. *Substructure*	25 149	4.83	5202 m^2	4.83	27 655	5.32	4.56
2. *Superstructure*							
2A. Frame	13 521	2.60	4459 m^2	3.03	14 869	2.86	
2B. Upper floors	482	0.09	97 m^2	4.97	530	0.10	
2C. Roof	27 555	5.30	5222 m^2	5.28	30 301	5.83	
2D. Stairs	–	–	–	–	–	–	
2E. External walls	14 333	2.75	2299 m^2	6.23	15 761	3.03	
2F. Windows and external doors	7437	1.43	487 m^2	15.27	8178	1.57	
2G. Internal walls and partitions	7731	1.49	2069 m^2	3.74	8501	1.63	
2H. Internal doors	3012	0.58	279 m^2	10.80	3312	0.64	
Group element total	74 071	14.24	–	–	81 453	15.66	13.26
3. *Internal finishes*							
3A. Wall finishes	2801	0.54	6898 m^2	0.41	3080	0.59	
3B. Floor finishes	1710	0.33	771 m^2	2.22	1880	0.36	
3C. Ceiling finishes	891	0.17	764 m^2	1.17	980	0.19	
Group element total	5402	1.04	–	–	5940	1.14	0.99
4. *Fittings and furnishings*	–	–	–	–	–	–	–
5. *Services*							
5A. Sanitary appliances	1244	0.24	–	–	1368	0.26	
5B. Services equipment	–	–	–	–	–	–	
5C. Disposal installations	392	0.08	–	–	431	0.08	
5D. Water installations	1269	0.24	–	–	1395	0.27	
5E. Heat source	–	–	–	–	–	–	
5F. Space heating and air treatment	–	–	–	–	–	–	
5G. Ventilating system	–	–	–	–	–	–	
5H. Electrical installations	4500	0.87	–	–	4949	0.95	
5I. Gas installations	–	–	–	–	–	–	
5J. Lift and conveyor installations	–	–	–	–	–	–	
5K. Protective installations	–	–	–	–	–	–	
5L. Communication installations	–	–	–	–	–	–	
5M. Special installations	–	–	–	–	–	–	
5N. Builder's work in connection with services	744	0.14	–	–	818	0.16	
5O. Builder's profit and attendance on services	70	0.01	–	–	77	0.02	
Group element total	8219	1.58	–	–	9038	1.74	1.51
Sub-total excluding external works, preliminaries and contingencies	112 841	21.69	–	–	124 086	23.86	20.32
6. *External works*							
6A. Site work	25 835	4.97	–	–	28 410	5.46	
6B. Drainage	12 484	2.40	–	–	13 728	2.64	
6C. External services	5974	1.15	–	–	6569	1.26	
6D. Minor building works	–	–	–	–	–	–	
Group element total	44 293	8.52	–	–	48 707	9.36	8.13
Preliminaries	15 659	3.01			–	–	–
TOTALS (*less* contingencies)	£172 793	£33.22			£172 793	£33.22	28.45

TABLE 8.3 (*continued*)

Element and design criteria	*A* Total cost of element £	*B* Cost of element per m² of gross floor area £	*C* Element unit quantity	$D = \frac{A}{C}$ Element unit rate £	Specification
1. *Substructure* Permissible soil loading: 157 kN/m² Nature of soil: Keuper marl. Bearing strata depth: 1 m. Site levels: undulating – 1:200 Water table depth: none Preliminaries 9.97% of remainder of contract sum.	25 149	4.83	5202 m² Area of lowest floor	4.83	Excavation and removal of 300 mm vegetable soil; excavation of average 300 mm to reduce levels over half the site; hardcore filling in making up levels over the other half of the site; reinforced concrete edge beam under external and internal load-bearing walls to factory area; traditional strip foundations and engineering brick walls under office, load bearing internal and external walls; 150 mm reinforced concrete slab to factory area; 100 mm reinforced concrete slab to office area; 1000 grade polythene damp-proof membrane; asbestos-based bituminous felt dampproof courses.
2A. *Frame* Grid pattern 4.60 m × 15.25 m. Preliminaries 9.97% of remainder of contract sum.	13 521	2.60	4459 m² Area of floors relating to frame	3.03	*Steel stanchions:* grid layout of 4.6 × 15.25 m; 300 × 165 stanchions along external walls; 250 × 145 stanchions along internal walls. *Steel portal frame:* 200 × 130 mm beams forming pitched roof, at 4.6 m centres. *Purlins:* 125 mm deep Z purlins at 1.67 m centres. No protective casing of steelwork.
2B. *Upper floors* Preliminaries 9.97% of remainder of contract sum.	482	0.09	97 m² Total area of upper floors	4.97	The area measured in this element comprises the covering to the office areas that protrude inside the factory. There are no stairways to these areas, and the construction is so light that it is unlikely that they will be utilised for storage. 50 × 125 mm sw impregnated joists, 25 mm Weyroc boarding.
2C. *Roof* Design loads: 0.75 kN/m². Spans: 15.24 and 4.5 m. Angle of pitch of sloping roof: 15°. Preliminaries 9.97% of remainder of contract sum.	27 555	5.30	5222 m² Area of roofs	5.28	*2C1. Roof structure* Offices: sw gangnail trusses; 200 × 75 mm sw plates. Factory: steel frame and steel Z purlins (see frame). *2C2. Roof coverings* Offices: 50 mm woodwool slabs; three-layer bituminous felt (incl. verges and gutters). Factory: Big Six asbestos sheeting, Nuralite flashings. *2C3. Roof drainage* Felt dressed into outlets in office roof. *2C4. Rooflights* North lights to factory area; patent glazing consisting of steel frames, and lead flashings. (see roof breakdown below)
2D. *Stairs*	–	–	–	–	Not applicable.

Roof	£	Area m²	All-in unit rate £
2.C.1 Roof structure	4343	5222	0.83
2.C.2 Roof coverings	19 662	5222	3.77
2.C.3 Roof drainage (factory roof drainage incl. in coverings PC sum)	15	5222	0.00
2.C.4 Rooflights	3535	5222	0.68

TABLE 8.3 (*continued*)

Element and design criteria	A *Total cost of element* £	B *Cost of element per m² of gross floor area* £	C *Element unit quantity*	$D = \frac{A}{C}$ *Element unit rate* £	*Specification*
2E. *External walls* $\frac{\text{External walls}}{\text{Gross floor area}} = \frac{2299\ m^2}{5202\ m^2}$ $= 0.442$ The approx. value of thermal conductivity: 1.70 W/m°C. Preliminaries 9.97 % of remainder of contract sum.	14 333	2.75	2299 m² Area of external walls	6.23	*Factory:* Outer skin of LBC heather facing bricks in half-brick wall; inner skin of common bricks, half-brick thick, increased to one-brick thick where required for strengthening purposes; 37½ mm cavity. *Offices:* Outer skin of LBC heather facings in half-brick wall; inner skin of columbia fair-faced blocks, 100 mm thick; 37½ mm cavity.
2F. *Windows and external doors* $\frac{\text{Windows}}{\text{Gross floor area}} = \frac{133\ m^2}{5202\ m^2}$ $= 0.026$ $\frac{\text{External doors}}{\text{Gross floor area}} = \frac{354\ m^2}{5202\ m^2}$ $= 0.068$ $\frac{\text{Area of opening lights}}{\text{Total window area}}$ $= \frac{27\ m^2}{133\ m^2}$ $= 0.203$ Preliminaries 9.97 % of remainder of contract sum.	7437	1.43	487 m² Total area of windows and external doors	15.27	*2F1. Windows* Module four standard metal windows, galvanised after manufacture, fixed to wood surround; 3.5 mm clear sheet glass; 5 mm thick drawn clear sheet glass; 6 mm georgian wired polished plate glass; softwood rebated and weathered sills; asbestos 152 mm × 102 mm based bituminous felt dampproof courses; eighteen no. full height softwood framed glazed screens to factory area. *2F2. External doors* Twelve no. sliding folding doors, each 4.27 m × 4.27 m; 50 mm two panelled doors, with glazed panels; 42 mm flush door, external quality plywood faced size 0.825 m × 2.00 m; 125 mm × 63 mm softwood frame; 100 mm pressed steel butts, mortice locks and furniture, mortice cylinder night latches; Dorman Long lintels; reveals as windows. Windows and external doors — £ — Area m² — All-in unit rate £ 2F1. Windows — 2599 — 133 — 19.54 2F2. External doors — 4838 — 354 — 13.67
2G. *Internal walls and partitions* $\frac{\text{Internal walls and partitions}}{\text{Gross floor area}}$ $= \frac{2069\ m^2}{5202\ m^2}$ $= 0.398$ Preliminaries 9.97 % of remainder of contract sum.	7731	1.49	2069 m² Total area of internal walls and partitions	3.74	*Offices:* Half-brick walls in common brickwork, fair-faced and flush pointed both sides; one-brick ditto; 100 mm wall of columbia fair-faced blocks flush pointed both sides. *Factory:* 225 mm wall of columbia fair-faced blocks flush pointed both sides. Internal walls and partitions — £ — Area m² — All-in unit rate £ Structural: One-brick wall in commons — 142 — 34 — 4.18 100 mm fair-faced blockwork — 1235 — 565 — 2.19 225 mm ditto — 5859 — 1286 — 4.56 Non-structural: Half-brick wall in commons — 353 — 150 — 2.35 One-brick wall ditto — 142 — 34 — 4.18

TABLE 8.3 (*continued*)

Element and design criteria	A Total cost of element £	B Cost of element per m² of gross floor area £	C Element unit quantity	$D = \frac{A}{C}$ Element unit rate £	Specification
2H. *Internal doors*	3012	0.58	279 m^2 area of internal doors	10.80	*Wrought softwood* *Doors:* Two-panelled, left open for glazing; hollow cored flush doors. *Fanlights:* 125 mm × 40 mm linings, all glazed with glazing beads over all doors. *Linings:* 125 mm × 40 mm and 125 mm × 35 mm on 100 mm × 12 mm or 115 mm × 12 mm grounds (sawn softwood); 40 mm × 20 mm stop planted on; 40 mm × 20 mm glazing beads; 50 mm × 25 mm splayed architrave. *Ironmongery:* 100 mm steel butts, mortice latch and furniture, mortice lock and furniture, hat and coat hooks, door closers, WC indicator bolts on WC doors (forty-eight no.). *Glazing:* Broad reeded glass and georgian wired polished plate glass. *Painting:* All doors – prepare, twice knot, touch up primer, two undercoats and one finishing coat gloss.

Internal doors	£	Area m^2	All-in unit rate £
Twelve no. two-panelled doors, fully glazed with georgian wired polished plate glass. 125 mm × 40 mm lining on 100 mm × 12 mm grounds. Fanlight over	375	80	4.68
Seventy-two no. 0.686 m × 1.981 m, 35 mm Fl. dr. lining, grounds, fanlight over	1752	128	13.69
Twenty-four no. 0.762 m × 1.981 m ditto	588	46	12.78
Twelve no. 0.838 m × 1.981 m ditto	297	25	11.88

Preliminaries 9.97% of remainder of contract sum.

Element and design criteria	A	B	C	D	Specification
3A. *Wall finishes*	2801	0.54	6898 m^2 total area of wall finishes	0.41	*Factory:* Two coats of emulsion paint on fair-faced brick or block walls. *Offices:* 12 mm render and set, three coats emulsion paint. White glazed wall tiles over basins in toilets.

Wall finishes	£	Area m^2	All-in unit rate £
Finishes to walls internally:			
12 mm render and set	1264	1469	0.86
Three coats emulsion on plaster	316	1469	0.21

(*contd.*)

Element and design criteria	A Total cost of element £	B Cost of element per m² of gross floor area £	C Element unit quantity	$D = \frac{A}{C}$ Element unit rate £	Specification
3A. *Wall finishes (contd.)* Preliminaries 9.97% of remainder of contract sum.					(see wall finishes breakdown below)
3B. *Floor finishes* Preliminaries 9.97% of remainder of contract sum.	1710	0.33	771 m² total area of floor finishes	2.22	150 mm × 150 mm quarry tiles to BS 1286 on 32 mm screeded bed; 225 mm × 225 mm thermoplastic tiling on 47 mm cement and sand trowelled bed; 50 mm granolithic paving, trowelled smooth; 3 mm × 38 mm galvanised division strip; 100 mm × 25 mm softwood skirting, and knot, prime, stop and two undercoats and one finishing coat. (see floor finishes breakdown below)
3C. *Ceiling finishes* Preliminaries 9.97% of remainder of contract sum.	891	0.17	764 m² total area of ceiling finishes	1.17	*3C2. Suspended ceilings:* 12 mm aluminium backed Gyproc lath scrimmed and skimmed with Thistle Board plaster; 26 gauge eml and Thistle plaster; three coats of emulsion paint
4. *Fittings and furnishings*	–	–	–	–	Not applicable.

Wall finishes	£	Area m²	All-in unit rate £
White glazed wall tiles	37	9	4.11
Two coats emulsion on fair-faced brick or block walls	461	2570	0.18
Finishes to inside face of ext. walls: 12 mm render and set	214	251	0.85
Three coats emulsion on ditto	64	251	0.26
Two coats emulsion on fair-faced brickwork or blockwork	466	2599	0.18

Floor finishes	£	Area m²	All-in unit rate £
32 mm screeded bed of cement and sand 1:3	10	20	0.50
47 mm ditto	427	625	0.68
50 mm granolithic paving trowelled smooth	163	126	1.29
150 mm × 150 mm quarry tiles, 12 mm thick	90	20	4.50
225 mm × 225 mm thermoplastic tiling	633	625	1.01
Division strip	28	32 lin. m.	0.88
Skirting and decoration	359	773 lin. m.	0.46

TABLE 8.3 (*continued*)

Element and design criteria	A Total cost of element £	B Cost of element per m² of gross floor area £	C Element unit quantity	$D = \frac{A}{C}$ Element unit rate £	Specification
5A. *Sanitary appliances*	1244	0.24	–	–	Low-level WC suites; two-gallon plastic cisterns; 560 mm × 200 mm vitreous china washbasins on brackets; light gauge 20 mm diameter copper tubing; 32 mm diameter 75 mm seal bottle traps with compression outlets.

Sanitary appliances type and quality	Number	Cost of unit £	Total £
Low-level WC suite with S trap, 400 mm medium duty pedestal plugged to concrete floor; black plastic dual flush cistern; CP 12 mm low-pressure ball valve and float; white plastic flush pipe and black plastic seat;	48	12.52	600.96
560 mm × 200 mm wash basin with cast iron brackets; one cp tap, one vitreous china tap stopper, 32 mm cp waste plug and chain	48	9.50	456.00

Element and design criteria	A	B	C	D	Specification
Preliminaries 9.97% of remainder of contract sum.					
5B. *Services equipment*	–	–	–	–	Not applicable.
5C. *Disposal installations*	392	0.08	–	–	*5C1. Internal drainage:* 32 mm and 40 mm PVC wastes and fittings; 100 mm diameter PVC S and V pipe; 75 mm diameter PVC rainwater pipes and fittings.
Preliminaries 9.97% of remainder of contract sum.					
5D. *Water installations* No. of cold water draw-off points: sixty. No. of hot water draw-off points: forty-eight.	1269	0.24	–	–	*5D1. Mains supply:* 20 mm diameter copper tubing to floor level; 12 mm and 20 mm diameter stainless steel tubing and fittings from floor level. *5D2. Cold water service:* 12 mm stainless steel tubing and fittings; 32 mm diameter overflow pipes; low pressure stop valves; cp bib taps; 910 litres mild steel cisterns to BS 417. *5D3. Hot water service:* heaters included in electrical installation (PC sum).

Water installations	Total £	Cost per m² of gross floor area £
5D1. Mains supply	431	0.08
5D2. Cold water service	838	0.16

Element and design criteria	A	B	C	D	Specification
Preliminaries 9.97% of remainder of contract sum.					
5E. *Heat source*	–	–	–	–	Not applicable.
5F. *Space heating and air treatment*	–	–	–	–	Not applicable.
5G. *Ventilating system*	–	–	–	–	Not applicable.

TABLE 8.3 (*continued*)

Element and design criteria	A Total cost of element £	B Cost of element per m^2 of gross floor area £	C Element unit quantity	$D = \frac{A}{C}$ Element unit rate £	Specification
5H. *Electrical installations* Total electric load not known – tenants to supply all electrical equipment. Total number of power outlets: 168. Preliminaries 9.97 % of remainder of contract sum.	4500	0.87	–	–	*5H2. Electric power supplies* PC sum.
5I. *Gas installations*	–	–	–	–	Not applicable.
5J. *Lift and conveyor installations*	–	–	–	–	Not applicable.
5K. *Protective installations*	–	–	–	–	Not applicable.
5L. *Communication installations*	–	–	–	–	Not applicable.
5M. *Special installations*	–	–	–	–	Not applicable.
5N. *Builder's work in connection with services* Preliminaries 9.97 % of remainder of contract sum.	744	0.14	–	–	NOTE: Gas services and communication services are to be provided by the tenant. The contractor has only to provide the ducts, holes, etc., in readiness for this installation. (see services table below)
5O. *Builder's profit and attendance on services*	70	0.01	–	–	(see services table below)

5N specification:

Services elements	£	Cost of BWIC per m^2 gross floor area £
5A. Sanitary appliances	–	–
5B. Services equipment	–	–
5C. Disposal installations	97	0.02
5D. Water installations	148	0.03
5E. Heat source	–	–
5F. Space heating and air treatment	–	–
5G. Ventilating system	–	–
5H. Electrical installations	454	0.09
5I. Gas installations	13	–
5J. Lift and conveyor installations	–	–
5K. Protective installations	–	–
5L. Communication installations	32	–
5M. Special installations	–	–

5O specification:

Services elements	£	Cost of BP and A per m^2 gross floor area £
5A. Sanitary appliances	–	–
5B. Services equipment	–	–
5C. Disposal installations	–	–
5D. Water installations	–	–
5E. Heat source	–	–
5F. Space heating and air treatment	–	–

Element and design criteria	*A* Total cost of element £	*B* Cost of element per m² of gross floor area £	*C* Element unit quantity	$D = \frac{A}{C}$ Element unit rate £	Specification		
5O. *Builder's profit and attendance on services* (*contd.*)					5G. Ventilating system	–	–
					5H. Electrical installations	70	0.01
					5I. Gas installations	–	–
					5J. Lift and conveyor installations	–	–
					5K. Protective installations	–	–
Preliminaries 9.97% of remainder of contract sum.					5L. Communication installations	–	–
					5M. Special installations	–	–
Sub-total excluding external works, preliminaries and contingencies £	112 841	21.69					

Element design criteria	Total cost of element £	Specification	Cost of subelement £	Element unit quantity	Element unit rate £
6A. *Site works*	25 835	6A1. Site preparation Preparatory earthworks to form new contours.	726	1933	0.38
Cost per m² gross floor area: £4.97		6A2. Surface treatment Roads and yards £13 416 Vehicle parks £ 5277 Paths and paved areas £ 1233 Retaining walls £ 502 Landscape work £ 2319	22 747	8968	2.54
		6A3. Site enclosure and division Gates and entrance £ 573 Fencing £ 1789	2362	138	17.12
Preliminaries 9.97% of remainder of contract sum.		6A4. Fittings and furniture	–	–	–
6B. *Drainage* Cost per m² gross floor area: £2.40.	12 484	*Surface water drainage* 100 mm, 150 mm, 225 mm diameter GVC BS pipes and fittings with flexible joints, laid and jointed in trenches up to 2 m total depth. 150 mm thick grade E *in situ* concrete bed and backfilling to pipes laid on granular bedding material. GVC mud and yard gullies, GVC rodding eyes, heavy duty road gullies. Eleven no. traditional construction manholes; Two no. precast concrete circular manholes.	7902	–	–
Preliminaries 9.97% of remainder of contract sum.		*Foul drainage* Pipes and bedding as surface water drainage. Twenty no. traditional construction manholes; One no. precast concrete circular manhole.	4582	–	–
6C. *External services* Cost per m² gross floor area: £1.15.	5974	6C1. Water mains	2300	–	–
		6C2. Fire mains	–	–	–
		6C3. Heating mains	–	–	–
		6C4. Gas mains	720	–	–
		6C5. Electric mains	2300	–	–
		6C6. Site lighting	500	–	–
		6C7. Other mains and services	–	–	–

TABLE 8.3 (*continued*)

Element and design criteria	*Total cost of element* £	*Specification*	*Cost of subelement* £	*Element unit quantity*	*Element unit rate* £
6C. *External services* (contd.)		6C8. Builder's work in connection with external services			
		6C1.	30	–	–
		6C2.	–	–	–
		6C3.	–	–	–
		6C4.	–	–	–
		6C5.	45	–	–
		6C6.	51	–	–
		6C7.	–	–	–
		6C9. Builder's profit and attendance on external services			
		6C1.	23	–	–
		6C2.	–	–	–
		6C3.	–	–	–
		6C4.	–	–	–
		6C5.	–	–	–
Preliminaries 9.97% of remainder of contract sum.		6C6.	5	–	–
		6C7.	–	–	–
6D. *Minor building work*	–		–	–	–
Preliminaries Cost per m² of gross floor area: £3.01.	15 659				
9.97% of remainder of contract sum.					
Total (*less* contingencies) £	172 793				

the cost of building so as to reduce the burden on the taxpayer, the ratepayer and the tenant and to ensure that the resources available for housing were used as effectively as possible'. The housing cost yardstick was detailed in appendix II of the circular[6] and this showed the limit of cost which would rank for subsidy for dwellings conforming to Parker Morris standards. For cost planning purposes the relevant figure in the cost tables is the total cost per person, broken down in the three main elements of superstructure, substructure and external works. Where schemes have to include provision for more than seventy-five cars per hectare of site, the yardstick for the dwellings is increased, based upon a notional equivalent higher density. Similarly, special extra allowances apply to dwellings designed for elderly persons subject to the provision of certain additional amenities for the residents. There is also a table of regional variations giving additions for certain parts of the country, rising to twelve-and-a-half per cent for inner London, and these have been increased in subsequent circulars.[6] The Minister was also empowered to apply a special *ad hoc* yardstick to meet 'an exceptional combination of adverse site characteristics' or 'very special local architectural requirements'. The method of operation of the housing cost yardstick at medium and high densities is well described in Design Bulletin 7.[26]

Nunn[7] emphasised how important it was in cost planning housing layouts, that the choice of dwelling plans and the pattern of site layout should remain fluid whilst cost was under consideration. Indeed, the first task after establishing the appropriate functional yardstick is to ensure that the selection of low, medium rise, or high building types, or a proportion of each, is such that an ultimate total cost within the yardstick is feasible. Potential savings within the buildings are small in comparison with economies that can be achieved by securing the best pattern of layout and building form. Table 1 of Ministry circular 1/68 shows the optimum proportions of accommodation in low, medium or high rise buildings for a large range of densities (in bed spaces/hectare) for average numbers of persons per dwelling varying from two to five.

As was emphasised in chapter 5 and in Ministry circular 36/67, two-storey houses are in general the least costly form of residential building and are preferred by most tenants; they are also less costly to manage and maintain than multistorey flats. As required densities are increased so it becomes necessary to introduce some three or four-storey blocks of flats or maisonettes, although these may be twenty to fifty per cent more expensive than two-storey houses, and so their numbers should be kept to a minimum. At even higher densities taller blocks of flats will have to be introduced and their costs may be in the order of fifty per cent more expensive for five-storey flats than two-storey houses, rising to eighty per cent for nine to twelve-storey flats. Budd[8] has pointed out that certain sections of the yardstick tables are more beneficial than others in that the yardstick figures and probable constructional costs are more closely related.

There has been extensive criticism of the level of yardstick in recent years, on the grounds that the yardstick was not revised sufficiently frequently to take account of the accelerating cost of building. It has been argued that this results in a lowering of housing standards and increased future maintenance costs, although Budd[8] believes that these suppositions are difficult to sustain. A RIBA deputation to the Ministry in 1969 reiterated these contentions and emphasised the difficulties of designing and redesigning to meet the yardstick with consequent increased design costs and delays; possible distortion of the desirable mix of dwelling types and sizes and the unfortunate effects of incorporating one-bedroom dwellings and in planning awkwardly shaped sites. Brett-Jones[27] has expressed a preference for a realistic yardstick which, while being tight enough to give an incentive for economical design, is not so tight that excessive mathematical ingenuity has to be used to keep within it. The ultimate criterion should be to try and provide the best value for money within the overall ceiling. A survey of local authorities by the National Housing and Town Planning Council in 1968 also showed that half the authorities responding to the survey were experiencing difficulties in keeping to the yardstick. The National Federation of Building Trades Employers asserted that because the yardstick has failed to match the rising price of building it has resulted in 'ill-judged and shortsighted reductions in specifications and amenities' and pressed for an annual review of the yardstick. Increases in the housing yardstick of between seven and twelve per cent were announced in April 1971, but these were considered inadequate by many local authorities, and in October 1971 further increases ranging between seven and eleven per cent were introduced.

Target prices for dwellings in early 1971 ranged from £1796 for one-bed space dwellings to £2624 for four-bed space dwellings in low rise development. The comparable figures for medium rise development were £2128 and £3109 and for high rise, £2617 and £3824. In 1970 average local authority housing costs were £63.30/m^2 in high developments, £53.40/m^2 in low-storey blocks of flats and £39.50/m^2 in houses and bungalows.

Other Cost Limits

Budd[9] has listed the levels of cost limits applied by various government departments over a four-year period, adopting as a base the second quarter of 1967, as compared with the relative cost of new construction. These statistics are scheduled in table 8.4.

Educational Buildings

Tayler[10] has described how since the early postwar years there has been in school building a steady process of refinement of layout, economy of specification, dual use of space and general elimination of waste. Architects, quantity surveyors, educationalists and others made a great contribution to the success of school building both at home and overseas, through ingenuity, innovation and co-operation, and much of the stimulus came from the application of school building cost limits. However, it will be seen from table 8.4 that there was no revision of these cost limits between the summer of 1966 and the early part of 1970, whereas building costs increased substantially throughout this period. Tayler asserted that, in consequence, by late 1969, the difficulty of maintaining standards consistent with good performance, low maintenance costs and the demands for increasing teaching space, resulted in delays in the commissioning and building of schools due to protracted and complicated negotiations to meet cost limits. Other unfortunate implications of unrealistic cost limits are the use of heating systems which have lower capital costs but are more expensive to operate; general disregard of costs in use; adoption of increasingly stereotyped architectural solutions; and strict limitation of variations in plan, shape, size, height and elevational treatment.

TABLE 8.4

TRENDS IN COST LIMITS: 1967 – 1971 (2nd quarter 1967 = 100)

Year	Quarter	Department of the Environment		Department of Education and Science		Department of Health and Social Security	
		Cost of new construction	Housing cost yardstick	Primary and special schools cost allowance per place	Secondary schools cost allowance per place	Residential accommodation for elderly people, clinics, junior training centres, hostels for mentally disordered, ambulance stations cost allowance	Adult training centres cost allowance
1967	2nd	100.0	100	100	100	100	100
	3rd	100.9	100	100	100	100	100
	4th	100.9	100	100	100	108	108
1968	1st	102.7	100	100	100	108	108
	2nd	104.4	100	100	100	108	108
	3rd	106.2	100	100	100	108	108
	4th	106.2	100	100	100	108	108
1969	1st	108.0	100	100	100	108	108
	2nd	108.8	107	100	100	108	108
	3rd	109.7	107	100	100	108	108
	4th	109.7	107	100	100	108	108
1970	1st	111.5	107	110	119	108	108
	2nd	115.9	107	110	119	108	108
	3rd	117.7	113	110	119	119	114
	4th	119.5	113	110	119	119	114
1971	1st	122.1*	113	110	119	119	114
	2nd	125.7*	122	125	125	119	114

Source: R. Budd. *The Chartered Surveyor* (June 1971) * Computed figures

Cost limits for educational buildings are prescribed by the Department of Education and Science. Schools are assessed by net costs per place, and in 1971 were £227 for primary schools and £424 for secondary schools (both with kitchens). In arriving at the number of cost places, an addition varying between five and twenty-five is made if the number of pupils in the school is less than 350 and a subtraction of between ten and forty where the number of pupils exceeds 350. Further small additions are made to the number of cost places where the percentage of pupils taking midday meals at the school exceeds eighty per cent. Special formulae have also been devised for determining the cost limits for remodelling of old schools.[11] Net cost limits in colleges of further education in 1971 were in the order of £50/m^2 for workshops, £60/m^2 for nonspecialised accommodation and £75/m^2 for specialised accommodation.

Hospital Buildings

The Department of Health has issued cost tables for establishing the cost limits of hospitals varying with the type of accommodation and number of beds. Allowances are added for communication space (notionally fifteen per cent), external works, height factor for buildings over four storeys, auxiliary buildings and abnormals, such as piling due to adverse site conditions and air-conditioning due to density of buildings on a restricted site.

BUILDING COST INDICES

Purpose of Cost Indices

The former RICS Cost Research Panel[12] in 1958 described how the cost of any building design is determined primarily by the cost of labour and material involved in its erection. Variation in the cost of either of these basic factors will influence the cost of an item of work, both absolutely and relative to the cost of the entire structure. The Panel also emphasised that the detailed labour and material content of every building differed but that these variations must be taken into account by the adjustment of cost data used for cost planning purposes. The compilation of indices of building costs is the most satisfactory method of approach and the Panel advocated the use of labour and material price variations, suitably weighted to take account of the quantities of each involved in a particular building type. There are however weaknesses in this approach which will be discussed later in this chapter.

The main function of building cost indices is to assess the differences in levels of tenders at varying dates. They may be used to compare building costs where tenders were obtained at different dates or to adjust analyses of the past costs of buildings to current prices. Future trends in price levels may be assessed, albeit rather imprecisely, by a study of cost indices and, having regard to possible future changes in labour and material prices, output of work and other related factors.

Source and Nature of Cost Indices

There are several published cost indices which are intended to provide an empirical guide to changes in building costs. Several of the better-known indices are listed with quarterly values over a six-year period in table 8.5. For comparison purposes, each of the indices has been adjusted to a 1963 base; they have been produced using different methods which are summarised below.

Department of Trade and Industry – price index of cost of construction. This is based on the statistics of earnings in the building industry compiled by the Department of Employment and Productivity adjusted to allow for changes in productivity, and the price index for construction materials computed by the Department of Trade and Industry. It thus reflects general changes in market conditions but does not take account of local conditions, variations for different building types or different types of contract.

Cost of building chart published in 'Building'. This is based upon an average of tenders received but it is impracticable in a single chart to make allowance for regional variations and differences in building types and contractual arrangements.

'Cubitt' index of construction costs. This is based upon average weekly earnings in the building industry, suitably adjusted for changes in productivity; and price trends in materials published by the Department of the Environment to which appropriate weighing factors have been applied.

Department of the Environment: Bulletin of Construction Statistics. This contains an index for a combination of new building and civil engineering work, which makes allowance for changes in productivity and wages rates on a countrywide basis but fails to take account of market conditions or different project types.

RICS *Building Cost Information Service.* This service is provided for the benefit of chartered surveyors and the indices are based upon basic hourly wages rates, holidays-with-pay, national insurance, etc. for building labour adjusted to allow for changes in productivity; materials prices based upon indices compiled by the Department of the Environment to which appropriate cost weighting factors have been applied; a fluctuations risk is built into the index by allowing the equivalent of a fixed prices tender for a period of eighteen months from the operative date of the index; and market conditions through an assessment of the cost effect of the prevailing tendering climate in the building industry. The BICS indices will be considered in greater depth a little later in the chapter.

An examination of table 8.5 shows that the various indices reflect broadly similar cost changes. Sweett[13] has described how he compiles his own price indices from all-in labour costs of building, electrical and heating and ventilating operatives, and plant and materials costs, all suitably weighted according to the type of job. For

TABLE 8.5

COMPARISON OF BUILDING COST INDICES (1963 = 100)

Year	Quarter	Department of Trade and Industry cost of construction	'Building' cost of building chart	'Cubitt' index of construction costs	RICS Building Cost Information Service Concrete frame construction
1963	*Average*	100	100	100	100
1965	1st	104	107	–	112
	2nd	106	108	–	112
	3rd	107	109	–	112
	4th	107	110	–	111
1966	1st	109	111	–	112
	2nd	111	112	–	114
	3rd	113	113	–	114
	4th	113	114	–	114
1967	1st	113	115	–	114
	2nd	113	116	–	117
	3rd	114	118	–	117
	4th	114	120	115	120
1968	1st	116	123	119	120
	2nd	118	125	120	121
	3rd	120	127	120	121
	4th	120	128	121	122
1969	1st	122	130	122	123
	2nd	123	131	123	125
	3rd	124	133	124	127
	4th	124	135	126	130
1970	1st	126	135	131	133
	2nd	131	137	133	136

Source: C. Sweett: *Building Cost Indices: Building* (23 April 1971)

instance, a typical office contract could be made up of thirty-three per cent building labour, two per cent electrical labour, three per cent heating and ventilating labour, five per cent plant and fifty-seven per cent materials. It is interesting to note that building labour costs increased by sixty per cent between January 1965 and December 1970, while over the same period electrical labour costs rose by eighty-three per cent and heating and ventilating labour by seventy-five per cent. Sweett[13] explains that his practice prepare their own indices to overcome the timelag which arises between the occurrence of price movements and the inclusion of their consequences in published building cost indices, and because it is much simpler to calculate the probable effect of specific cost factors on total costs from a detailed knowledge of the basis of the index.

Forecasting of future building costs is likely to be somewhat hazardous even with the benefit of building cost indices. Price movements can be very erratic on occasions. From 1965–9 building costs rose at a rate of about five per cent per annum, but both labour costs and materials prices rose much more sharply during 1970 than in the previous years. Sweett[13] suggests that in the foreseeable future building costs may rise by about seven-and-a-half per cent per annum based on a consideration of current trends in industrial relations and price movements, and even this prediction may now prove inadequate.

BCIS Cost Indices

Building cost indices compiled by the RICS Building Cost Information Service are scheduled in table 8.6. Indices are provided for four different classes of buildings – steel frame, concrete frame, brick and light frame. Hence

TABLE 8.6

BCIS BUILDING COST INDICES 1963–1970
1958 = 100

	Index figures								*Percentage addition for adjusting tenders to levels ruling fourth quarter 1970*							
Steel frame construction: class A buildings																
Year	1963	1964	1965	1966	1967	1968	1969	1970	1963	1964	1965	1966	1967	1968	1969	1970
Quarter																
1st	131	140	149	150	153	160	163	177	45	36	28	27	24	19	17	7
2nd	133	142	149	152	155	161	165	180	43	34	28	25	23	18	15	6
3rd	133	142	149	153	156	161	170	185	43	34	28	24	22	18	12	3
4th	136	145	148	152	159	162	173	190	40	31	28	25	19	17	10	0
Concrete frame construction: class B buildings																
Year	1963	1964	1965	1966	1967	1968	1969	1970	1963	1964	1965	1966	1967	1968	1969	1970
Quarter																
1st	130	139	148	148	151	159	163	176	45	35	27	27	25	18	15	7
2nd	132	141	148	151	154	160	165	179	42	33	27	25	22	18	14	5
3rd	133	142	148	151	155	160	168	183	41	32	27	25	21	18	12	3
4th	135	144	147	151	158	161	172	188	39	31	28	25	19	17	9	0
Brickwork construction: class C buildings																
Year	1963	1964	1965	1966	1967	1968	1969	1970	1963	1964	1965	1966	1967	1968	1969	1970
Quarter																
1st	132	142	152	153	156	164	169	179	46	36	27	26	24	18	14	8
2nd	134	144	152	154	159	165	170	182	44	34	27	25	21	17	14	6
3rd	135	145	152	155	159	166	173	188	43	33	27	25	21	16	12	3
4th	138	148	152	155	162	166	176	193	40	30	27	25	19	16	10	0
Light frame construction: class D buildings																
Year	1963	1964	1965	1966	1967	1968	1969	1970	1963	1964	1965	1966	1967	1968	1969	1970
Quarter																
1st	131	140	149	149	153	160	164	176	45	36	28	28	24	19	16	8
2nd	133	142	149	151	155	162	167	179	43	34	28	26	23	17	14	6
3rd	134	143	149	152	156	162	170	185	42	33	28	25	22	17	12	3
4th	136	146	148	152	159	163	173	190	40	30	28	25	19	17	10	0

changes in prices of materials may have quite different effects on the various classes of building according to the extent that the particular materials are used. A steep rise in brick prices bears more heavily on class C buildings than on class A. In addition to the cost indices the tables include percentage additions to be added to the costs of past jobs to bring them up to date. Figure 8.1 shows in graph form the relative costs of brick buildings over the period 1963–70.

In addition BCIS circulate subscribing members with tables of quarterly cost indices relating to various groups of elements, namely foundations, superstructure, internal finishes, fittings, services, and external works. Supplementary information is also supplied on labour costs, productivity levels and materials price changes (DOE index of construction materials). The Service attempts to explain the underlying influences behind price movements and the following extract relating to the fourth quarter of 1970 will serve to illustrate this aspect.

> Cost indices continue to show a rising trend, mainly due to increased material prices, the steadily increasing influence of longterm wage agreements, contractors' more realistic view of competitive tendering and an increasing doubt over future cost increases in assessing the fluctuations risk. Generally, the 'whole building' index series has risen by five points from the revised third quarter figure (about two-and-three-quarter per cent).

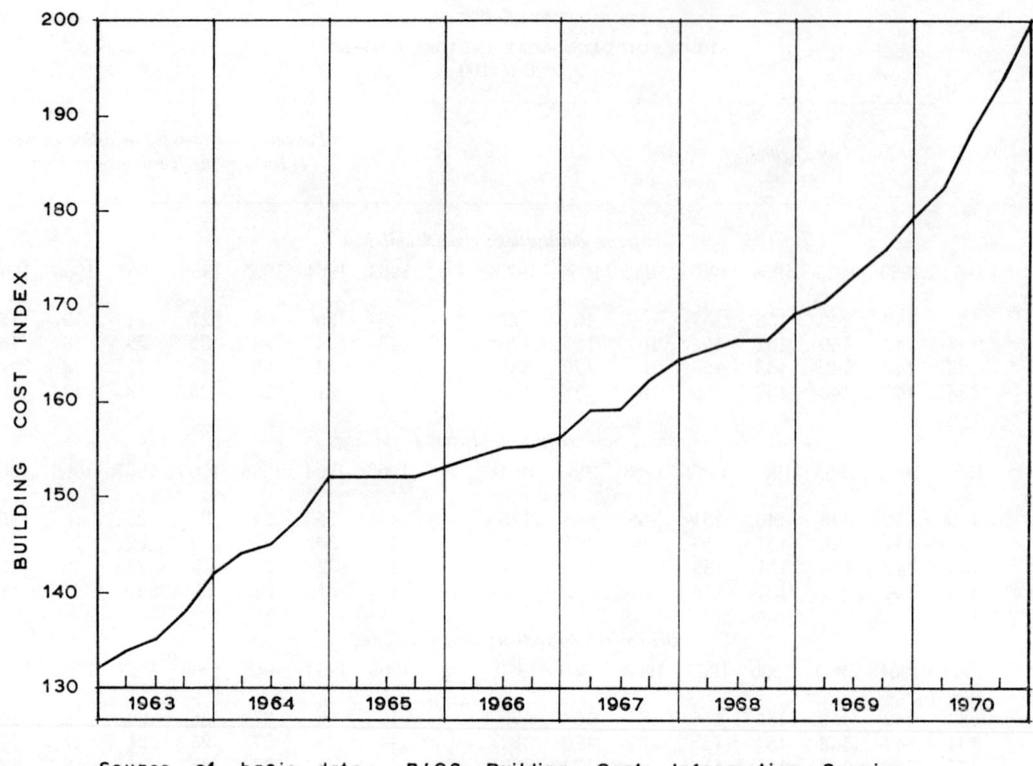

Figure 8.1 Brick building costs 1963–1970 base year: 1958=100

Trends in Building Prices and Validity of Building Cost Indices

For some years there has been a growing concern with regard to the degree of reliability of available building price indices. This concern stems partly from the lack of consistency between various published indices and partly by their failure to indicate the movement in building prices that persons in the building industry felt, from their own experience, to have occurred. An additional and significant reason was the wide coverage of the majority of indices, without regard to specific locality or building type. The former RICS Cost Research Panel formed a working party in 1963 to examine the problems associated with the construction of price indices and this showed the need for formal research.

In 1965 the RICS and the former Ministry of Public Building and Works jointly sponsored a three-year study at University College, London whose objective was

> to make a study of the movement of tender prices of buildings of all types over the previous ten years and to isolate the effects of variations in the design and type of building, the region of the country and the form of contract; to relate the movements in tender prices with the changes in building costs; and to consider critically suitable methods of constructing indices of building costs and prices and whether separate indices ought to be set up for different conditions.

Jupp[14] has described how preliminary investigations of the study group centred around three main concepts of building costs or prices: firstly, the cost to the builder of carrying out a specified job; secondly, the cost to the purchaser or client of the building of a similar specified amount of work of constant quantity and quality; and thirdly, the cost to the client measured in terms of functional units. The conclusion of the study group[15] was that most users of indices require the second approach – indices reflecting market prices paid by the client. Furthermore, a series of index numbers covering prices of different categories of building seemed to be more useful for the majority of purposes than one general index relating to all building and construction. It might be desirable to have separate index numbers for different localities, although more data would be required with increased stratifications, and further work might be advisable to establish a balance between the extent of stratification and the cost of collecting and analysing the data.

The study group also concluded that there are serious problems associated with the construction of index numbers based on labour and material costs, notably concerning productivity and plant, overhead costs and profit. Regarding indices based on full repricing of bills, quantity differences and variability of pricing between bills could lead to high coefficients of variation, although price variability might be reduced by stratification into geographical areas. The cost of constructing indices by full repricing of bills would be dependent upon the size of sample chosen and would probably be very high. Jupp[14] has, however, pointed out that the use of computers for repricing and the use of full bill repricing for cost control purposes might reduce the cost considerably. Furthermore, it seems likely that indices based on short lists of items selected from the bills reflect reasonably well the trends in prices derived from indices based on a full bill. The use of tender prices for index construction involves isolation of the effects of quantity and quality.

The study group's investigation of housing tender prices produced some interesting cost relationships:

(1) Two-storey houses have a lower cost per square metre than all other types of dwelling. Other types in ranking order are: bungalows +£4.5/m^2; two-storey flats +£7/m^2; three-storey maisonettes +£15.5/m^2; three-storey flats +£15.7/m^2; four-storey maisonettes +£20.2/m^2; and four-storey flats +£22.3/m^2, with standard errors of £0.65–£1.30.

(2) The approximate amounts by which regional costs per square metre differ from the average for all regions are: south-east +£4.3; east and south +£8.2; south-west +£7.3; Midlands −£3; north Midlands −£4.1; East and West Riding −£3.65; north-west +£0.65; north −£6.5; Wales +£2.7.

(3) An upward trend of £2/m^2 per annum (3.2 per cent) is indicated.

(4) Two-bedroom houses have the lowest cost per square metre, followed by: one-bedroom +£2/m^2; three-bedroom +£2.30/m^2; four-bedroom + £5.55/m^2; and five-bedroom +£7.8/m^2, after eliminating the effects of all other variables.

Application of Building Cost Indices

Building cost indices are often used to bring a tender figure for a previous similar job up to current prices, so that it can be used for comparison purposes as part of cost planning. An example may serve to illustrate this use.

An original firm price contract for a steelframed office building had a tender figure of £160 000. The BCIS index at the date of tender was 153, and this included the assessed value of fluctuation risks on the contract. The present index is 190, and an estimate is required for a building of similar design. It is proposed to use the tender figure for the earlier contract as a basis and it will be necessary to update that figure as follows:

$$\text{current value of contract} = £160\,000 \times \frac{190}{153} = £198\,500$$

It will also be necessary to make some allowance for increasing costs during the period between the date of preparation of the estimate and the probable tender date. This sum is often referred to as *design risks* and will be calculated on the basis of the likely time period and current price trends, including market conditions.

On occasions it will be necessary to assess the effect of price increases in materials on the estimated cost of a project, particularly where the rate of increase is substantial. In order to do this it is necessary to know the approximate breakdown of the cost of the job into labour and materials, and a further breakdown of t he cost o materials so that the proportion by value of each major material or component to be used on the job can be assessed. This cannot be done with any great precision at the approximate estimate stage and would involve a large amount of work at the bill stage. It is therefore customary to use analyses already prepared for previous jobs of a similar type. The reader may find the following analyses useful as a general guide.

TABLE 8.7

ANALYSIS OF GROUPS OF ELEMENTS IN TYPICAL SCHOOLS

Group of elements	*Class A school (steelframed)* %	*Class C school (brick)* %
Foundations	10	7
Superstructure	50	50
Fittings	11	14
Internal finishes	4	3
Services	19	14
External works	6	12
Total	100	100

TABLE 8.8

ANALYSIS OF LABOUR AND MATERIALS IN TYPICAL SCHOOLS

Class A schools		*Class C schools*	
Labour and materials	*Percentage (cost)*	*Labour and materials*	*Percentage (cost)*
Heavy steel sections	12.9	Bricks	20.6
Steel sheets	10.3	Softwood	16.4
Metal windows	9.0	Roofing tiles	6.0
Bricks	8.6	Partition blocks	4.6
Mild steel bars	5.4	Sand	3.8
Stramit	5.1	Metal windows	3.8
Softwood	3.0	Cement	2.7
Sand	2.8	Ironmongery	2.1
Marble	2.6	Roofing felt	1.4
Cement	2.5	Glass	0.4
Ironmongery	0.7	General labour	38.2
Glass	0.4		
Concrete products	0.3	Total	100.0
Cast-iron pipes and fittings	0.3		
Asphalt	0.3		
General labour	23.4		
Steel erector	12.4		
Total	100.0		

TABLE 8.9

ANALYSIS OF MATERIALS IN GENERAL CONSTRUCTION AND HOUSING

Material	*Construction percentage (cost)*	*Housing percentage (cost)*
Building stone	0.6	0.1
Roadmaking materials	3.1	–
Roofing slates	0.3	0.7
Sand and gravel	6.7	8.2
Bricks	7.3	12.8
Clay floor tiles	0.2	0.3
Clay roofing tiles	0.9	1.0
Pipes and conduits (vitrified)	1.6	2.4
Fire clay sanitary ware	1.0	1.2
Clay partition blocks	0.2	0.1
Sanitary earthenware	0.5	0.6
Glazed wall tiles	1.6	1.0
Sheet glass	0.4	0.5
Plate glass	0.3	0.1
Cement	7.0	6.4
Cast stone and concrete products including clinker blocks	5.1	4.1
Asbestos cement products	1.7	0.9
Plaster (gypsum)	1.2	1.8
Lime	0.3	0.5
Roofing felt	0.8	1.2
Asphalt	0.9	0.4
Paint for building and decoration	7.1	4.8
Steel	14.5	3.1
Cast-iron pipes and fittings	1.2	2.0
Solid fuel appliances and radiators	2.4	3.3
Cast-iron baths and cisterns	2.8	4.3
Copper tube	0.9	0.8
Copper sheet	0.4	0.5
Aluminium sheet	0.4	0.1
Lead (pig and pipe)	1.1	0.6
Zinc	0.2	–
Domestic gas appliances	1.3	1.3
Domestic electrical appliances	1.2	1.6
Electrical installation materials	4.2	2.5
Thermal insulation materials	1.0	0.2
Fencing	0.2	0.1
Plumbers' brassware	1.3	1.1
Ironmongery	1.0	1.7
Metal windows	1.4	1.2
Joinery	5.1	12.0
Imported hardwood	1.1	2.5
Imported softwood	7.3	10.9
Imported plywood	0.7	0.8
Railway sleepers	0.9	–
Wallpaper	0.3	0.1
Polythene tubing	0.1	0.1
Plastic household mouldings	0.2	0.1
Total	100.0	100.0

A further example may serve to illustrate the use of these schedules in assessing the probable cost effects of increases in the price of materials. A house is estimated to cost £9500 and the following price increases have been notified. It is required to determine the probable effect of these increases on the estimated cost of the house.

Bricks +0.6 per cent; cement +8.4 per cent; softwood +0.7 per cent;
paint +4.7 per cent; glazed wall tiles +8.3 per cent; steel plate +6.3 per cent.

The additional material costs will consist of

Bricks	0.6% of 12.8 =	0.08
Cement	8.4% of 6.4 =	0.54
Softwood	0.7% of 10.9 =	0.08
Paint	4.7% of 4.8 =	0.23
Glazed wall tiles	8.3% of 1.0 =	0.08
Steel plate	6.3% of 3.1 =	0.20
	Total	1.21 per cent

This represents a percentage addition of 1.21 on the materials part of the contract. Assuming that labour costs account for thirty-eight per cent of the total contract price, the overall percentage addition resulting from the increased material costs will be 1.21 × 62 per cent = 0.74 per cent, and the probable total addition to cost is £9500 × 0.74 = £70.2 (to the nearest pound, £70).

APPLICATION AND USE OF COST ANALYSES

The purpose of elemental cost analysis is to show the distribution of the cost of a building among its elements or main functional parts in terms which are meaningful to both designers and building clients and so to permit the costs of two or more buildings to be compared.[16] The cost analyses can be used to fulfil four main purposes.

(1) To enable designers and building clients to appreciate how cost is distributed among the functional components of a building.

(2) To help designers and building clients to develop ways in which costs could be allocated to produce a better balanced design.

(3) To allow prompt remedial action on receipt of high tenders, by indicating the components where reductions in cost are possible.

(4) To assist in the cost planning of future projects by comparison of cost analyses.

In comparing cost analyses for different buildings, variations in the cost of specific elements between the different projects often have to be analysed and explained. If, for instance, internal doors cost £1.80/m² of gross floor area in one office project and £3.10/m² on another; then reasons would have to be sought for this considerable variation in cost. Possible reasons are that there are more internal doors in relation to floor area on the second project (quantity); that the doors are of higher quality in the second project (quality); and that the second project relates to a more recent tender and so all the prices in this analysis are higher than those for the first project (price level).

When adjusting element rates in a cost analysis of an existing building as a basis for a preliminary cost plan for a new project, it is necessary to consider each of these three factors. Adjustments for period increases between the tender date of the previous job and the present day will apply to the rates for all the elements, and will probably be made on the basis of the appropriate BCIS building cost indices. Quantity and quality factors will vary from one element to another and each item will have to be considered individually.

For instance, supposing that the element cost of external walls in the analysis of the previous building was £8.53/m² and the wall/floor ratio was 1.080. The new building has a wall/floor ratio of 0.650 and it can be assumed that the wall construction will be similar in both cases. The external wall rate for the new building will be much lower as there is much less wall in relation to the enclosed floor area, and the wall/floor ratios can be used to determine the external wall rate for the new building.

$$\text{New element rate for external walls} = £8.53 \times \frac{0.650}{1.080} = £5.90/\text{m}^2.$$

A further example relates to internal doors.

	Previous building	*Proposed building*
Floor area	12 000 m²	30 000 m²
Number of internal doors	25	80
Cost of internal doors/ m² of gross floor area	£1.20	$£1.20 \times \frac{12\,000}{25} \times \frac{80}{30\,000} = £1.54$

Adjustments frequently have to be made to the elemental cost rates to take account of differences in quality. All-in unit rates of amplified cost analyses, priced approximate quantities or price books may be used individually or collectively to determine differences in cost stemming from variations in the quality of elements. For instance, assuming that the previous building contained 45 mm softwood doors and the proposed building is to have 45 mm veneered flush doors, then the doors in the new building would be about one-third more costly than those in the existing building and a further adjustment will be needed to the new internal door element rate

$$£1.54 \times \frac{4}{3} = £2.05$$

It is generally considered preferable to update element rates before adjusting them for quantity or quality. This sequence of events has the main advantage of producing, at the intermediate stage, a current element unit rate which could prove useful in preparing cost plans for other buildings.

COST DATA

Type of Information Required at Different Stages of Design

The cost information required by designers differs in its form and general characteristics according to the stage of cost planning at which it is required. The cost information required at each stage is now considered.

Preliminary estimate. This is required at the inception of a scheme, or more probably at the feasibility stage. Only the broadest of information is available such as the location and type of building, approximate amount of accommodation required and possibly a general quality standard and an outline drawing. The estimate is likely to be prepared on a floor area basis or unit method by comparisons of cost data available for previous similar jobs, possibly subdivided into groups of elements. The data used will be obtained from within the quantity surveyor's own organisation and from published sources such as BCIS and various technical journals.

Comparative costings. These are necessary when alternative building designs or building techniques are under consideration. A variety of estimating methods and sometimes a combination of them are used – in particular, a comparison of costs of past buildings, floor area and approximate quantities are popular methods. The quantity surveyor's own cost records, priced bills and published cost data will form the principal sources of cost information at this stage.

Initial cost plan. The earlier figures must be confirmed as soon as possible since they are influenced considerably by plan, shape, storey height, use of building and other factors.

Elemental approximate estimates operate on the basis of the synthesis of cost by functional elements (synthesis being the composition, putting together or building up of separate elements and is the opposite of analysis).

The elemental rates may be computed from rates for the same functional elements extracted from cost analyses of previous jobs and it will be advisable to use all-in unit rates for some elements, such as internal partitions, windows, doors, finishings and sanitary appliances, where the type and amount of the component parts can vary so widely. Adjustments must also be made for variations in design criteria, including soil conditions, loads and spans. Once again the quantity surveyor's own cost records, published data from a variety of sources, but particularly the BCIS and approximate quantities, will all play their part. On occasions it is good policy to use one method to check costs obtained by another.

Cost checks. As the working drawings are developed cost checks are made to ensure that the design is keeping within the cost budget. As the design is finalised, comparisons of smaller and smaller parts of the design in one or more alternative constructions may be required so that the architect can control the design within the overall target and give value for money. Alternative designs will require costing in more detail to compile a more precise cost plan. Whether these costs are submitted as a cost market of alternatives from which selections can be made, or whether they are used to make comparisons with elemental cost targets, they require careful and realistic assessment. In general, the measurement and costing of all-in rates for general contractor's work and specialist quotations for specialists' work will form the basis of the computation. The same sources of cost information will be relevant as in the previous stage. Finally, the priced bill of quantities from the accepted tender is analysed in the same form as the cost plan, both as a check of it and as the recorded analysis of tender cost. Final accounts are rarely used as a basis for cost analyses as the information is too dated.

It will be appreciated that to obtain reliable cost information from as many sources as possible entails standardisation of elements and cost analysis format, and this is now being achieved through the use of the BCIS standard form of cost analysis.[3] It is generally recognised that the proper purpose of cost analysis is to isolate and give a value to each element which performs the same function, irrespective of the building in which it appears. Only then can comparisons be made between one building and another in a meaningful way, and elemental targets have any real value.

One of the critiques on *Building Cost Information*[25] postulates that within the context of precontract estimating three stages can be defined, each of which requires a different kind of cost data.

Order of cost estimate. The initial estimate of cost which may be prepared from schematic diagrams and area schedules. At this point the client wishes to know the range of cost which a particular project may incur, and this information may normally be wanted in hours rather than days.

Outline cost plan. An estimate prepared at initial sketch plan stage which indicates both an accurate total target cost which the client can reasonably insist upon his professional advisers to maintain, and a broad elemental division of this target sum for the guidance of architects and engineers. This will normally be prepared in a matter of days.

Full cost plan. Prepared in parallel with the preliminary working drawings, and giving an itemised breakdown and outline specification of the cost allocation to design elements. This will normally be prepared in a matter of weeks.

Sources of Cost Information

It might be helpful at this stage to consider some of the principal sources of cost information available to quantity surveyors.

Ministry of Education (now Department of Education and Science). The Ministry of Education in the early nineteen-fifties, faced with an urgent school building programme and a limited allocation of funds, formed a development group which, amongst other things, evolved a system of cost planning and control, details of which were published in Building Bulletin 4.[2] This document laid down standards and provided cost and other information of considerable assistance to education authorities and their architects in the rational design and construction of a substantial number of efficient and architecturally pleasing schools.

The Wilderness Group. The Wilderness Group, composed of ex-members of the RICS Junior Organisation Quantity Surveyors' Committee formed a Cost of Building Study Working Party which published a paper on the storey-enclosure method of approximate estimating in 1954, and this was followed by an investigation into building cost relationships of various design variables,[17] which has proved to be of considerable assistance in costing alternative designs of steelframed buildings.

Technical journals. The *Architects' Journal* commenced the publication of cost analyses for building projects in 1954, based on the Ministry of Education elemental approach. Sweett[18] has described how these articles provoked some criticism on the grounds that the information provided might be dangerously misleading in inexperienced hands. There can be no doubt that these analyses provided some useful information for architects and quantity surveyors and stimulated considerable interest in the costs of buildings. Other journals subsequently published information on building costs and *Building*, in particular, has produced many informative and well-illustrated cost analyses covering a wide and diverse range of building types over the years.

RICS *Cost Research Panel/Building Cost Information Service.* The Cost Research Panel was formed in 1955 by the Council of the Royal Institution of Chartered Surveyors on a recommendation of the Quantity Surveyors' Committee. The Panel undertook a vast amount of research work into many aspects of building costs and the results of many of the investigations were the subject of papers in *The Chartered Surveyor*, many of which are referred to in this book as they are still valuable sources of cost information. Cost planning conferences and courses were organised jointly with the RIBA and Brixton School of Building (now South Bank Polytechnic) and these helped to maintain and encourage interest in this field. Branches of the Royal Institution were also encouraged to undertake specific cost research projects, usually concerned with a certain building element or subelement.

The QS Research and Information Committee Cost Research Panel was superseded by the BCIS Management Committee and the Building Cost Information Service was introduced in 1965. This service is available to subscribing members, about 750 in number in 1971, who are chartered quantity surveyors in practice or heads of departments in the public service. Members receive batches of cost analysis forms which they complete and return as they are able for contracts selected by them as suitable. These returns when edited are published as they stand as part of the Building Cost Information Service, to provide a library of cost data for comparison purposes, and are made available to subscribers and bona fide researchers. This service also includes information on cost indices, cost trends, relevant publications and other matters of interest. This constitutes useful background information and the cost analyses are of particular value where they cover special type buildings with which the quantity surveyor concerned has had no previous direct experience. The circulation BCIS was widened in 1972.

Building Maintenance Cost Information Service. This service which operated initially from the University of Bath was introduced in 1971 to assist its subscribers in obtaining value for money from efficient property occupancy and economic building maintenance. The service makes a regular distribution of up-to-date data on cost indices, labour and materials costs, maintenance techniques, legislation, statistics, publications, case studies, occupancy cost analyses, design/performance data, desk design appraisals and research and development papers. Further reference will be made to the work of this service in chapter 11.

Other sources. Building price books[19,20] provide useful information on current measured rates in London, together with other helpful data such as all-in or comprehensive rates and constants and costs of labour and materials. This data forms a useful guide but needs adjusting to local conditions.

James[21] has described how quantity surveyors can assemble cost data to good advantage. The office copy of the priced bill can have cube, floor area, storey-enclosure and similar rates inserted in it. The preliminaries bill can contain a full description of the construction of the building with its overall dimensions, cubic content, floor area, site restrictions, order of works and other peculiarities. Final accounts should be filed with a set of small-scale drawings and the accepted estimates and final accounts of all the specialist subcontractors. The author also sees considerable merit in the preparation of cost analyses, amplified where possible, for each job passing through the quantity surveyor's office, as part of his own library of cost information.

Reliability of Cost Information

The main source of cost information, contractors' prices, is itself subject to considerable variation – from contractor to contractor, district to district, job to job and over periods of time. This does not invalidate its use but does mean that the user must interpret the cost information in its appropriate context and adjust it as necessary before reuse. Computing average costs from a number of jobs has little relevance and it is better to base an assessment on a limited number of suitable jobs where extensive background information is available. It is necessary to know the conditions surrounding the prices which are to be used for cost planning, and how these conditions differ from those applicable to the new project.

Many factors should be considered when collecting and analysing cost data. These factors include the size of project; climate of building industry and relative keenness of tendering; stability of prices and availability of labour and materials; whether fixed price tender; special requirements as to speed or order of completion; method of pricing preliminaries; location of project; type of contract; and time of year when executed.

Much of the cost information used by the quantity surveyor is obtained from priced bills of quantities and it is advisable to use average prices for a particular job rather than those associated with the lowest tender. All quantity surveyors are fully aware of the wide variations in billed rates for even the most commonly encountered items of building work, sometimes referred to as the vagaries of tendering. One has only to examine the price per cubic metre for excavating a foundation trench not exceeding 1.50 m deep or the price per linear metre for providing, laying and jointing 100 mm British Standard glazed vitrified clay pipes, in half-a-dozen priced bills for the same job, to appreciate how great these differences can be. Strangely enough, although wide differences may occur in the rates submitted for individual billed items, the differences between the various tender figures are often quite small. Wales[22] in describing cost investigations undertaken by the former Ministry of Public Building and Works (now Department of the Environment), instanced one case study where the priced bills submitted by the two lowest tenders for a £110 000 job were examined. The two tenders differed by less than two-and-a-half per cent, yet the variations in price of works sections were: excavation – two-and-a-half per cent; concrete work – twelve-and-a-half per cent; glazed-ware drains – two-and-a-half per cent; cast-iron drains – thirty per cent; paving – fifteen per cent; plasterwork – twenty per cent; steel reinforcement – twenty-five per cent; and brickwork – fifteen per cent.

Skilled judgement is therefore necessary when extracting and using cost information from priced bills of quantities, obtained as a result of competitive tendering. Price differences in billed rates are not necessarily due to mistakes in estimating, but are more likely to stem from the use of different techniques in pricing and in performing the work on site. The treatment of preliminaries varies considerably between contractors; some cover them in the preliminaries bill, some spread them over the billed rates, while others adopt a combination of both approaches. If the preliminaries are spread then billed rates for like items are comparable. Preliminaries may however embrace much more than the standing overheads of the contractor's establishment and varying overheads such as foreman, use of plant, site huts, etc. They may reflect major cost factors such as access conditions, restricted site working, phasing of the work, labour shortage and similar matters, and these are just the sort of items for which allowance should be made by the quantity surveyor. One of the major processes in cost planning is estimating the anticipated cost of a project, and separation of the cost of preliminaries is probably beneficial in this context.

Furthermore, Southwell[25] has described how the builder's price is the building owner's cost, and when we are using data provided by bills of quantities for cost analysis, it is cost-to-client in respect of the elements of design rather than the cost-to-contractor, with which we are concerned. Southwell has emphasised that the distinction between cost and price is not merely an academic terminological issue. The fact that there is a lack of coincidence between the unit of production cost and unit of design cost accounts for some part of the wide discrepancy between unit rates in bills of quantities. Other causes include the difficulty of predicting operational times, alternative operational methods, and all other situational aspects, such as site peculiarities, regional differences, tendering conditions, market levels, etc. Southwell has made a study of the reliability of the unit rate as an indicator of the true costs of the specification. He asserts that the unit cost is made up of three main components:

firstly, standard costs within determinable parameters; secondly, additional costs due to specific design requirements; and thirdly, abnormal requirements due to factors which are indeterminate at forecasting stage, but which can be expressed as probabilities.

Sweett[28] emphasises the point that cost data should never be transferred indiscriminately from a bill to an estimate. Some process of modification and adjustment is needed before it can be applied to other projects, and this involves both the study of the factors affecting the original pricing and of the characteristics of the new project. Allowances must be made for differences in tender dates, contract conditions, regional differences, type of contractor, site conditions, availability of local labour and materials, any abnormalities in pricing, etc. Indeed it is not always possible to determine whether billed rates have been subject to abnormal pricing.

Southwell[25] has also referred to the need to make adjustments for differences in tendering keenness and price levels generally. The only possible indication we have of the general keenness in tendering is the closeness of the tender figures. In some cases a very low tender is so exceptionally low that it throws doubt upon its use as a basis for cost analysis. For these reasons, probably the most reliable information for cost planning and cost investigations is that based on approximate quantities prepared for each design and priced by an experienced quantity surveyor, who will make allowance for relevant abnormal and local conditions. The pricing of approximate quantities should represent an intelligent forecast of the level of prices contractors may offer for that work. Furthermore, when estimating future costs it should be borne in mind that building costs sometimes increase at a faster rate than rises in price of labour and materials. This is difficult to explain but may be due to various factors such as higher oncosts and profit margins, higher additional payments to operatives, or even lower rates of output.

Some quantity surveyors are opposed to the centralised processing of cost information as through the Building Cost Information Service, using the argument that even detailed cost analyses are insufficient and that all the necessary background information can only be secured through close familiarity with a project and access to the contract documents. If this view is accepted then the only place for a library of cost information is within the quantity surveyor's own organisation.

There is little doubt that the quantity surveyor must be prepared to build up extensive cost records of jobs passing through his office and to prepare cost analyses tabulated in a suitable form. Without this information he cannot really prepare reliable estimates of cost of future projects or give the wide range of cost information which building clients require. Different classes of work call for different methods of cost planning and analysis, and quantity surveyors will often need to exercise considerable skill and ingenuity in developing systems which are particularly suited to the type of work in which they tend to specialise. Thus quantity surveyors engaged mainly on educational buildings are likely to use quite different costing techniques to those concerned with power station contracts. In all cases, however, the cost planning service can be regarded as a natural extension of the quantity surveying function.

Farrar and Malthouse[23] assert that in order to provide the data needed in cost planning, a store of cost information must be built up from analyses of completed work, and that there is some benefit to be derived from presenting bills in a form that can be more readily analysed than the conventional arrangement by works sections, or from preparing bills by some process that permits a conventional presentation to be easily re-sorted for the purpose of analysis.

Cost Information Service for Engineering Systems

In 1965 the Heating and Ventilating Contractors Association authorised an investigation into the setting up of a cost information service for engineering systems for buildings. It was considered that 'the growing application of cost control techniques to buildings, the greater sophistication of engineering systems and their ever-increasing proportion of building cost make it essential that the engineering services industry should have up-to-date and detailed information about costs upon which it can budget and cost plan'. The investigations have taken place and questionnaires covering the cost of engineering services were circulated to engineering service offices in 1970.

The main problems have been described by Cox[24] as the collection of sufficient data and, in some cases, lack of design information and incomplete price breakdown. It is envisaged that designers and designer-contractors, for whom the service is intended, should subscribe towards the operating costs. Unfortunately, Cox[24] believes that the cost planning of engineering services is the responsibility of the engineer and not the quantity surveyor. This attitude may prevent beneficial co-operation and co-ordination between the engineering and quantity surveying professions which seems so desirable in the sphere of cost planning.

COST RESEARCH

One of the primary weaknesses in the building industry is undoubtedly that of communication. Cost planning improves direct communication between the building client, architect and quantity surveyor, and on occasions valuation and building surveyors also join the team. There is often, however, a wide gap between the designers and the contractor, as the contractor is frequently not appointed until the design is complete and the design team lacks the contractor's expertise on site problems and the practical application of constructional techniques. Hence information on important aspects such as the use of heavy plant and likely costs of new or unusual forms of construction may not be available. It has also been found that building clients often possess a considerable fund of knowledge of costs in use but that this is not always made available to the design team.

Co-ordinated programmes of research can assist in bridging this gap and in making the cost control mechanism more effective. Mention has already been made of the considerable volume of research undertaken by the Royal Institution of Chartered Surveyors; and various Government departments, including the Department of the Environment and the Department of Education and Science, are conducting extensive research programmes into various aspects of building costs. The Department of the Environment is particularly concerned with pricing and design economics and in 1971 instituted, in conjunction with many universities and polytechnics, some useful case studies concerned with costs in use and building performance. Further research has been carried out at London University into building price trends and at Bath University into building cost forecasting. The need for a much enlarged research programme into many aspects of building economics is evident and it is to be hoped that sufficient funds will be made available to permit its implementation on an adequate scale.

REFERENCES

1. C. SWEETT. Cost analysis – its application to cost planning and cost control techniques. *The Chartered Surveyor* (June 1959)
2. MINISTRY OF EDUCATION. Building bulletin No. 4 – cost study. HMSO (1957) (latest edition issued by DES, 1972).
3. ROYAL INSTITUTION OF CHARTERED SURVEYORS – BUILDING COST INFORMATION SERVICE. Standard form of cost analysis (1969)
4. INSTITUTE OF QUANTITY SURVEYORS. A research and development report: design cost planning. *The Quantity Surveyor*, **27.3** (1970)
5. D. POLDING. RIBA gives cool reception to revised standard cost analysis form. *Building Design* (27 February 1970)
6. MINISTRY OF HOUSING AND LOCAL GOVERNMENT. Circulars 36/67, 1/68, 31/69 and 56/69. HMSO
7. D. W. NUNN. The housing cost yardstick. *The Chartered Surveyor* (February 1968)
8. R. BUDD. Housing cost yardstick. *The Chartered Surveyor* (February 1971)
9. R. BUDD. Cost yardsticks. *The Chartered Surveyor* (June 1971)
10. J. TAYLER. School building cost limits. *The Architect and Building News* (15 January 1970)
11. DEPARTMENT OF EDUCATION AND SCIENCE. Building code. HMSO (September 1962 and various subsequent amendments)
12. ROYAL INSTITUTION OF CHARTERED SURVEYORS: COST RESEARCH PANEL. Indices of building costs by trades: multistorey and traditional housing. *The Chartered Surveyor* (April 1958)
13. C. SWEETT. Building cost indices: the measurement and forecasting of changes in building costs. *Building*, (23 April 1971)
14. B. C. JUPP. Trends in building prices. *The Chartered Surveyor* (January 1971)
15. M. E. A. BOWLEY and W. J. CORLETT. Report on the study of trends in building prices. HMSO (1970)
16. MINISTRY OF PUBLIC BUILDING AND WORKS, RESEARCH AND DEVELOPMENT. *Building Management Handbook 4: Cost Control in Building Design*. HMSO (1968)
17. THE WILDERNESS COST OF BUILDING STUDY GROUP. An investigation into building cost relationships of the following design variables: storey heights, floor loadings, column spacings and number of storeys. Royal Institution of Chartered Surveyors (1964)
18. C. SWEETT. Building economics and the chartered quantity surveyor. *The Chartered Surveyor* (August 1961)
19. SPON'S ARCHITECTS' AND BUILDERS' PRICE BOOK. Edited by Davis, Belfield and Everest. Spon (1972)
20. LAXTON'S BUILDING PRICE BOOK. Kelly's Directories (1971/72)
21. W. JAMES. The quest for building design – cost data. *The Chartered Surveyor* (November 1957)
22. C. A. WALES. Cost information. *The Chartered Surveyor* (April 1962)
23. C. H. FARRAR and R. F. W. MALTHOUSE. Quantity surveying aspects of coding and data co-ordination for the construction industry. *The Chartered Surveyor* (May 1969)

24. W. R. COX. Cost information service for engineering systems for buildings. *Journal of Institution of Heating and Ventilating Engineers* (February 1971)
25. J. SOUTHWELL. Building cost forecasting. Quantity Surveyors' Research and Information Committee. Royal Institution of Chartered Surveyors (1971)
26. MINISTRY OF HOUSING AND LOCAL GOVERNMENT. Housing cost yardstick for schemes at medium and high densities: Design Bulletin 7. HMSO (1964)
27. A. T. BRETT-JONES. Public accountability and the quantity surveyor. *The Chartered Surveyor* (March 1970)

9 PRACTICAL APPLICATION OF COST CONTROL TECHNIQUES

IN THIS CHAPTER cost planning techniques are applied to a variety of practical situations involving the computation of building costs for a range of building projects.

EXAMPLE 1

A preliminary estimate of cost is required for a two-storey factory building to be erected in a small East Anglian town, as an extension to an existing light engineering factory. It is envisaged that work on site will commence in January 1972 and that the contract period will be twelve months. A cost analysis on the new BCIS format of a Midlands factory (table 9.1) is to be used as a basis for the estimate, as the form and construction of the proposed factory is likely to be very similar. A drawing of the Midlands factory is provided in figure 9.1.

The following factors are to be taken into account in the preparation of the estimate.

(1) The location of the new factory is rather remote and it is anticipated that the contractor's employees will have to travel 30 km to the site. There is a shortage of contractors in the region due to the operation of a number of expanding town schemes.

(2) The site is flat and of filled ground of many years standing, but with a low bearing value.

(3) The external dimensions of the new building are to be 50 m × 16 m and the storey heights 5 m on the ground floor and 3.75 m on the first floor.

(4) External works are estimated at £14 000 and fittings at £1200.

(5) No office accommodation is required as the present accommodation is adequate.

Solution
The first step is to assess the probable cost effects of the different features of the two projects.

Time and situational factors. The BCIS building cost index for steelframed buildings for the third quarter of 1970 was 185, and if the price trends for the succeeding eighteen months are similar to the preceeding eighteen months (difference of twenty-two points or thirteen per cent) then the price index at January 1972 is likely to be about 207. Further adjustment is also needed to take account of the lack of competition in the area and the greater distance that building operatives will have to travel to the job, resulting in increased travel costs. An overall weighting of $16\frac{2}{3}$ per cent ($\frac{1}{6}$) would seem appropriate to meet all these additional costs.

Substructure. It can be reasonably anticipated that the ground floor will need to be a fully suspended reinforced concrete slab and that stanchion bases will require support from piles about 6 m long. It would therefore be prudent to increase the substructure cost to £8/m² of gross floor area.

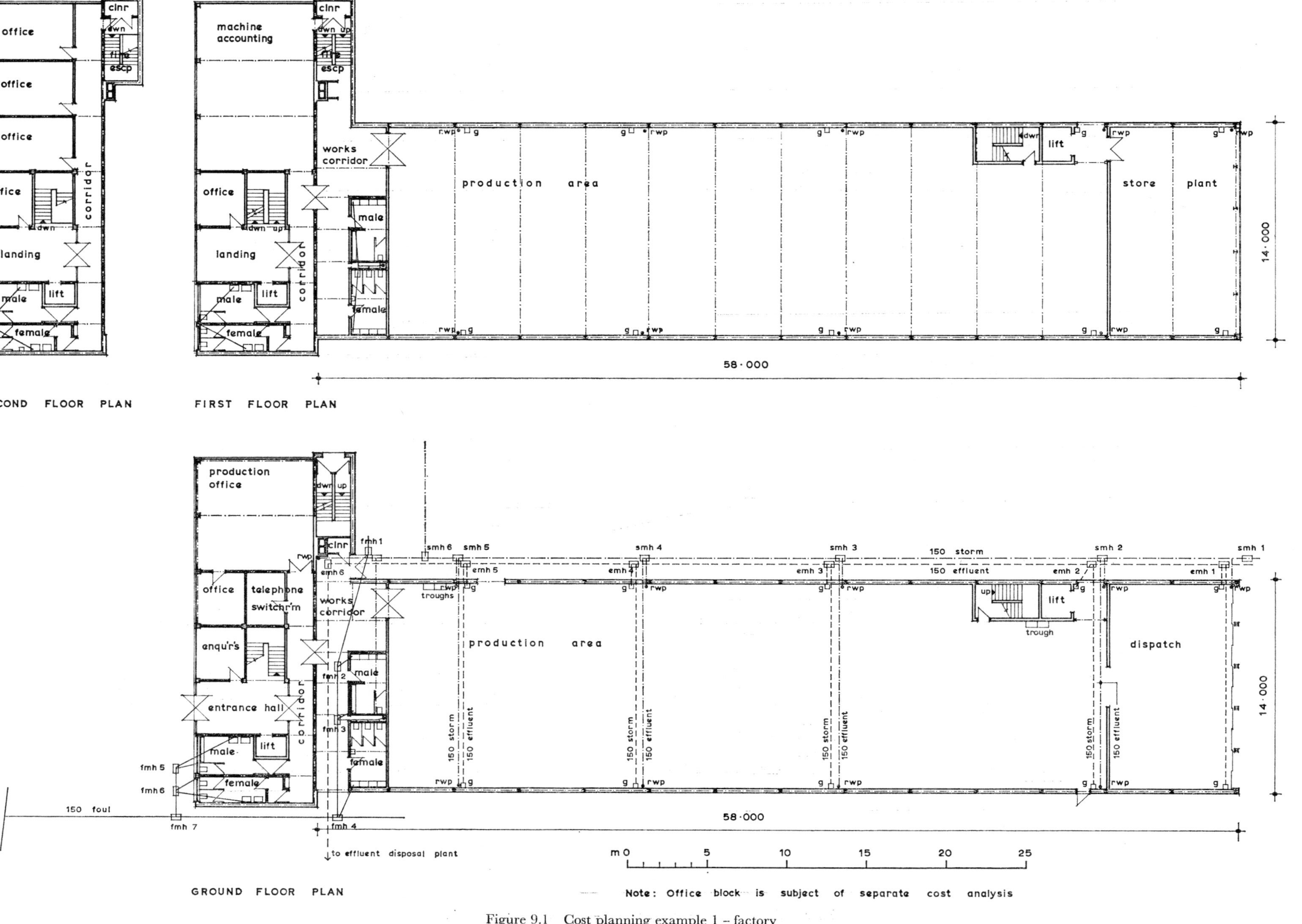

Figure 9.1 Cost planning example 1 – factory

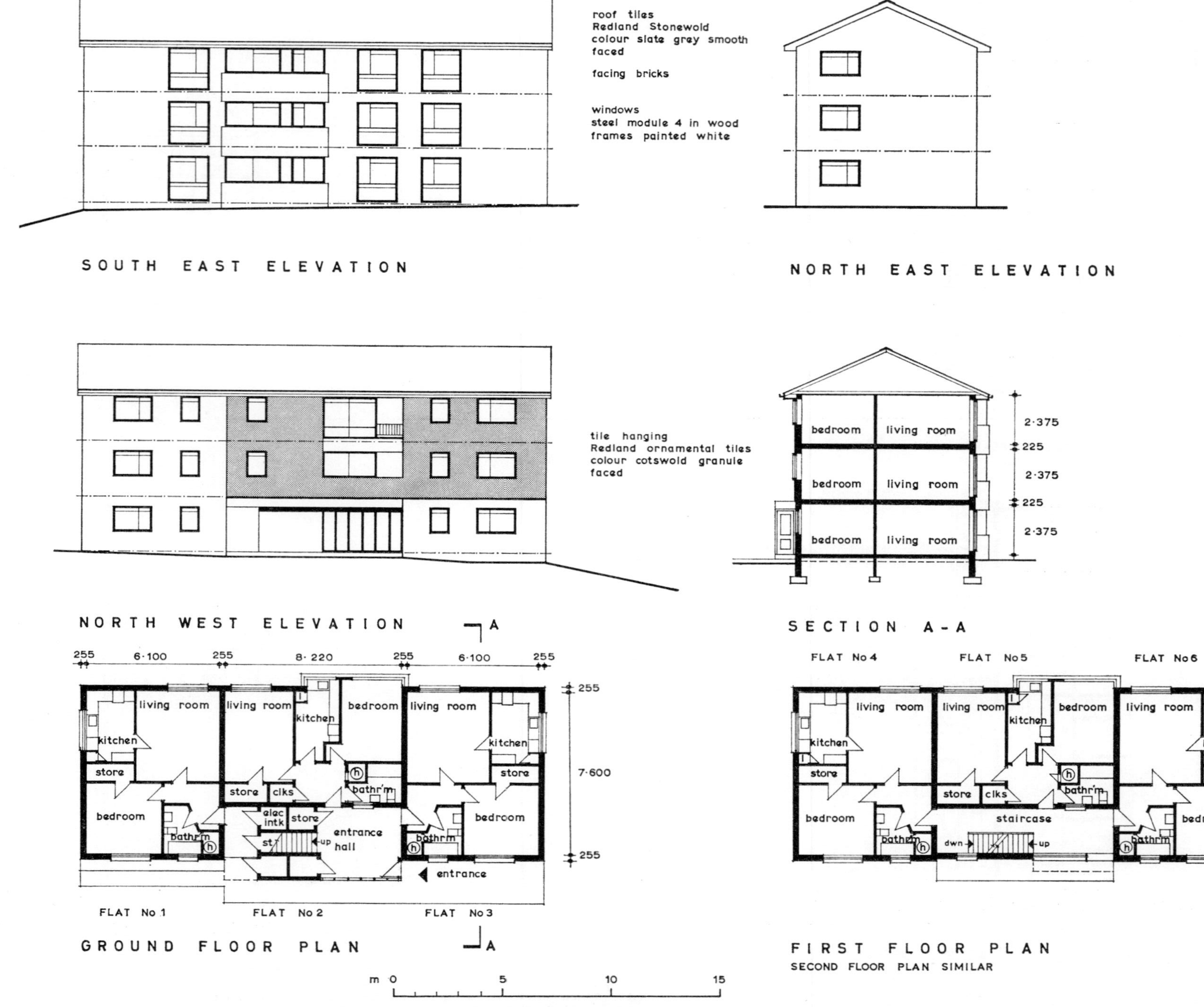

Figure 9.2 Cost planning example 2 – three-storey block of flats

TABLE 9.1

COST PLANNING EXAMPLE 1

Brief cost analysis of factory

Code: A – 2 – 1602

Job title: Factory
Location: Midlands

Client: Owner occupier
Tender date: July 1970

Information on total project

Project and contract information

Project details and site conditions
Two-storey factory producing meat products with lift motor rooms at roof level and attached to three-storey office block (analysed separately). Site adjacent to existing factory and office block. Access to site from main road, no standing water encountered. Site previously access road to existing factory and row of terraced houses and rear gardens.

Contract
Standard Form of Building Contract, private edition with quantities; firm price selected tenders, six issued, six received, all local contractors, lowest accepted, second lowest approx. two per cent above lowest; contract period stipulated by client – fourteen months.

Market conditions
Successful contractors headquarters within $1\frac{1}{2}$ km of site. Contractors asked to tender were within a radius of 25 km of site.

Contract particulars
Type of contract: Standard Form of Building Contract with quantities
Basis of tender: ~~Open~~/Selected competition ☒
Bill of quantities ☒ Negotiated ☐
Bill of approx. quantities ☐ Serial ☐
Schedule of rates ☐ Continuation ☐
Contract period stipulated by client: fourteen months.
Contract period offered by builder: fourteen months.
Number of tenders issued: six
Number of tenders received: six

Cost fluctuation: YES ☐ NO ☒
LABOUR ☐
MATERIALS ☐
Adjustments based on formula: YES ☐ NO ☐

Provisional sums	£ 4530
Prime cost (PC) sums	£ 117 130
Preliminaries	£ 9930
Contingencies	£ 4550
Contract sum	£ 209 850

Competitive tender list

£	N/L
209 850	L
214 000	L
214 111	L
216 987	L
219 750	L
228 860	L

Analysis of single building

Design/shape information

Accommodation and design features
Two-storey food factory with production space on both floors. Despatch on ground floor and store and plant room at first floor level.

Areas

Basement floors	– m²
Ground floor	801 m²
Upper floors	801 m²
Gross floor area	1602 m²
Usable area	1145 m²
Circulation area	146 m²
Ancillary area	272 m²
Internal division	39 m²
Gross floor area	1602 m²
Floor spaces not enclosed	– m²
Roof area	787 m²

Functional unit: usable floor space

$$\frac{\text{External wall area}}{\text{Gross floor area}} = \frac{1206}{1602} = 0.75$$

Internal cube = 6464 m³

Storey heights

Average below ground floor	– m
at ground floor	4.800 m
above ground floor	3.600 m

Design/shape:
Percentage of gross floor area

(a)	Below ground floor	%
(b)	Single-storey construction	%
(c)	Two-storey construction	100.0%
(d)	-storey construction	%
(e)	-storey construction	%

Brief cost information

Contract sum	£ 151 308	
Provisional sums	£ 4100	
Prime cost sums	£ 84 680	
Preliminaries	£ 7208	being 5.08% } of remainder of contract sum.
Contingencies	£ 2275	being 1.60% }
Contract sum less contingencies		£ 149 033

Functional unit cost excluding external works:

Tender	£ 118.85
Base date	£ 104.72

TABLE 9.1 – *(contd.)*

Summary of element costs

Gross internal floor area: 1602 m²

Tender date: July 1970

Element	*Preliminaries shown separately* Total cost of element £		Cost per m² gross floor area £		Element unit quantity	Element unit rate £	*Preliminaries apportioned amongst elements* Total cost of element £		Cost per m² gross floor area £		Cost per m² gross floor area at 1st quarter 1969 £
1. *Substructure*		9958		6.22	801 m²	12.43		10 464		6.53	5.79
2. *Superstructure*											
2A. Frame	16 498		10.30		1602 m²	10.30	17 336		10.82		
2B. Upper floors	4599		2.87		801 m²	5.74	4833		3.02		
2C. Roof	2073		1.29		787 m²	2.63	2178		1.36		
2D. Stairs	377		0.23		–	–	396		0.25		
2E. External walls	6825		4.26		1206 m²	5.66	7172		4.48		
2F. Windows and external doors	582		0.36		32 m²	18.19	612		0.38		
2G. Internal walls and partitions	2462		1.54		576 m²	4.27	2587		1.61		
2H. Internal doors	826		0.52		47 m²	17.57	868		0.54		
Group element total		34 242		21.37				35 982		22.46	19.75
3. *Internal finishes*											
3A. Wall finishes	2157		1.35		1554 m²	1.39	2267		1.42		
3B. Floor finishes	7453		4.65		1512 m²	4.39	7832		4.89		
3C. Ceiling finishes	2794		1.74		1406 m²	1.99	2936		1.83		
Group element total		12 404		7.74				13 035		8.14	7.34
4. *Fittings and furnishings*	–	–	–	–	–	–	–	–	–		–
5. *Services*											
5A. Sanitary appliances	1335		0.83				1403		0.87		
5B. Services equipment	–		–				–		–		
5C. Disposal installations	3042		1.90				3197		2.00		
5D. Water installations 5E. Heat source 5F. Space heating and air treatment 5G. Ventilating system	44 594		27.84				46 860		29.25		
5H. Electrical installations	17 704		11.05				18 604		11.61		
5I. Gas installations	–		–				–		–		
5J. Lift and conveyor installations	5534		3.45				5815		3.63		
5K. Protective installations	–		–				–		–		
5L. Communication installations	59		0.04				62		0.04		
5M. Special installations	–		–				–		–		
5N. Builder's work in connection with services 5O. Builder's profit and attendance on services	Included with service elements						–		–		
Group element total	–	72 268	–	45.11			–	75 941	–	47.40	42.27
Sub-total excluding external works, preliminaries and contingencies	128 872			80.44				135 422		84.53	75.15
6. *External works*											
6A. Site work	9079		5.67				9540		5.96		
6B. Drainage	3874		2.42				4071		2.54		
6C. External services	–		–				–		–		
6D. Minor building works	–		–				–		–		
Group element total	–	12 953	–	8.09			–	13 611	–	8.50	7.55
Preliminaries	–	7208	–	4.50						–	–
TOTALS (*less* contingencies)		£149 033		93.03				£149 033		93.03	82.70

TABLE 9.1 (*contd.*)

Specification and design notes

1. *Substructure*	Reinforced concrete stanchion bases and floor slab; concrete casing to steel ground beams.
2. *Superstructure*	
2A. Frame	Steelframe fire cased with metal lathing and plaster or render or Sprayed Limpet asbestos.
2B. Upper floors	Precast prestressed concrete beam floors.
2C. Roof	Pitched roof asbestos cement double-six sheets; flat roof Holorib steel decking covered with fibreboard and three layers felt roofing; galvanised mild steel pressed gutters; three no. Duplex domelights.
2D. Stairs	Precast reinforced concrete, mild steel handrail, granolithic finish to treads and risers.
2E. External walls	Cavity walls facings externally, Aglite blocks internally; exposed Colerclad aluminium cladding on timber battens to small areas in lieu of facings.
2F. Windows and external doors	Standard metal windows; softwood purpose-made matchboard doors, Gliksten Mark 9 flush doors.
2G. Internal walls and partitions	Engineering bricks, class B, Aglite blocks, Venesta Plymel WC compartments.
2H. Internal doors	Softwood matchboard and panelled doors, Gliksten flush doors.
4. *Fittings and furnishings*	
5. *Services*	
5A. Sanitary appliances	Armitage ware fittings lavatory basins (sixteen no.) WC suites (eight no.), urinal ranges (two no.), stainless steel sinks (five no.) mirrors (eight no.).
5C. Disposal installations	PVC rainwater soil and overflow pipe, polypropylene waste pipes, cast-iron soil pipes, asbestos cement incinerator flues, stoneware drainpipe encased in concrete.
5D. 5E. 5F. 5G. Heating and mechanical installations	Air-conditioning, refrigeration, special services, tank and mains pressure, cold and hot water services. (NOTE: boiler house cost not included as served by boiler house in office block which is subject to a separate analysis.) Builder's work in connection.
5H. Electrical installations	Lighting points 134 no., fitting outlet points fifty-two no., socket outlet points fourteen no.
5J. Lift and conveyor installations	Two no. goods lifts serving two floors.
5N. Builder's work in connection with services; 5O. Builder's profit and attendance on services	Included with service elements.
6. *External works*	
6A. Site works	Demolition, access road, temporary access roads, carriage crossings, landscaping, screen walls, boundary fencings, entrance gates.
6B. Drainage	Storm, foul and effluent drains.
Preliminaries	

Superstructure.

$$\text{gross floor area} = 2 \times (50.000 - 510) \times (16.000 - 510)$$

where 510 is twice the external wall thickness.

$$= 2 \times 49.490 \times 15.490$$
$$= 1533 \text{ m}^2$$

$$\text{external wall area} = 2/(50.000 + 16.000) \times (5.000 + 3.750)$$
$$\text{(measured on external face)} = 2/66.000 \times 8.750$$
$$= 132.000 \times 8.750$$
$$= 1155 \text{ m}^2$$
$$\text{wall to floor ratio} = \frac{1155}{1533} = 0.753$$

Hence the wall-to-floor ratios of both buildings are practically identical, the improvement in plan shape of the new building being entirely offset by the increased height of the external wall. The external wall cost will not vary but the cost of the frame and internal walls will increase due to the rise in storey heights. The percentage increase in storey heights is 8.750/8.400 = four per cent and the frame, internal walls and wall finishes should be increased in price by this amount. Hence the costs/m^2 will be

$$\text{frame} = 10.82 + 0.43 = £11.25$$
$$\text{internal walls and partitions} = 1.61 + 0.06 = £\ 1.67$$
$$\text{wall finishes} = 1.42 + 0.06 = £\ 1.48$$

The stairs will also need increasing in the same ratio and the cost will be 0.25 + 0.01 = £0.26.

External works are estimated at £14 000 and fittings at £1200 and the costs/m² of gross floor area will be

$$\text{external works} = \frac{£14\,000}{1533} = £9.13$$

$$\text{and fittings} = \frac{£1200}{1533} = £0.78$$

The Midlands factory is used for the manufacture of food products and contains air-conditioning, refrigeration and special services, which will not be required in the new factory. £9/m² should be adequate for water, heating and ventilating systems.

It is now possible to build up a total building cost/m² of gross floor for the new factory and to adjust this total by an addition of one-sixth to cover time and situational or context factors.

Groups of elements	*Cost/m² of gross floor area* (including preliminaries)
Substructure	8.00
Superstructure (£22.46 + 0.43 (frame) + 0.06 (internal walls and partitions) + 0.01 (stairs)	22.96
Internal finishes (£8.14 + 0.06)	8.20
Fittings	0.78
Services (£47.40 – 20.25)	27.15
External works	9.13
Total	£76 22

Total building cost/m² based on cost analysis of previous factory	£76.22
add 16⅔ per cent (time and situational factors)	12.70
Adjusted current rate	£ 88.92
Estimated cost of new factory = 1533 × £88.92 =	£136 314
add contingencies (1.60%)	2 186
Total estimated cost	£138 500

EXAMPLE 2

The second cost planning example covers the preparation of a first cost plan for a four-storey block of flats, to be built in London, from sketch designs (figure 9.3), and with the help of an amplified cost analysis (orginal BCIS format but suitably metricated in table 9.2) and drawings (figure 9.2) relating to a three-storey block of flats in south-west England. Basic design information and brief specification notes covering the new project follow.

Basic Information

The block of flats is to be built for a housing trust on a steeply sloping site on the outskirts of London. It is anticipated that construction will commence in early January 1972 and a contract period of twelve months will be specified with competitive firm price tenders. The proposed accommodation consists of fourteen dwellings made up of one-bedroom flats and bedsitters, with access from a central common landing, and with six garages on the ground floor.

Brief Specification

Substructure: Concrete strip foundations, 375 mm deep and 750 and 900 mm wide; 255 mm hollow wall 900 mm high; 150 mm hardcore bed, 50 mm concrete blinding, polythene damp-proof membrane and 100 mm concrete bed with steel fabric reinforcement.

Upper floors: Reinforced concrete – 175 mm thick to living areas and 150 mm thick to circulation areas.

Roof: Built-up felt roofing and prescreeded wood-wool slabs on 50 × 175 mm and 75 × 225 mm softwood joists.

Rooflights: Allow pc sum of £200.

Staircases: In situ reinforced concrete; 3.2 mm felt-backed PVC treads, risers and strings; mild steel handrail with hardwood handrail and kneerail.

External walls: 255 mm brick and block cavity wall faced externally.

Windows: EJMA softwood standard section windows.

External doors: 50 mm hardwood glazed door and entrance screen (two no.); up-and-over metal garage doors with timber frames.

Internal load-bearing walls: One-brick solid walls.

Partitions: 100 and 75 mm hollow block partitions.

Internal doors: 35 mm hollow core flush doors in softwood linings.

Ironmongery: Allow pc sum of £700.

Wall finishes: 12 mm gypsum plaster to brick and block walls; 50 m^2 of 100 × 100 mm white glazed tiles.

Floor finishes: 3.2 mm felt-backed PVC paving.

Ceiling finishes: 12 mm gypsum plaster.

Decoration: Plastic emulsion on plastered ceilings and walls; three coats oil paint on wood and metal work; polyurethane lacquer on hardwood.

Fittings: Kitchen fittings of standard base unit, under sink unit and wall units; wardrobe fronts, shelving and small cupboard doors.

Sanitary appliances: 600 × 450 mm white vitreous china lavatory basins and valves; 1050 × 450 mm stainless steel sinks, drainers and valves; 1650 mm porcelain enamelled baths, fittings and valves; low level white vitreous china WC suites.

Waste, soil and overflow pipes: PVC soil, ventilating and waste pipes; plastic overflow pipes.

Hot and cold water services: 15 and 22 mm copper tubing; ninety-eight draw-off points; immersion heaters; tanks of 3600 litres capacity.

Ventilation: Allow pc sum of £340.

Electrical and heating services: Allow pc sum of £3800.
Drainage: Allow pc sum of £1200.

External works: Allow pc sum of £5000.

SOLUTION

Time and Situational Factors

In making comparisons with element rates in the cost analysis of the three-storey block of flats, allowance must be made for the three-year difference in contract dates (January 1969 to January 1972), any regional price differences and any likely variation in the market situation.

The building cost index for brick buildings is 169 for January 1969 and is estimated at 210 for January 1972. An examination of housing cost yardsticks and building price trends indicate that building prices in outer London are likely to be about twelve per cent higher than in south-west England. The market situation in the south-west is quite keen as there is not an excess of work in the region, and there is also ample evidence of very keen competitive tendering in outer London with very small differences between tenders. It does not therefore seem that any adjustment of prices would be needed to cover varying market conditions. The total percentage addition to the tender rates for the existing block of flats would be 210/169 = 24 + 12 = thirty-six per cent.

Quantity Factors

Gross floor area

	Length	*Width*
	17.100	13.400
less twice external wall thickness 2/255	510	510
	16.590	12.890

gross floor area = 4 × 16.590 × 12.890 = 855 m² made up of 213.75 m² on each floor.

Roof area. The roof area is measured to the inside face of the parapet walls and will also be 213.75 m².

External walls. The external wall area is measured on the outer face of the external walls without any adjustments for windows or doors. The perimeter girth is 2 × (17.100 + 13.400) × 10.700 = 652.7 m². The wall to floor ratio $= \dfrac{\text{external wall area}}{\text{gross floor area}} = \dfrac{652.7}{855} = 0.764.$

Windows. The total window area calculated from the plans and elevations is 174.3 m², and the window to floor ratio is 174.3/855 = 0.204.

Internal load-bearing walls.

Lengths:		
ground floor	=	43.500
first floor	=	45.700
second floor	=	37.800
third floor	=	37.800
total length	=	164.800

Area of internal load-bearing walls = 164.800 × 2.375 = 391.5 m².

$$\frac{\text{Area of internal load-bearing walls}}{\text{Gross floor area}} = \frac{391.5}{855} = 0.457.$$

TABLE 9.2

COST PLANNING EXAMPLE 2

Amplified cost analysis of block of flats

(pre-1970 format)

Code: C – 3 – 499

Type: Flats
Region: south-west England

Client: Local authority
Tender date: 15 January 1969

Brief description

Building and site conditions: Three-storey residential block with load-bearing brick walls, reinforced concrete first and second floors, well pitched roof containing nine one-bedroom flats built on site 800 metres from city centre.

Contract: Competitive firm price tenders; six tenders received, lowest accepted, second lowest approx. five per cent above lowest; stipulated contract period twelve months.

Accommodation and design/shape information: Three-storey block size 21.440 × 8.130 × 9.750 to ridge with projections from each of long faces, one size 5.150 × 0.600 and one size 8.380 × 1.150, containing nine one-bedroom flats.

No. of storeys: three

Storey heights: 2.600, 2.600 and 2.375

$$\frac{\text{Wall}}{\text{Floor}} = \frac{410}{499} = 0.82$$

Ground floor:	171 m²
Other floors:	328 m²
Total gross floor area	499 m²

Roof area: 206 m²

Brief cost summary

Preliminaries 8.59% } of remainder of
Contingencies 0.53% } contract sum.

Elements	*Preliminaries and contingencies shown separately* Totals £	Tender £/m²	*Preliminaries and contingencies apportioned amongst elements* Totals £	Tender £/m²
Substructure	1193	2.44	1302	2.66
Superstructure	8182	16.70	8928	18.23
Internal finishes	3031	6.18	3307	6.74
Fittings	749	1.53	817	1.67
Services	4481	9.14	4890	9.97
Total excluding external works	17 636	35.99	19 244	39.27
External works	1378	2.81	1504	3.06
Preliminaries	1634	3.33	–	–
Contingencies	100	0.20	–	–
TOTALS	20 748	42.33	20 748	42.33

TABLE 9.2 (*contd.*)

Summary of amplified analysis

Gross internal floor area: 499 m²

Date of tender: 15 January 1969

Element	*Total cost of element*		*Cost of element per m² of gross floor area* £		*Element unit quantity*	*Element unit rate* £
1. Work below lowest floor finish	£1215	£1215	2.435	2.435	171 m²	7.105
2. Frame	–		–		–	–
3. Upper floors	1138		2.281		344 m²	3.308
4. Roof	1340		2.685		237 m²	5.654
5. Rooflights	–		–		–	–
6. Staircases	586		1.174		7 m²	83.714
7. External walls	1910		3.828		310 m²	6.161
8. Windows	1015		2.034		88 m²	11.534
9. External doors	75		0.150		11 m²	6.818
10. Internal load-bearing walls	1243		2.491		503 m²	2.471
11. Partitions	96		0.192		52 m²	1.846
12. Internal doors	742		1.487		122 m²	6.082
13. Ironmongery	185	£8330	0.371	16.693	–	–
14. Wall finishes	796		1.595		1207 m²	0.659
15. Floor finishes	965		1.934		456 m²	2.116
16. Ceiling finishes	258		0.517		436 m²	0.592
17 Decoration	1067	£3086	2.138	6.184	–	–
18. Fittings	762	£ 762	1.527	1.527	–	–
19. Sanitary fittings	468		0.938		–	–
20. Waste, soil and overflow pipes	228		0.457		–	–
21. Cold water services	309		0.619		Twenty-seven draw-off points	11.444
22. Hot water services	505		1.012		Twenty-seven draw-off points	18.704
23. Heating services	1239		2.483		–	–
24. Ventilation and air-conditioning	–		–		–	–
25. Gas services	132		0.265		Twenty-seven draw-off points	4.888
26. Electrical installation	1379		2.763		187 points	7.374
27. Passenger and goods lifts	–		–		–	–
28. Fire-fighting equipment	–		–		–	–
29. Sprinkler installation	–		–		–	–
30. Special installations	–		–		–	–
31. Drainage	302	£4562	0.605	9.142	–	–
Totals excluding external works		£17 955		35.981	–	–
32. External works		1403		2.812	–	–
33. Preliminaries		1663		3.333	–	–
34. Contingencies		102		0.204	–	–
TOTALS		£21 123		£42.330	–	–

TABLE 9.2 (*contd.*)

Amplified cost analysis

	A	B	C	$D = \frac{A}{C}$	
Element and design criteria	*Total cost element*	*Cost of element per m² of gross floor area*	*Element unit quantity*	*Element unit rate*	*Specification*
	£	£		£	
1. *Work below lowest floor finish*	1215	2.44	171 m²	7.11	Concrete strip foundations; concrete block walls; 150 mm hardcore bed; 150 mm concrete bed.
2. *Frame*	–	–	–	–	Not applicable.
3. *Upper floors*	1138	2.28	344 m²	3.31	150 mm reinforced concrete first and second floor slabs.
4. *Roof*	1340	2.69	237 m²	5.65	Pitched tiled roof of concrete interlocking tiles on battens and felt; timber roof construction with roof trusses. 150 mm reinforced concrete canopy to front entrance area covered with 25 mm two-coat mastic asphalt.
5. *Rooflights*	–	–	–	–	Not applicable.
6. *Staircases* Two straight flights each flight 2.600 rise.	586	1.17	7 m²	83.71	Precast concrete steps faced with terrazzo on exposed surface and with mild steel balustrades
7. *External walls* $\frac{\text{External wall}}{\text{Gross floor area}} = \frac{455}{499} = 0.912$	1910	3.83	310 m²	6.16	(a) 255 mm cavity wall – half-brick skin in Lungfield Multi Rustic facings, 50 mm cavity, half-brick skin in commons. (b) 252 mm cavity wall – half-brick skin in Lungfield Multi Rustic facings, 50 mm cavity, 100 mm Lignacite block skin. (c) 250 mm cavity wall – concrete arrow-head tile hanging on battens, two 100 mm Lignacite block skins, 50 mm cavity. (d) 250 mm cavity wall – concrete arrow-head tile hanging on battens, 100 mm Lignacite block skin, half-brick skin in Phorpres Dapple Light facings, 38 mm cavity. (e) 252 mm cavity wall – half-brick skin in commons, 50 mm cavity, 100 mm Lignacite block skin. (f) 367 mm cavity wall – one-brick skin faced one side, 50 mm cavity, half-brick skin in commons. (g) 365 mm cavity wall – one-brick skin faced one side, 50 mm cavity, 100 mm Lignacite block skin.

Roof	£	*Area in m²*	*All-in unit rate* £
Tiled roof	1230	226	5.44
Concrete canopy	110	9	12.22

Walls	£	*Area in m²*	*All-in unit rate* £
(a)	512	93	5.50
(b)	798	150	5.32
(c)	196	20	9.80
(d)	229	25	9.16
(e)	161	18	8.95
(f)	4	1	4.00
(g)	10	3	3.33

TABLE 9.2 (*contd.*)

Element and design criteria	*A* Total cost of element £	*B* Cost of element per m² of gross floor area £	*C* Element unit quantity	$D = \frac{A}{C}$ Element unit rate £	Specification
8. *Windows* $\frac{\text{Area of windows}}{\text{Gross floor area}} = \frac{88}{499} = 0.18$	1015	2.03	88 m²	11.53	Standard metal windows to wood surround; fixed lights with clear sheet glazing; fixed lights with polished plate glazing; fixed lights with Armour-clad toughened plate glazing.
9. *External doors*	75	0.15	11 m²	6.82	Softwood glazed screen incorporating one no. 50 mm thick panelled softwood glazed door (screen 4.800 girth × 2.300 high).
10. *Internal load-bearing walls* $\frac{\text{Area of internal load-bearing walls}}{\text{Gross floor area}} = \frac{503}{499} = 1.01$	1243	2.49	503 m²	2.47	(a) 255 mm cavity wall – two half-brick skins in commons, 50 mm cavity. (b) 260 mm cavity wall – two 100 mm Lignacite block skins, 60 mm cavity. (c) 255 mm cavity wall – half-brick skins in Dapple Light facings, 50 mm cavity, half-brick skin in commons. (d) 252 mm cavity wall – half-brick skin in Dapple Light facings, 50 mm cavity, 100 mm Lignacite block skin. (e) One-brick wall faced in Dapple Light facings one side. (f) Half-brick wall in commons. (g) Half-brick wall in Dapple Light facings. (h) 100 mm Lignacite block wall.
11. *Partitions* $\frac{\text{Area of partitioning}}{\text{gross floor area}} = \frac{52}{499} = 0.10$	96	0.19	52 m²	1.85	50 mm thick solid Lignacite block wall. 75 mm ditto.

Windows	£	*Area in m²*	*All-in unit rate £*
Metal windows	865	75	11.52
Fixed lights – clear glazing	22	3	7.34
Fixed lights – with polished plate glazing	102	8	12.73
Fixed lights with Armour-clad glazing	26	2	13.00

Internal load-bearing walls	£	*Area in m²*	*All-in unit rate £*
(a)	105	29	3.62
(b)	170	48	3.54
(c)	33	8	4.12
(d)	67	17	3.95
(e)	244	58	4.20
(f)	229	124	1.85
(g)	23	9	2.56
(h)	372	210	1.77

Partitions	£	*Area in m²*	*All-in unit rate £*
50 mm Lignacite	5	4	1.25
75 mm ditto	91	48	1.90

TABLE 9.2 (*contd.*)

Element and design criteria	*A* Total cost of element £	*B* Cost of element per m² of gross floor area £	*C* Element unit quantity	$D = \frac{A}{C}$ Element unit rate £	Specification
12. *Internal doors*	742	1.49	122 m²	6.08	(a) 25 mm thick flush blockboard door. (b) 45 mm thick flush softwood skeleton core flush door with 6 mm plywood facings both sides. (c) 35 mm thick flush softwood skeleton core flush door to BS 459 part 2. (d) 35 mm thick flush softwood skeleton core flush door to BS 459 part 2, with 3 mm hardboard facings both sides. (e) 45 mm thick flush softwood skeleton core door to BS 459 part 2, with 6 mm plywood facings both sides. (f) 45 mm thick fire check flush door to BS 459 part 3 half-hour type with softwood core, plasterboard protective infilling.
13. *Ironmongery*	185	0.37	–	–	*External doors: £*18. Satin chromium-plated door handles; silver anodised aluminium kicking plates; floor spring door closer. *Internal doors: £*167. Silver anodised aluminium letter plates; silver anodised aluminium kicking plates; pressed steel butt hinges and rising butt hinges; brass barrel bolts; nylon mortice catches; satin chromium-plated mortice latches; mortice dead lock; mortice lock; plastic name plates.
14. *Wall finishes*	796	1.60	1207 m²	0.66	12 mm one-coat work of cement and sand (1:3) externally to concrete or brickwork base. 15 mm two-coat plaster to brickwork or blockwork base first coat of cement, lime and sand (1:1:6) and finishing coat of gypsum plaster. White glazed ceramic tiles and bedding in approved adhesive in narrow widths; white glazed ceramic tiles and bedding in cement mortar (1:3) to sills.

Internal doors	£	*Area in m²*	*All-in unit rate £*
(a)	181	29	6.25
(b)	9	1.5	6.00
(c)	250	27	9.26
(d)	180	46.5	3.87
(e)	15	3	5.00
(f)	107	15	7.13

Wall finishes	£	*Area in m²*	*All-in unit rate £*
Cement and sand	27	41	0.66
Plaster	736	1150	0.64
Ceramic tiles	21	–	–
Ditto sills	12	16 (lin. m)	–

TABLE 9.2 (*contd.*)

Element and design criteria	*A* Total cost of element £	*B* Cost of element per m² of gross floor area £	*C* Element unit quantity	$D = \frac{A}{C}$ Element unit rate £	Specification
15. *Floor finishes*	965	1.93	456 m²	2.12	40 mm granolithic, cement, sand and lime, chippings 6 mm single sized, BS 1201 (1:1:2). 100 × 100 × 10 mm fully vitrified ceramic floor tiles on cement and sand bed; vinylised domestic tiles size 225 × 225 × 2 mm and fixing with adhesive on 40 mm cement and sand bed; ditto on 40 mm cement and sand bed on sand deadening quilt, reinforced with chicken wire.

Floor finishes	£	*Area in m²*	*All-in unit rate* £
Granolithic	9	13	0.69
Vitrified tiles	188	42	4.48
Vinylised tiles	229	135	1.70
Ditto on sand deadening quilt	539	266	2.03

Element and design criteria	*A*	*B*	*C*	*D*	Specification
16. *Ceiling finishes*	258	0.52	436 m²	0.59	12 mm cement and sand (1:3) finishing coat to concrete base; 5 mm finishing coat of gypsum plaster on plasterboard base; 15 mm two-coat plaster to concrete or wood-wool base, first coat of cement, lime and sand (1:1:6) and finishing coat of gypsum plaster.

Ceiling finishes	£	*Area in m²*	*All-in unit rate* £
Cement and sand	2.5	4	0.62
5 mm plaster	86.0	142	0.61
15 mm plaster	169.5	290	0.59

Element and design criteria	*A*	*B*	*C*	*D*
17. *Decoration*	1067	2.14	–	–

Decoration	*Total £*	*Floor area £/m²*
External	140	0.28
Internal	927	1.86

Element and design criteria	*A*	*B*	*C*	*D*
18. *Fittings*	762	1.53	–	–

Fittings	*Quantity*	*Cost* £
25 mm blockboard cupboard fronts, backs, sides or divisions	35 m²	300
Curtain and hat and coat rails	77 lin. m	25
Slatted shelving	14 m²	55
12 mm blockboard shelving	18 m²	81
Standard kitchen units	18 no.	193
19 mm Formica covered blockboard worktops	16 m²	108

TABLE 9.2 *(contd.)*

Element and design criteria	A *Total cost of element* £	B *Cost of element per m² of gross floor area* £	C *Element unit quantity*	$D = \frac{A}{C}$ *Element unit rate* £	*Specification*
19. *Sanitary fittings*	468	0.94	–	–	*Type and quality of fittings* — *Number* — *Cost* £ Porcelain enamelled cast-iron baths including panels — 9 — 187 550 × 400 mm vitreous china lavatory basins — 9 — 59 1050 × 525 mm stainless steel sinks — 9 — 115 Low level vitreous china WC suites — 9 — 94 Vitreous china toilet paper boxes — 9 — 11 Builder's work — – — 2
20. *Waste, soil and overflow pipes*	228	0.46	–	–	Overflow in copper to BS 659 (10 lin. m) £14 Wastes in polypropylene (60 lin. m) £70 Soil pipe in rigid PVC (30 lin. m) £59 Builders' work £85
21. *Cold water services* Number of cold water draw-off points: twenty-seven nine sinks nine lavatory basins nine baths	309	0.62	Twenty-seven draw-off points	11.44	Supply pipework in copper to BS 659 with compression fittings to BS 864 type A (180 lin. m) £279 Builder's work £30
22. *Hot water services*	505	1.01	Twenty-seven draw-off points (including water heater points)	18.70	Service pipework in copper to BS 659, with compression fittings to BS 864 type A (830 lin m): £90 Instantaneous water heaters (8790 W): £202 Combination hot water storage unit (115 litres hot and 45 litres cold actual capacity) with immersion heater (nine no.): £184 Thermal insulation £18 Testing £10 Builder's work £1
23. *Heating services*	1239	2.48	65 925 W		Nine no. gas-fired warm air units (7325 W) £730 Sheet metal ducting with fibreglass insulation and registers and grilles (24 lin. m of ducting) £176 Asbestos cement flue pipes (24 lin. m) £58 Electrical work in connection with gas units £130 Builder's work £145
24. *Ventilation and air-conditioning services*	–	–	–	–	Not applicable.

TABLE 9.2 (*contd.*)

	A	B	C	$D = \frac{A}{C}$	Specification
Element and design criteria	*Total cost of element*	*Cost of element per* m^2 *of gross floor area*	*Element unit quantity*	*Element unit rate*	*Specification*
	£	£		£	
25. *Gas services* Number of draw-off points: twenty-seven nine fridges nine cookers nine gas heater units	132	0.27	Twenty-seven draw-off points	4.89	Service pipework in copper to BS 659 with compression fittings to BS 864, type A. Builder's work £7.
26. *Electrical services* Consumer distribution points: eighteen Power points: ninety-nine Lighting points: sixty-one Television points: nine	1379	2.76	187 points	7.37	*Installation* / *Cost* / *Cost per* m^2 / *Cost per point* (£ / £ / £): Main switchgear and distribution 207 0.41 11.50 Power 630 1.26 6.37 Lighting 271 0.54 4.45 Television 176 0.35 19.55 Builder's work 95 – –
27. *Passenger and goods lifts*	–	–	–	–	Not applicable.
28. *Fire-fighting equipment*	–	–	–	–	Not applicable.
29. *Sprinkler installations*	–	–	–	–	Not applicable.
30. *Special installations*	–	–	–	–	Not applicable.
31. *Drainage*	302	0.61	–	–	*Surface water* (£52); 100 mm diameter pitch fibre pipes to four no. soakaways. *Foul drains* (£250); 100 mm diameter pitch fibre main runs between manholes with 100 mm diameter clay pipe and pitch fibre branches; four no. engineering brick manholes.
32. *External works*	1403	2.81	–	–	Site works £242 Two detached stores and lines enclosure £214 Retaining walls and screen walls £253 Fencing £100 chain link £26 close boarded £46 post and rail £28 Pavings and steps £279 Parking bay £283 Main water services £ 32
33. *Preliminaries and insurances* 8.54% of remainder of contract sum.	1663	3.33	–	–	
34. *Contingencies* 0.48% of remainder of contract sum.	102	0.20	–	–	
Totals:	£21 123	£42.33			

Partitions.

		100 mm	75 mm
Lengths:	ground floor	21.500	33.000
	first floor		64.600
	second floor		68.800
	third floor		68.800
	Totals	21.500	235.200

Areas of partitioning: 100 mm $= 21.500 \times 2.375 = 51\ m^2$

75 mm $= 235.200 \times 2.375 = 558\ m^2$

$$\frac{\text{area of partitioning}}{\text{gross floor area}} = \frac{609}{855} = 0.713.$$

Costs of Elements

The next step is to calculate the estimated costs of each of the elements expressed in terms of cost per square metre of gross floor area. The costs can in some cases be computed from the costs of the same element shown in the cost analysis of the existing block of flats in table 9.2, with adjustments as necessary for time and situational, quantitative and qualitative factors. In other cases it may be simpler and more realistic to compute the costs from approximate quantities (element unit quantities) priced at current rates. On occasions it can be a valuable exercise to calculate the cost by both methods, with one acting as a check on the other.

(1) *Substructure.* The element unit rate in table 9.2. should be increased by thirty-six per cent to cover time and situational factors and a further fifteen per cent to cover the more expensive construction stemming from thicker strip foundations and reinforced slab: £7.11 + 36 per cent + 15 per cent = £11.12.

$$\text{Cost of element} = 213.75 \times 11.12 = £2378$$

(ground floor area)

$$\text{Cost of element/m}^2 \text{ of gross floor area} = \frac{2378}{855} = £2.78$$

(2) *Frame.* Not applicable.

(3) *Upper floors.* The element unit in table 9.2. can be adjusted by the addition of thirty-six per cent (time and situational factors) and ten per cent (increased thickness of part of floor slabs from 150 to 175 mm and increased hoisting costs): £3.31 + 36 per cent + 10 per cent = £4.95.

$$\text{Cost of element} = 641.25 \times 4.95 = £3176$$

(floor area of three upper floors)

$$\text{Cost of element/m}^2 \text{ of gross floor area} = \frac{3176}{855} = £3.71$$

(4) *Roof.* The roof construction for the existing block of flats is not comparable and it is therefore necessary to calculate an all-in rate for the built-up felt roofing on wood-wool slabs and timber joists. A suitable all-in rate would be £6.20/m² of roof area plus £90 for rainwater disposal.

$$\text{Element cost then becomes } 213.75 \times 6.20 = 1324 + 90 = £1414$$

$$\text{Cost of element/m}^2 \text{ of gross floor area} = \frac{1414}{855} = £1.65$$

(5) *Rooflights.* A pc sum of £200 is to be provided.

$$\text{Cost of element/m}^2 \text{ of gross floor area} = \frac{200}{855} = \underline{£0.23}$$

(6) *Staircases.* The form of construction covered by the cost analysis in table 9.2 is rather different from that specified for the new project. Hence it is more satisfactory to compute a new element unit rate from first principles, possibly using other cost analyses and price books as a guide. The cost is estimated at 8 m² @ £55 = £440

$$\text{Cost of element/m}^2 \text{ of gross floor area} = \frac{440}{855} = \underline{£0.52}$$

(7) *External walls.* The external walls in both projects are of similar construction but it is necessary to make adjustment for the time and situational factors, the changed wall-to-floor ratio and the extra cost of building brickwork to the fourth storey. Thus the all-in unit rate of 5.32/m² of net wall area can be taken as a basis with suitable adjustments.

(i) Add 36 per cent to cover time and situational factors

5.32 + 36 per cent = £7.24

(ii) There is a considerable improvement in the wall to floor ratio and this will be reflected when the element cost/m² of gross floor area is calculated.

(iii) Increase the unit rate by a further five per cent to take account of the additional cost of building the brickwork to the fourth storey

£7.24 + 5 per cent = £7.59

The net area of external brickwork is calculated as follows

Gross area of external wall		652.7 m²
less window area:	174.3 m²	
door area:	32.9 m²	207.2 m²
Net external wall area		445.5 m²

Cost of element = 445.5 × 7.59 = £3382

$$\text{Cost of element/m}^2 \text{ of gross floor area} = \frac{3382}{855} = \underline{£3.96}$$

(8) *Windows.* The cost analysis in table 9.2 covers metal windows whilst the specification of the new project relates to wood windows. It is therefore advisable to build up a new unit rate. The cost of the element can then be estimated as

174.3 m² × £7.00 = £1220

$$\text{Cost of element/m}^2 \text{ of gross floor area} = \frac{1220}{855} = \underline{£1.43}$$

(9) *External doors.* Variations in specification again make it advisable to compute the new element rate from all-in rates, and to break down the element into its two main constituent parts, namely (1) two glazed doors with screens (2) metal up-and-over garage doors.

(1) 50 mm hardwood doors and screens (two no.) = 5.6 m² @ £22 = £123
(2) Metal up-and-over garage doors = 27 m² @ £4.50 = £122

Total cost of element £245

$$\text{Cost of element/m}^2 \text{ of gross floor area} = \frac{245}{855} = £0.29$$

(10) *Internal load-bearing walls.* There is a wide variety of walls used in the existing block of flats and so once again it is advisable to calculate the new rate independently. A suitable all-in rate for one-brick internal load-bearing walls would be £3.70/m².

Cost of element then becomes 391.5 × 3.70 = £1448

$$\text{Cost of element/m}^2 \text{ of gross floor area} = \frac{1448}{855} = £1.69$$

(11) *Partitions.* Variations in specification and quantity make separate computation desirable, although with the partition/floor ratios it is possible to use the existing element rate as a check. The element cost can be built up from the quantities already calculated.

75 mm hollow block partition = 558 m² @ £1.06 = £592
100 mm hollow block partition = 51 m² @ £1.18 = £ 60

Total cost of element £652

$$\text{Cost of element/m}^2 \text{ of gross floor area} = \frac{652}{855} = £0.76$$

This rate checks quite well with the rate of £0.19 for the previous job, having regard to the much lower partition/floor ratio but more expensive type of partition, and the time and situational factors.

(12) *Internal doors.* It is possible to adjust the element rate for the previous job, by making allowance for the time and situational factors and the varying qualities and thickness of door.

Previous rate: £1.49 + 36 per cent (time and situational factors) = £2.02

Deduct twenty per cent to allow for variations in quality of doors

£2.02 − 20 per cent = £1.62

Cost of element = £1.62 × 855 = £1385

(13) *Ironmongery.* A pc sum of £700 is to be provided.

$$\text{Cost of element/m}^2 \text{ of gross floor area} = \frac{700}{855} = £0.82$$

(14) *Wall finishes.* The specification for both jobs is very similar for this element and the simplest approach would be merely to adjust the previous rate for time and situational differences. The new rate then becomes £1.60 + thirty-six per cent = £2.18

The total element cost is £2.18 × 855 = £1862

(15) *Floor finishes.* On account of the rather wide variations in the types of floor finishes, it is advisable to calculate the new rate independently. A suitable all-in rate for 3.2 mm felt-backed PVC paving would be £2.50/m^2.

To arrive at the net area of flooring, it will be necessary to deduct the space occupied by internal load-bearing walls and partitions and the floor area of the garages.

Gross floor area		855 m^2
less floor area of garages	87 m^2	
Space occupied by internal walls and partitions, say 10% of 855–87 =	77 m^2	164 m^2
Net floor area		691 m^2

Cost of element = 691 × £2.50 = £1728

$$\text{Cost of element/m}^2 \text{ of gross floor area} = \frac{1728}{855} = £2.02$$

(16) *Ceiling finishes.* It will be in order merely to adjust the previous rate for time and situational differences. New rate then becomes

£0.52 + 36 per cent = £0.71

Cost of element = 855 × £0.71 = £606

(17) *Decoration.* Once again there is not likely to be any significant difference in the relative quantity or quality of the work. Hence the previous rate can be used and adjusted by the combined time and situational factor.

New rate = £2.14 + 36 per cent = £2.92

Cost of element is 855 × £2.92 = £2500

(18) *Fittings.* Due to varying specification particulars it is advisable to build up a rate for this element from the three component parts:

(i) Seventy no. kitchen fittings (five per dwelling)	£ 850
(ii) Shelving, say, 20 sq m @ £6	£ 120
(iii) Wardrobe fronts = twenty-eight no. @ £17 each	£ 476
Total cost of element	£1446

$$\text{Cost of element/m}^2 \text{ of gross area} = \frac{1446}{855} = £1.69$$

(19) *Sanitary appliances.* The specification particulars are identical for the two jobs and the element cost in the previous job can be used as a basis for the new rate. The number of appliances is directly proportional to the number of dwellings and so the previous element cost should be adjusted accordingly. The new element cost then becomes £468 × 14/9 = £729 + 36 per cent (time and situational factors)

Total element cost = £990

$$\text{Cost of element/m}^2 \text{ of gross floor area} = \frac{990}{855} = \underline{£1.16}$$

Alternatively, the cost could be calculated by pricing up the appliances to be provided.

Fourteen no. lavatory basins @ £12	£ 168
Fourteen no. sinks and drainers @ £18	£ 252
Fourteen no. baths @ £26	£ 364
Fourteen no. WC suites @ £16	£ 224
Total cost	£1008

The alternative method of costing the element provides a useful check on the first method of computation.

(20) *Waste, soil and overflow pipes.* The rate in table 9.2 will need adjusting on two counts; firstly time and situational factors and secondly, the extra height of the new building increases the amount of branch pipework more than proportionately, as appliances on ground floor are connected direct to drains. Allow an extra ten per cent to cover the second aspect.

New rate then becomes £0.46 + 36 per cent + 10 per cent = £0.69

Total cost of element is 855 × £0.69 = £590

(21) *Hot and cold water services.* In the new project elements 21 and 22 have been combined and the proposals include for ninety-eight draw-off points with a similar specification. It would seem reasonable to average the cost of the cold and hot water draw-off points in the previous job to give an average cost per draw-off point, and then to increase this rate to take account of time and situational differences. Average cost per draw-off point on previous job is

$$\frac{£11.44 + £18.70}{2} = \frac{£30.14}{2} = £15.07$$

Adjusted rate = £15.07 + 36 per cent = £20.50

Total cost of element then becomes 98 × £20.50 = £2009

$$\text{Cost of element/m}^2 \text{ of gross floor area} = \frac{2009}{855} = \underline{£2.35}$$

(22) *Hot water services.* Included with element 21.

(23) *Heating services.* Included with element 26.

(24) *Ventilation.* A pc sum of £340 is to be provided.

$$\text{Cost of element/m}^2 \text{ of gross floor area} = \frac{340}{855} = \underline{£0.40}$$

(25) *Gas services.* Not applicable.

(26) *Electrical and heating services.* A pc sum of £3800 is to be provided.

$$\text{Cost of element/m}^2 \text{ of gross floor area} = \frac{3800}{855} = \underline{£4.45}$$

(27) *Passenger and goods lifts.* Not applicable.

(28) *Fire-fighting equipment.* Not applicable.

(29) *Sprinkler installations.* Not applicable.

(30) *Special installations.* Not applicable.

(31) *Drainage.* A pc sum of £1200 is to be provided.

$$\text{Cost of element/m}^2 \text{ of gross floor area} = \frac{1200}{855} = \underline{£1.40}$$

(32) *External works.* A pc sum of £5000 is to be provided to cover extensive earthworks, parking areas, access road, paths, landscaping and fencing.

$$\text{Cost of element/m}^2 \text{ of gross floor area} = \frac{5000}{855} = \underline{£5.85}$$

(33) *Preliminaries and insurances.* To be calculated at ten per cent of the remainder of the contract sum.

(34) *Contingencies.* To be calculated at one per cent of the remainder of the contract sum.

The estimated elemental costs and costs of elements per square metre of gross floor area will now be summarised and totalled and in doing this the first or initial cost plan will also be produced.

Initial cost plan: four-storey block of flats, outer London

Gross floor area: 855 m²

Element	*Total cost of element* £	*Total cost of groups of elements* £	*Cost of element per m² of gross floor area* £	*Cost of groups of elements per m² of gross floor area* £
1. Substructure	2378		2.78	£ 2.78
	—	£ 2378	—	
2. Frame	—		—	
3. Upper floors	3176		3.71	
4. Roof	1414		1.65	
5. Rooflights	200		0.23	
		c.f. £ 2378		£ 2.78

Initial cost plan: four-storey block of flats, outer London

Gross floor area: 855 m²

Element	*Total cost of element*	*Total cost of groups of elements*	*Cost of element per m² of gross floor area*	*Cost of groups of elements per m² of gross floor area*
	£	£	£	£
		b.f. £ 2378		b.f. £ 2.78
6. Staircases	440		0.52	
7. External walls	3382		3.96	
8. Windows	1220		1.43	
9. External doors	245		0.29	
10. Internal load-bearing walls	1448		1.69	
11. Partitions	652		0.76	
12. Internal doors	1385		1.62	
13. Ironmongery	700		0.82	
	——	£14 262	——	16.68
14. Wall finishes	1862		2.18	
15. Floor finishes	1728		2.02	
16. Ceiling finishes	606		0.71	
17. Decoration	2500		2.92	
	——	£ 6696	——	7.83
18. Fittings	1446		1.69	
	——	£ 1446	——	1.69
19. Sanitary appliances	990		1.16	
20. Waste, soil and overflow pipes	590		0.69	
21. Cold water services } 22. Hot water services }	2009		2.35	
23. Heating services (see element 26)	—		—	
24. Ventilation	340		0.40	
25. Gas services	—		—	
26. Electrical and heating services	3800		4.45	
		c.f. £ 24 782		£28.98

Initial cost plan: four-storey block of flats, outer London

Gross floor area: 855 m²

Element	Total cost of element	Total cost of groups of elements	Cost of element per m² of gross oor area	Cost of groups of elements per m² of gross floor area
	£	£	£	£
		b.f £24 782		b.f. £28.98
27. Passenger and goods lifts	—		—	
28. Fire-fighting equipment	—		—	
29. Sprinkler installations	—		—	
30. Special installations	—		—	
31. Drainage	1200		1.40	
	—	£ 8929	—	10.45
Totals excluding external works		£33 711		39.43
32. External works		5000		5.85
33. Preliminaries		3871		4.52
34. Contingencies		387		0.45
Totals		£42 969		£50.25

NOTE: No sum has been included to cover price and design risks as the element rates have been increased to cover probable cost increases up to the date of tender. An alternative approach would have been to work on price levels current at the date of the estimate and to have included a percentage (possibly about three per cent) to cover possible increases in building prices during the design period.

EXAMPLE 3

The third example in this chapter is concerned with one of the later cost planning processes in the design stage of a building project. The project is a social club building and is illustrated in figure 9.4. It is a single-storey building with a gross floor area of 470 m². The initial cost plan for the scheme is illustrated in table 9.3, together with the entries relating to the first set of cost checks undertaken as the design developed. The building client has requested that the scheme should desirably be kept within a budget figure of £20 000, which is equivalent to £42.60 per square metre of gross floor area, but does not favour any reduction in the size of the building. The examples will endeavour to show the way in which costs of elements are investigated and steps taken to reduce the estimated cost of the project, together with the appropriate communications to the architect.

TABLE 9.3

COST PLANNING EXAMPLE 3

Initial cost plan and record of cost checks of social club

	Initial cost plan		*Cost check 1*		*Cost check 2*	
Element	*Total cost of element* £	*Cost of element/m² of gross floor area* £	*Date*	*Cost of element/m² of gross floor area* £	*Date*	*Cost of element/m² of gross floor area* £
1. Substructure	1900	4.04	2.9.71	4.02		
2. Frame	178	0.38	3.9.71	0.39		
3. Upper floors	110	0.23	3.9.71	0.22		
4. Roof	4540	9.66	6.9.71	9.64	4.10.71	6.44
5. Rooflights	180	0.38	6.9.71	–		
6. Staircases	60	0.13	6.9.71	0.13		
7. External walls	1086	2.31	17.9.71	2.30	4.10.71	2.03
8. Windows	550	1.17	17.9.71	1.48		
9. External doors	190	0.41	17.9.71	0.40		
10. Internal load-bearing walls	446	0.95	10.9.71	0.92	4.10.71	0.77
11. Partitions	25	0.05	10.9.71	0.05		
12. Internal doors	225	0.48	10.9.71	0.39		
13. Ironmongery	334	0.71	10.9.71	0.69		
14. Wall finishes	523	1.11	14.9.71	1.10		
15. Floor finishes	1323	2.81	14.9.71	2.80	4.10.71	2.56
16. Ceiling finishes	420	0.89	14.9.71	1.06		
17. Decoration	510	1.08	14.9.71	1.08		
18. Fittings	500	1.06	16.9.71	0.28		
19. Sanitary appliances	340	0.73	21.9.71	0.73		
20. Waste, soil and overflow pipes	40	0.08	21.9.71	0.08		
21. Cold water services	195	0.41	21.9.71	0.42		
22. Hot water services	–	–		–		
23. Heating services	2245	4.78	21.9.71	4.78		
24. Ventilation	480	1.02	21.9.71	1.02		
25. Gas services	20	0.04	21.9.71	0.04		
26. Electrical installation	1988	4.23	22.9.71	4.21		
27. Passenger and goods lifts	–	–		–		
28. Fire-fighting equipment	165	0.35	22.9.71	0.06		
29. Sprinkler installation	–	–		–		
30. Special installations	250	0.53	23.9.71	0.21		
31. Drainage	886	1.89	23.9.71	1.88		
32. External works	2040	4.35	24.9.71	3.26		
33. Preliminaries	755	1.61		1.60		
34. Contingencies	250	0.53		0.53		
35. Price and design risk	500	1.06		1.06		
Totals	£23 254	£49.46		£46.83		

Note: The arrangement of elements in this cost plan follows the pre-1970 BCIS cost analysis format

As the design of the social club has developed and working details and detailed specification requirements have been prepared, cost checks have been carried out to confirm that it is feasible to provide the elements within the cost targets listed in the initial cost plan and, in view of the building client's request, to look for economies at the same time. Slight adjustments have been made to the costs of a number of elements and their consequences, in relation to the costs per square metre of gross floor area, are duly recorded in table 9.3 under the heading of cost check 1. Significant changes have been made to the costs of rooflights, windows, internal doors, ceiling finishes, fittings, fire-fighting equipment, special installations and external works, and details of the changes are now shown in the form of a report to the job architect.

To: Architect
From: Quantity surveyor

27.9.71

Social Club

Cost checks have been made of all elements of the above project and a summary sheet (table 9.3) shows the current position. It has been possible to reduce the total cost expressed in terms of the square metre of gross floor area from £49.46 to £46.83, but this is still outside the client's budget limit. Details of the cost changes arising from the amendments to design, as agreed with you, are now listed.

	Savings per m² of gross floor area £	*Excess per m² of gross floor area* £
(5) *Rooflights.* These have been omitted with a consequent saving of £180. There will be no corresponding increase in the cost of the roof element, as the cost of the additional area of roof is offset entirely by the saving in cost resulting from the omission of forming and waterproofing the roof around the rooflights.	0.38	
(8) *Windows.* The purpose-made windows are proving to be more expensive than was originally anticipated and the specialist quotation now received shows an increase of £145, raising the total element cost to £695.		0.31
(12) *Internal doors.* The sliding hardwood doors have been replaced by side-hung softwood glazed doors, resulting in a saving of £45. The total element cost now becomes £180.	0.09	
(16) *Ceiling finishes.* The use of acoustic ceiling tiles to the hall has resulted in an increase of cost of £70, raising the total element cost to £490.		0.17
(18) *Fittings.* A substantial saving has resulted from the proposal to adapt and fix second-hand fittings from the present club, instead of purchasing new fittings. The total element cost is reduced from £500 to £130.	0.78	
(28) *Fire-fighting equipment.* The extensive fire-fighting equipment originally proposed has now been confined to three portable fire extinguishers, reducing the total element cost to £30.	0.29	
(30) *Special installations.* As a result of the simpler burglar system, the total element cost has been reduced to £100.	0.32	
(32) *External works.* A length of 50 m of one-brick thick boundary wall has been omitted permitting a cost reduction of £500.	1.09	
Major cost changes	£2.95	£0.48

In view of the client's expressed wish for the total cost of the job not to exceed £20 000, further design/cost investigations are proceeding with particular reference to the elements of roof, external walls, internal load-bearing walls and floor finishes, as being the elements where significant cost savings may still be possible without seriously impairing the functional and aesthetic characteristics of the building.

These four elements are now investigated in detail with a view to reducing the total cost per m² of gross floor area to £42.60.

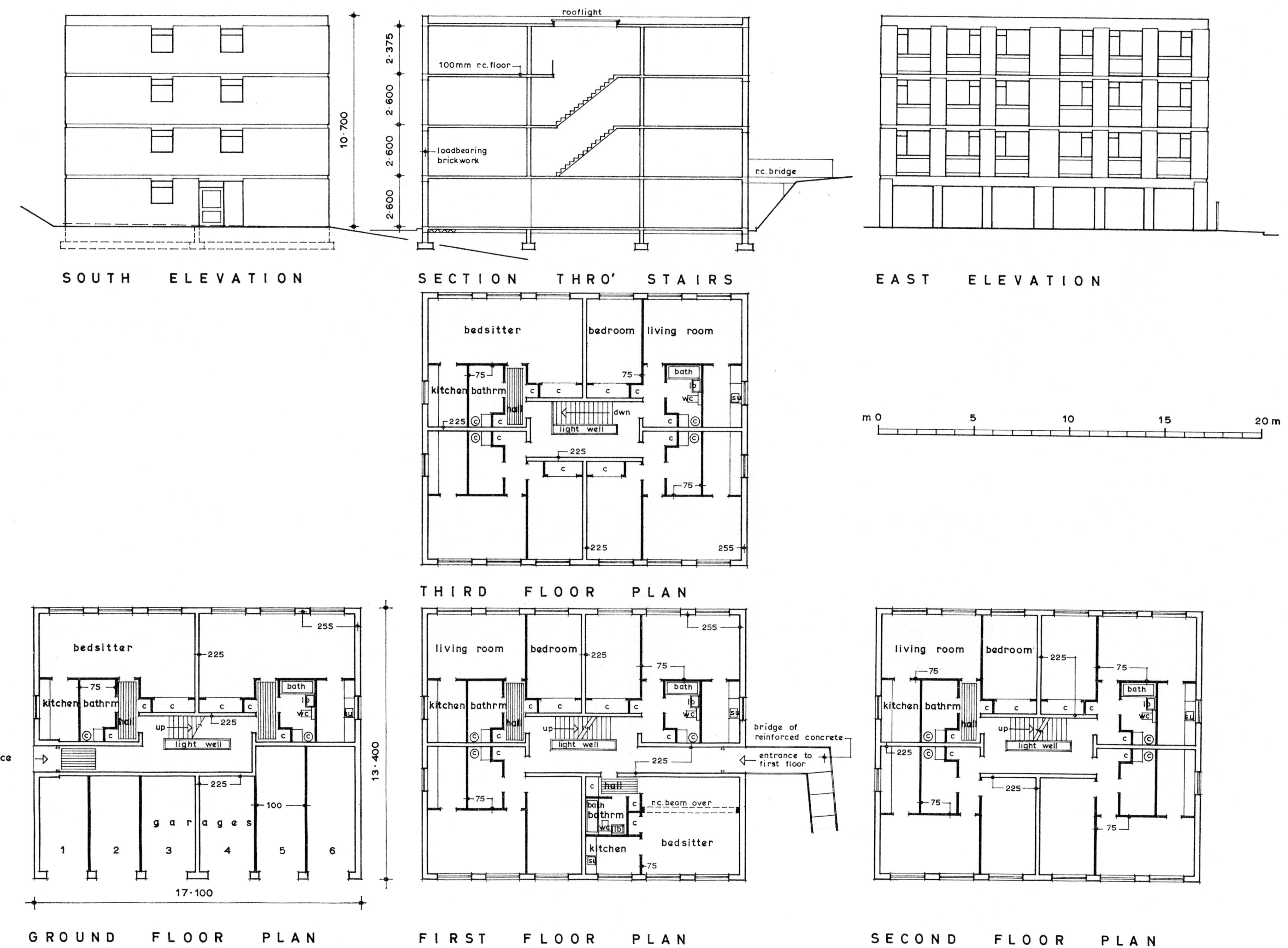

Figure 9.3 Cost planning example 2 – sketch design of four-storey block of flats

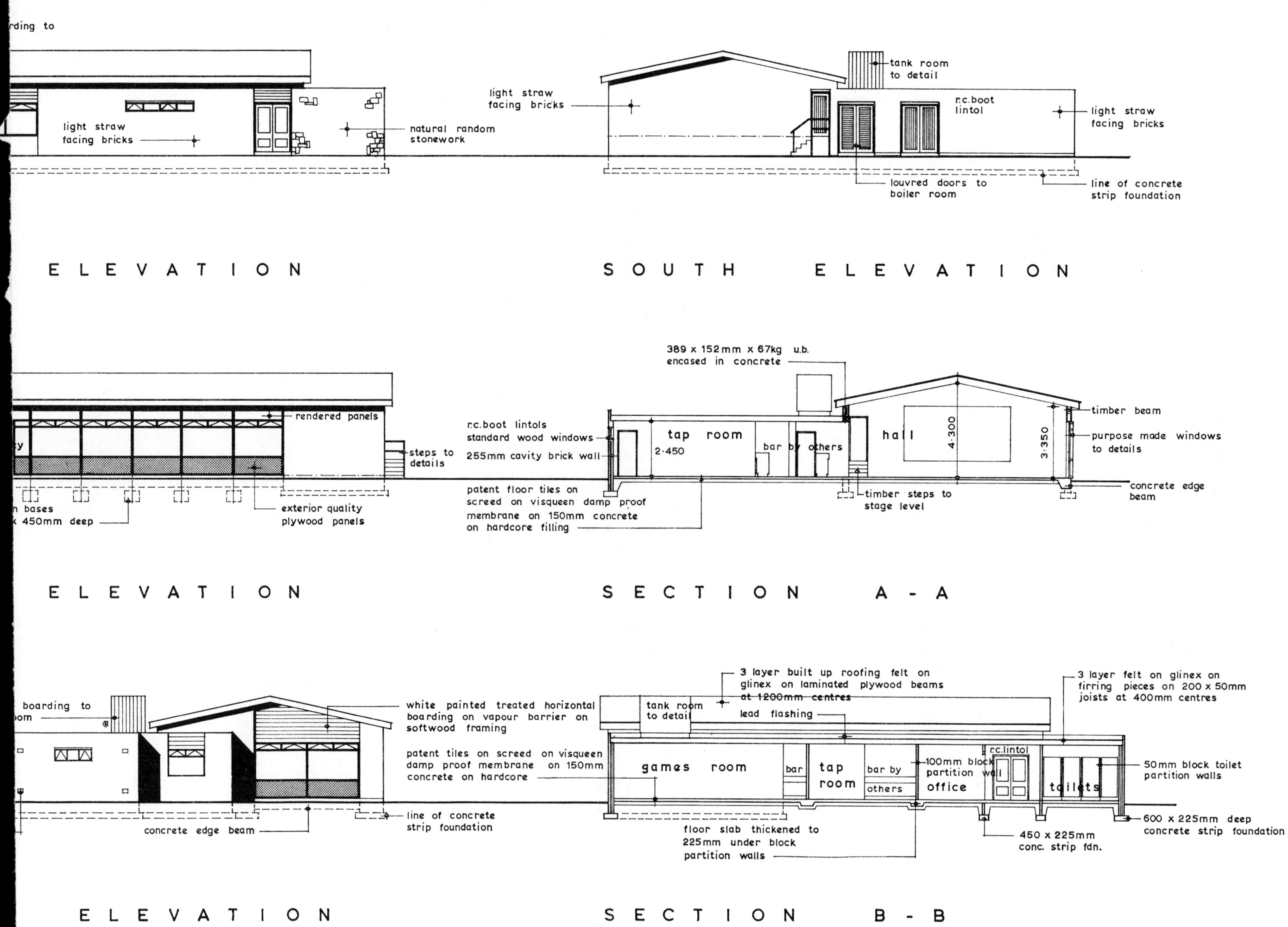

rding to
light straw
facing bricks
natural random
stonework
ELEVATION
light straw
facing bricks
tank room
to detail
r.c. boot
lintol
light straw
facing bricks
louvred doors to
boiler room
line of concrete
strip foundation
SOUTH ELEVATION
rendered panels
steps to
details
bases
450mm deep
exterior quality
plywood panels
ELEVATION
389 x 152mm x 67kg u.b.
encased in concrete
r.c. boot lintols
standard wood windows
255mm cavity brick wall
tap room
2·450
bar by others
hall
4·300
3·350
timber beam
purpose made windows
to details
patent floor tiles on
screed on visqueen damp proof
membrane on 150mm concrete
on hardcore filling
timber steps to
stage level
concrete edge
beam
SECTION A-A
boarding to
om
white painted treated horizontal
boarding on vapour barrier on
softwood framing
patent tiles on screed on visqueen
damp proof membrane on 150mm
concrete on hardcore
line of concrete
strip foundation
concrete edge beam
3 layer built up roofing felt on
glinex on laminated plywood beams
at 1200mm centres
3 layer felt on glinex on
firring pieces on 200 x 50mm
joists at 400mm centres
tank room
to detail
lead flashing
games room
bar
tap
room
bar by
others
100mm block
partition wall
office
r.c. lintol
toilets
50mm block toilet
partition walls
600 x 225mm deep
concrete strip foundation
floor slab thickened to
225mm under block
partition walls
450 x 225mm
conc. strip fdn.
ELEVATION
SECTION B-B

Figure 9.4 Cost planning example 3 – social club

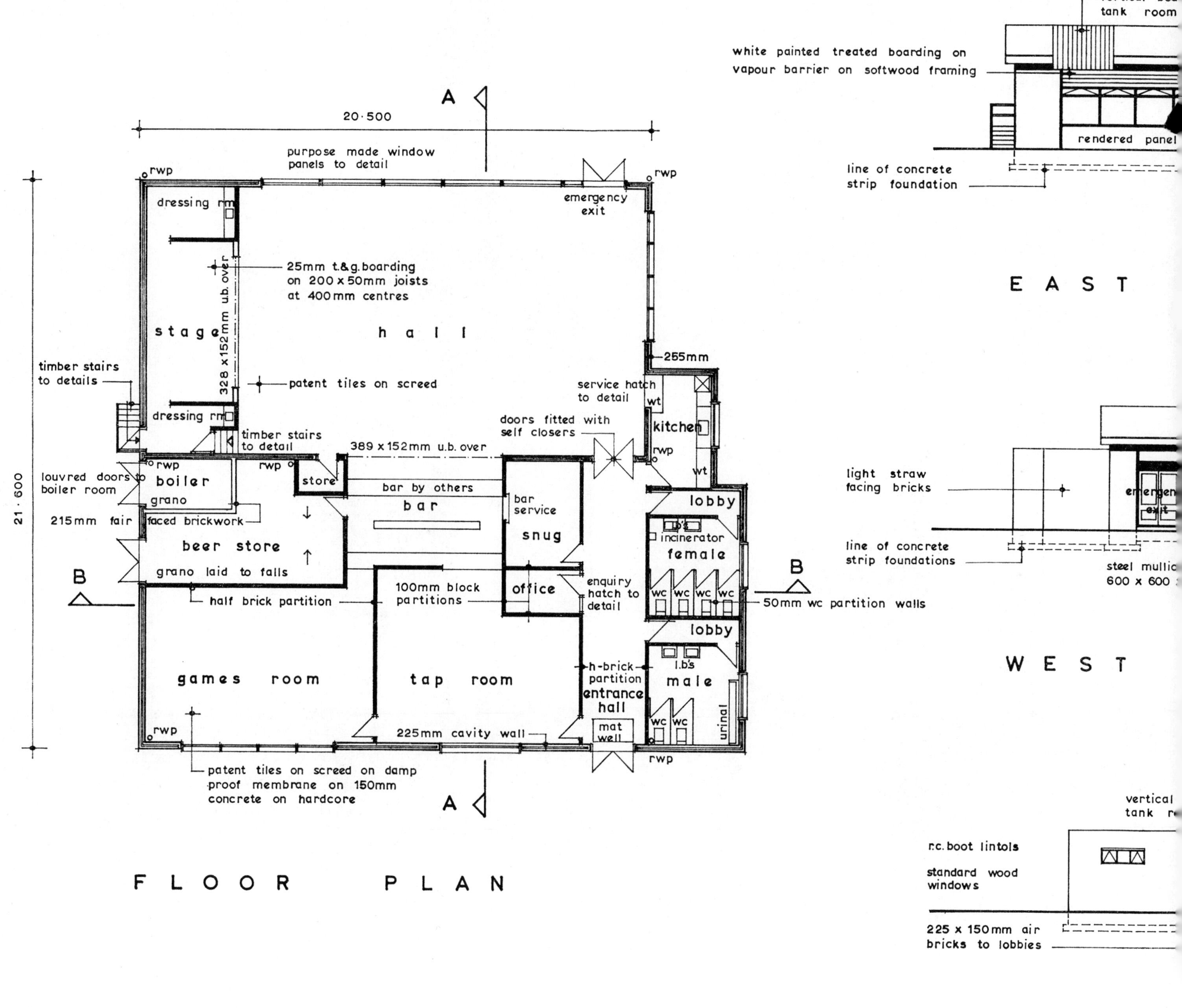

A
20·500
purpose made window panels to detail
rwp
dressing rm
25mm t.&g. boarding on 200 x 50mm joists at 400mm centres
328 x 152mm u.b. over
stage
hall
emergency exit
timber stairs to details
patent tiles on screed
255mm
service hatch to detail
wt
kitchen
dressing rm
timber stairs to detail
389 x 152mm u.b. over
doors fitted with self closers
louvred doors to boiler room
boiler
grano
store
bar by others
bar
bar service
snug
lobby
l.b's
incinerator
female
215mm fair faced brickwork
beer store
grano laid to falls
office
enquiry hatch to detail
B
half brick partition
100mm block partitions
wc
50mm wc partition walls
lobby
h-brick partition
male
games room
tap room
entrance hall
urinal
225mm cavity wall
mat well
patent tiles on screed on damp proof membrane on 150mm concrete on hardcore
21·600
FLOOR PLAN
m 0
5
10
15
20
25
vertical boa
tank room
white painted treated boarding on vapour barrier on softwood framing
rendered panel
line of concrete strip foundation
EAST
light straw facing bricks
emergen exit
line of concrete strip foundations
steel mullic
600 x 600
WEST
vertical
tank
r.c. boot lintols
standard wood windows
225 x 150mm air bricks to lobbies
NORTH

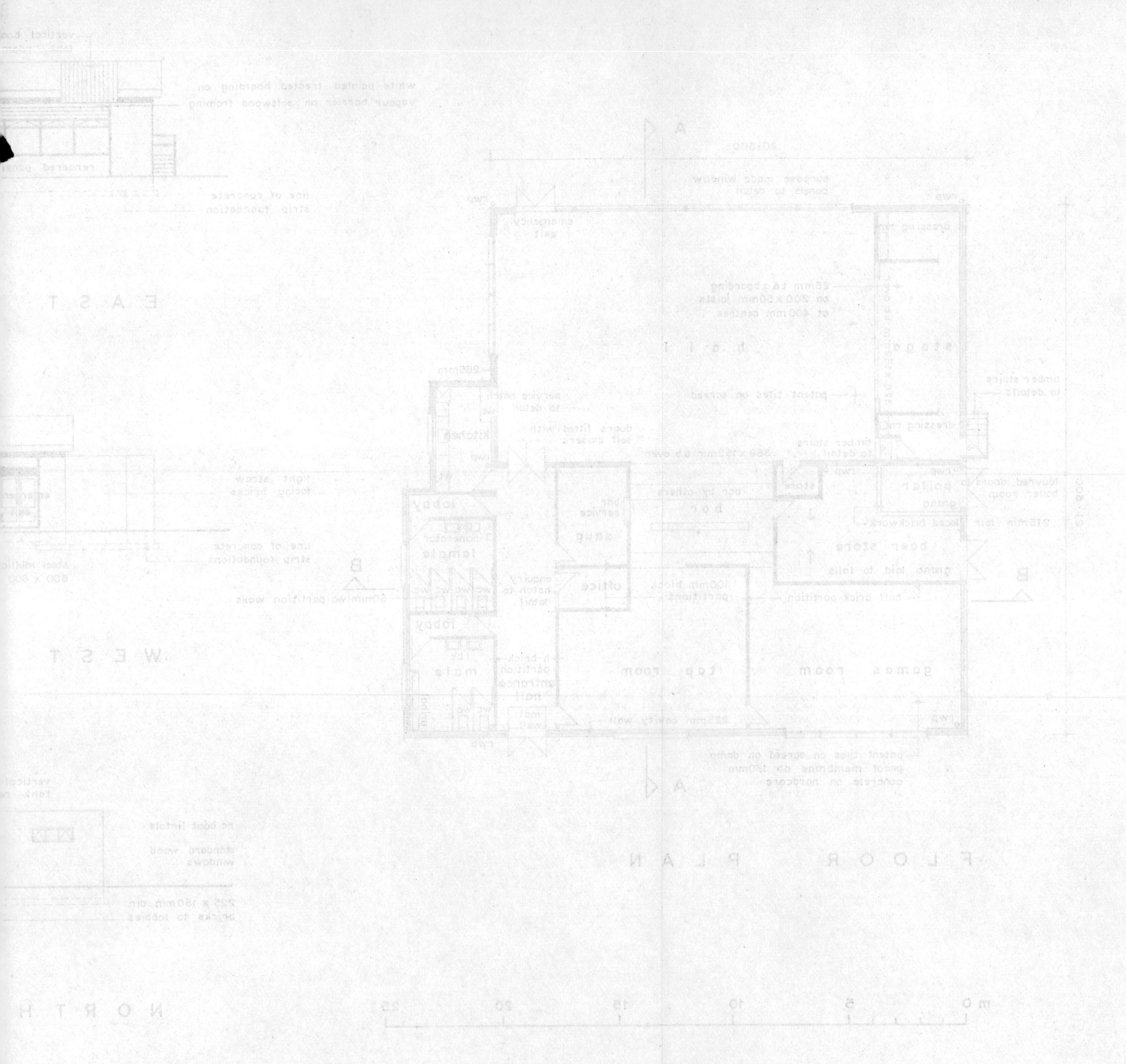

(4) *Roof.* It is agreed with the architect to replace the felt-covered plywood boarding on laminated beams with felt covered Weyroc boards supported on timber joists.

The cost of the substituted construction is

$$496 \text{ m}^2 \times £6.10 = £3026\ (£6.44/\text{m}^2)$$

(7) *External walls.* Lightweight concrete blocks, 100 mm thick, are to replace the half-brick inner skin of the external walls. The saving in cost is 190 m^2 × £0.70 = £133, and the total element cost becomes £953 (£2.03/m^2).

(10) *Internal load-bearing walls.* 100 mm block walls are to be substituted for half-brick walls and 75 mm block walls for 100 mm block walls.

The revised estimate of cost is

20 m^2 of one-brick wall @ £3.40 =	£ 68
132 m^2 of 100 mm block wall @ £1.40 =	£185
100 m^2 of 75 mm block wall @ £1.10 =	£110
Total cost	£363

This is equivalent to £0.77/m^2.

(15) *Floor finishes.* Substitution of granolex vinyl bonded block floor tiles on cement screed for the quarry tile flooring originally proposed for the entrance hall, toilets and kitchen. The cost variation is

62 m^2 of quarry tile paving @ £5.40 =	£335
62 m^2 of granolex paving @ £3.50 =	£217
Saving in cost	£118

The revised element cost then becomes £1205 and the cost/m^2 of gross floor area is £2.56.

A further letter to the architect summarises the current cost situation.

To: Architect
From: Quantity surveyor 5.10.71

Social Club

The cost consequences of the latest design/cost investigations are now listed.

Element	*Saving in cost/m^2 of gross floor area* £
(4) Roof (£9.64 – £6.44)	3.20
(7) External walls (£2.30 – £2.03)	0.27
(10) Internal load-bearing walls (£0.92 – £0.77)	0.15
(15) Floor finishes (£2.80 – £2.56)	0.24
Total saving in cost	£3.86

These further amendments to design reduce the total cost per square metre of gross floor area to £42.97 and the total cost of the project to £20 190. This is £190 in excess of the client's budget limit but could possibly be accommodated in the £500 allocated to price and design risk. Another alternative would be to consider the use of a cheaper floor finish to the hall and games room. I would welcome your comments on this latter suggestion.

CONCLUSIONS

It will be noted that all three examples have been based on the elemental cost planning approach, although it would have been quite conceivable to have developed example 2 on the comparative method, indicating a selection of choices with their appropriate costs for each of the major elements and the cost consequences of each of the choices. However, the prime aim of this chapter has been to help examination candidates in the compilation of cost plans or part of cost plans, from sketch drawings, cost analyses of similar buildings and other relevant information. It is believed that the examples will have shown suitable methods of approach to this type of problem.

It might be argued that it is not the best practice to restrict the basis for the new cost plan to a published cost analysis of one building, with which the surveyor is not directly familiar. At the very least, a number of cost analyses should be taken and average costs computed. Limitations of space have prevented the cost examples being more comprehensive and general principles are just as well illustrated on a small job as on a large project, and the smaller scheme has the benefit of greater simplicity and clarity.

Another aspect which may leave room for doubt in the reader's mind is that of the application of published building cost indices. There are very real dangers in applying them too rigidly and indiscriminately without regard to local circumstances. The type of building, its location and the tendering climate must each have a significant effect on price levels and all published cost data must therefore be used with the greatest discretion and be suitably adjusted, wherever necessary, to take account of local conditions. In practice the quantity surveyor will have the benefit of a vast amount of cost data of his own and published data will be largely used to form broad guidelines and for comparative purposes. Published cost data will be of particular value when dealing with special type buildings subject to the provisos already given.

An examination of BCIS cost analyses, all with prices adjusted to the first quarter of 1969, show wide price ranges for similar building types and this again reinforces the problems and dangers inherent in the application of even the best published cost data. The 1969 price ranges of a selection of common building types are

factories: £16.96 to £37.19/m^2
primary schools: £55.70 to £79.76/m^2
churches: £70.73 to £96.90/m^2
libraries: £47.69 to £124.02/m^2
flats: £31.15 to £80.13/m^2
houses: £25.25 to £43.49/m^2
residential blocks: £53.56 to £92.37/m^2
fruit markets: £29.67 to £83.33/m^2
offices: £26.24 to £66.06/m^2
old person's homes: £52.02 to £74.27/m^2

Libraries, churches and universities are buildings with high unit costs. All these buildings have large storey heights, are constructed to high standards of quality and contain expensive fittings. It may not however be quite so readily recognised that public conveniences are also very costly buildings with high wall-to-floor ratios and expensive fittings and finishings. In consequence they frequently cost more than £100/m^2 of gross internal floor area.

COST CONTROL OF ENGINEERING SERVICES

Berryman has made suggestions as to how the cost of heating and ventilating engineering services might be controlled during both design and postcontract stages. He admits that the cost control functions will have to be shared and believes that the arrangements listed in table 9.4 constitute an acceptable and workable arrangement.

TABLE 9.4

SUGGESTED PROCEDURE FOR COST CONTROL OF ENGINEERING SERVICES IN BUILDINGS

Contractor design stage	*Quantity surveyor*	*Contractor*
Precontract	(1) Initial budget and cost plan	*(1) Cost studies before design
	(2) Tender enquiries	(2) Design
	(3) Tender report	*(3) Cost checking during design
	(4) Cost analysis and budget reconciliation	(4) Priced bills of quantities or cost sheets
		(5) Precontract variations when required
Postcontract	(5) Budget control	(6) Design and detailing
	(6) Variations or remeasurement	(7) PC, daywork, increased or decreased costs and overtime accounts

Consultant design stage	*Consulting engineer*	*Quantity surveyor*	*Contractor*
Precontract	(1) Initial budget and cost plan	(1) Initial budget and cost plan	(1) Cost sheets (if bills of quantities not required)
	*(2) Cost studies before design	(2) Bills of quantities when required	(2) Pricing
	(3) Design	(3) Tender enquiries	
	*(4) Cost checking during design		
	(5) Tender report	(4) Tender report	
		(5) Precontract variations when required	
		(6) Cost analysis and budget reconciliation	
Postcontract	(6) Postcontract design and supervision	(7) Budget control	(3) Detailing
		(8) Variations or remeasurement	(4) PC, daywork, increased or decreased costs and overtime accounts

NOTE

(1) All postcontract duties are subject to negotiation, and consequently shared between both parties to the contract, irrespective of who is to take the initiative.

(2) The same applies to bills of quantities or cost sheets or precontract variations when these also are subject to negotiation.

(3) It may be helpful or necessary to delegate duties marked * to the quantity surveyor, especially where detailed measurement, analysis or integration with building costs are involved.

Source: A. Berryman, *Controlling the costs of engineering services in buildings. The Chartered Surveyor* (August 1971)

10 VALUATION PROCESSES

THIS CHAPTER IS concerned with the nature of value and investment, the construction and use of valuation tables and methods of valuation. Until quite recent years these matters were considered to be solely the province of the valuation surveyor, but it has now become apparent that the quantity surveyor also needs to be familiar with some of the valuation techniques and certain of the valuation tables, in order to be able to make feasibility studies and to deal satisfactorily with future costs.

THE CONCEPT OF VALUE AND INVESTMENT

Value

The cornerstone of the economic theory of value is that an object must be scarce relative to demand to have a value. Where there is an abundance of a particular object and only limited demand for it, then the object has little or no value in an economic sense. Value constitutes a measure of the relationship between supply and demand. An increase in the value of an object is obtained either through an increase in demand or a decrease in supply. Value also measures the usefulness and scarcity of an object relative to other objects or commodities.

The degree of response of supply and demand to price changes is referred to as the elasticity of supply or demand. Where a small change in price causes a large change in demand, then the demand is elastic, but if a large change in price leaves the demand virtually unchanged, then the demand is inelastic. The elasticity of demand is very much influenced by the availability of suitable substitutes. There are also the shortterm and longterm requirements of changes in supply to be considered.

Surveyors are primarily concerned with the value of landed property and this embraces all forms of building and land which may be put to a wide variety of uses. The market value of an interest in landed property will be the amount of money which can be obtained from a willing purchaser at a specific point in time, and is generally determined by the interaction of the forces of supply and demand. It will be appreciated that the supply of land as a whole is fixed but the quantity of various types of landed property is variable, as land and buildings can be transferred from one use to another, existing buildings destroyed and new ones built. The value of a specific form of landed property will be influenced by the amount coming onto the market at a particular time rather than the total stock in existence. It takes time to transfer one form of landed property to another use and to erect buildings to meet an increased demand, and so the supply of landed property is generally regarded as inelastic.

The demand for any particular type of landed property is influenced by a number of factors, such as:

(1) population changes;
(2) changes in the standard of living or in taste or fashion;
(3) changes in society;
(4) population movement;
(5) changes in social services (shops, schools, cinemas, etc.);
(6) changes in communications; and
(7) changes in statutory requirements, such as the Town and Country Planning Acts.

Each property is unique, with its own specific location and characteristics and no one property is a perfect substitute for another. It is these factors which make the valuation of landed properties so difficult.

Investment

In a capital investment project there is an outlay of cash in return for an anticipated flow of future benefits. The consequences of capital investment extend into the future and may involve decisions as to the type or quality of a new building and its best location. Buildings cannot always be readily adapted to other uses, so wrong development decisions can result in heavy losses to investors. In addition, future benefits are always difficult to evaluate. When comparing alternative building solutions it is essential that total costs are used. In this situation it is necessary to compare both present and future costs on a common basis with the help of valuation tables, which will be described later in this chapter.

Landed property has a basic characteristic of durability and can be used over lengthy periods of time. It is accordingly capable of yielding an income as individuals will be prepared to make periodic payments for its use. When an investor purchases an interest in landed property, he is tying up a certain amount of capital in the property and will expect a reasonable return comparable with what he might have received had he invested it elsewhere. The amount of yield or rate of interest will vary with the degree of security, regularity of payment, period of investment, ease of convertibility of capital and cost of acquiring or disposing of the asset. Inflationary tendencies and taxation arrangements also have a bearing on interest rates and the relative desirability of the investment. Changes in rates of income tax, profits tax, capital gains tax, and investment grants and allowances will influence interest rates. Nevertheless, interest rates on landed property tend, after a suitable lapse of time, to be similar to the yields of the nearest substitute in the capital market. There are, however, essential and significant differences between landed property and other forms of investment, as shown below.

(1) There is no central market for the comparison of prices as with the Stock Exchange. The transfer of landed property by conveyance is both costly and time-consuming.

(2) It is not possible to divide property into small units like shares on the stock market, and it is therefore difficult for an investor to invest small sums in landed property.

(3) The management of landed property creates problems which do not arise with other forms of investment.

(4) The income or rate of return from landed property can normally only be varied at the end of comparatively long leases, whereas the income from ordinary shares can be varied annually.

METHODS OF VALUATION

The main function of the valuation surveyor or valuer is to assess the value of any type of landed property under any set of conditions. Valuations are required for a variety of purposes – for sale, for purchase for occupation or investment, for determining auction reserves, mortgage loans, estate duty, or for income tax or rating purposes. Property values vary considerably from one district to another and so a valuer needs to have extensive experience of values in the area in which he is practising. It is a specialised function involving its own particular expertise and the quantity surveyor would be wise to consult a valuer whenever valuation of property is concerned.

A number of methods may be used to assess the market value of an interest in landed property.

Comparative Method

This method is a popular valuation technique and consists of making a direct comparison with the prices paid in the open market for other similar properties, where reasonably close substitutes are available and transactions occur quite frequently. Its prime use is for residential properties where there is likely to be a greater similarity between different properties. Difficulties do, however, frequently arise in the use of this method as it is unusual to find two entirely similar type properties – differences occur in size, amount of accommodation, quality and extent of finishings and fittings, condition of property and its situation. For instance, the price paid for one block of offices may not be a very good indicator of the value of an adjacent office building which may differ considerably in room sizes, internal layout, type of finishes and in many other ways. Furthermore, prices may vary appreciably over relatively short periods of time and so the valuer must also have regard to current trends.

The valuer generally finds it helpful to break down the property into suitable units for comparison purposes.

Land can conveniently be priced per hectare or possibly per metre of frontage in the case of building land, and buildings might be reduced to the price per square metre of total floor area. It is also advisable to have regard to the underlying economic factors influencing the prices as well as the prices themselves.

Contractor's Method

The basis of this approach is that the value of the land and buildings is equivalent to the cost of erecting the buildings plus the value of the site. This is usually an unsound assumption as the value of a property is determined not by what it cost to build but by the amount which purchasers in the open market are prepared to pay for it in relation to the price the seller is prepared to take. Its main use is for buildings which rarely change hands, such as hospitals, schools and town halls, where there is little or no evidence in the form of sale prices.

When applying this method it is necessary to make allowance for depreciation in older buildings, as a building which is sixty years old cannot have the same value as a similar type of building of comparable size and construction which is only five years old. Some buildings may be excessively ornate or extravagant in their construction and finishings and the value of these properties may not necessarily be increased in proportion to the additional expense incurred. A house specially designed to meet the needs of a particular occupant may not suit the requirements of prospective purchasers.

Residual Method

This is a valuation method which is sometimes used where the value of the property can be increased by carrying out certain works. A large house could, for instance, be profitably converted into flats when its potential will be exploited to the full. The building could be valued by taking its value after conversion and deducting the cost of conversion plus an allowance for developer's risk and profit. The residual figure will indicate the value of the property in its existing state but with a potential for development.

Profits Method

This is sometimes described as the accounts method and is used where the value is largely dependent upon the earning capacity of the property, as is the case with hotels, cinemas and dance halls. The usual approach is to estimate the gross earnings and to deduct from them the working expenses, interest on capital and tenant's remuneration. The balance represents the amount that is available for the payment of rent. It is an exceedingly indirect approach and is best checked by some other method, such as the value per seat. It has, however, been found useful in rating valuations for the classes of property previously described.

Investment Method

This method can be used where the property produces an income, as there will be a direct relationship between the income accruing and the capital value of the property. The income must show a reasonable return compatible with the interest which could be earned by investing the capital elsewhere. An example will serve to illustrate this aspect.

If an investor purchased a freehold property at £12 000 and required a six per cent rate of interest on his capital, he will only secure the required return if the net income accruing from the property is £720 per annum (£12 000 × 6/100). When the position is reversed, it is possible to calculate the capital value from the return and required rate of interest, thus

$$\pounds 720 \times \frac{100}{6} = \pounds 12\,000.$$

Years' Purchase

The multiplier used in the last example – 100/6 may be described as *years' purchase* (YP) or the *present value of £1 per annum*. Net income *x* years' purchase = capital value.

With a perpetual income, years' purchase or YP can be obtained by dividing 100 by the interest rate, thus

$$\text{YP in perpetuity} = \frac{100}{\text{rate of interest}}$$

$$\text{and YP in perpetuity @ 5 per cent} = \frac{100}{5} = 20$$

The following examples will illustrate its use.

Example (a). Value a freehold interest in a shop producing a net income (after deduction of all outgoings) of £1200 per annum. It can be assumed that a purchaser will require a return of six per cent on this capital.

Net income	£1200 pa
YP in perpetuity @ 6 per cent (100/6)	16.67
Capital value	£20 004
Rounded off to	£20 000

Example (b). Value a house capable of producing a net income of £400 per annum. This investor requires a six-and-a-half per cent rate of interest.

Net income	£400 pa
YP in perpetuity @ 6½ per cent (100/6½)	15.38
Capital value	£6152
Rounded off to	£6150

An analysis of each of these examples shows that the net income from the property represents the required rate of interest on the capital value or purchase price, for example

$$\text{six per cent on £20 004 or £20 004} \times \frac{6}{100} = \text{£1200 pa}$$

The figure of years' purchase (YP) varies considerably with the type of property and is largely dependent upon the degree of risk involved. Thus properties involving greater risk require higher rates of interest and lower YPs. The following figures are typical for the various classes of property.

Cottages – 12 YP – 8½% interest
Houses – 14 YP – 7 % interest
Shops – 16 YP – 6 % interest

Sinking Funds

Leasehold properties reduce in value throughout the duration of the lease, until finally at the termination of the lease, they cease to have any value. It is customary, therefore, for a leaseholder to provide for a sinking fund which will recoup the initial capital sum by the end of the lease. The interest on the capital and that on the sinking fund are generally at different rates (say seven per cent and three per cent). For this reason it is usual to use the dual-rate tables provided in valuation tables when valuing leasehold interests. Where a leasehold interest is sublet and the net rent so received exceeds the rent paid to the landlord, then a profit rent exists (the difference between the two).

An example follows to show the method of valuing leasehold interests.

Example. X owns the freehold interest of a house which is let on an annual tenancy at a net rent of £300 pa. Y has a lease of a similar house with forty years to run at a ground rent of £60 per annum. Value the interests of X and Y.

Freehold interest of X	
Net rent received	£300 pa
YP in perpetuity @ seven per cent	14.3
Capital value	£4290
Leasehold interest of Y	
Net rent received	£300 pa
less ground rent	60
	£240
YP forty years @ eight per cent and three per cent	10.72
(figure obtained from valuation tables)	
Capital value	£2573
Rounded off to	£2570

NOTE: Dual-rate tables have been used to value the leasehold interest. The remunerative rate of interest has been taken at one per cent above the corresponding freehold rate to allow for the greater risk involved, and an accumulative interest rate of three per cent for the sinking fund.

VALUATION TABLES

The quantity surveyor needs to be able to use certain valuation tables in connection with some of his cost planning calculations. Where returns are spread over a number of years, or the calculation involves both initial capital costs and annual running and maintenance costs, with possibly replacement costs at intervals throughout the life of the building, then the calculations become more complicated and valuation tables will assist in the computations. It may be desirable to obtain the *present value* (PV) of future expenditure or to convert from present and future costs to *annual equivalent* costs, and examples of both processes in costs in use calculations are given in chapter 11.

The best source of valuation tables is the current edition of *Parry's Valuation Tables* (*Estates Gazette*). Students are advised to examine a copy of these tables in order to become familiar with their general form, layout and contents. Some abridged valuation tables appear in appendices 1 to 5 at the back of this book to assist the reader in working through the various costs in use and other worked examples. The nature, construction and application of some of the more important valuation tables will now be described.

Amount of £1 Table

Extracts from the above table are shown in appendix 1 at the back of this book. It indicates the amount to which a sum of £1 will accumulate if invested at compound interest over a certain period of years. The table is based on the assumption that if £1 is invested for a given number of years at a specific rate of interest, at the end of the period the investor will receive his original £1 together with the compound interest which has accumulated on it. It is an important table, as it forms the basis for many of the other valuation tables.

This table is constructed in the following way.

	Amount invested at start of year	*Interest payable*	*Amount invested plus interest*	*Amount owed to investor at end of year*
Year 1	1	i	$1 + i$	$= 1 + i$
Year 2	$1 + i$	$(1 + i)i$	$(1 + i) + (1 + i)i$	
			$= 1 + i + i + i^2$	
			$= 1 + 2i + i^2$	$= (1 + i)^2$
Year 3	$1 + 2i + i^2$	$(1 + 2i + i^2)i$	$(1 + 2i + i^2) + (1 + 2i + i^2)i$	
			$= 1 + 2i + i^2 + i + 2i^2 + i^3$	
			$= 1 + 3i + 3i^2 + i^3$	$= (1 + i)^3$

where $i =$ the interest payable on £1.

A relatively simple example will serve to illustrate its application.

Example (*a*). Calculate the amount of £1 in three years at five per cent.

$$i = \frac{R}{100} \text{ where } R \text{ is the rate of interest (per cent)}$$

$$i = \frac{5}{100} = 0.05$$

Amount of £1 $= (1 + i)^n$ where $n =$ number of years for which the sum is invested.

$$\text{Amount of £1 in three years at five per cent} = (1 + 0.05)^3 = 1.05^3 = 1.05 \times 1.05 \times 1.05 = 1.158$$

A much quicker approach, particularly where a large number of years is involved, is to obtain the appropriate multiplier from the amount of £1 table.

Capital	£1.00
Amount of £1 in three years at five per cent	1.158
Capital plus interest	£1.158

Two further examples illustrate practical applications of this particular table.

Example (*b*). To what sum will £1000 accumulate if invested for twenty years at six per cent compound interest?

Capital	£1000
Amount of £1 in twenty years at six per cent	3.207
Accumulated amount	£3207

Example (*c*). An investor pays £30 000 for a building site and it remains undeveloped for five years. Calculate the cost of the land to him at the end of this period, assuming a six per cent rate of interest.

Purchase price	£30 000
Amount of £1 in five years at six per cent	1.338
Total equivalent cost	£40 140

This assumes that if the investor had not purchased the building site, he would have invested the money elsewhere and obtained interest on it.

Present Value of £1 Table

This table shows the capital sum that must be invested now to accumulate to £1 at the end of a certain period at a specific rate of compound interest. Whereas the amount of £1 table commences with a capital sum of £1 and ends with £1 plus compound interest, the present value of £1 table is based on the assumption that if a sum less than £1 is invested for a certain number of years at a given rate of interest, it will accumulate with compound interest to equal £1. Expressed in another way, if a person is going to receive £1 after a given number of years, how much could he sell this for today in order that the purchaser will be able to receive the capital of £1 in the future but without interest? A part of this table is illustrated in appendix 2. Hence

$$\text{the present value of £1} = \frac{1}{\text{amount of £1}}$$

If the present value of £1 is represented by V and and amount of £1 by A, then $V = \dfrac{1}{A}$

As $A = (1 + i)^n$, therefore $V = \dfrac{1}{(1 + i)^n}$

Some examples will help to indicate the practical applications of the present value of £1 table.

Example (*a*). What capital sum must be invested today to accumulate to £10 000 in eight years' time at six per cent compound interest?

Capital in eight years' time	£10 000
Present value (PV) of £1 in eight years at six per cent	0.627
Present value (amount to be invested now)	£6270

Example (*b*). A developer has been given the option to purchase a building site which, it is estimated, will be worth £45 000 in five years' time. The provision of public services will delay its use until that time. What sum could the developer be expected to pay now assuming an interest rate of six per cent?

Estimated value of site in five years' time	£45 000
Present value (PV) of £1 in five years at six per cent	0.7473
Price the developer could be expected to pay now	£33 629
Rounded off to	£33 600

The sum of £33 600 can be described as the *present value* or alternatively the *deferred value* of the £45 000. This method of making allowance for the receipt of a sum at some future date is often described as deferring or discounting the sum. The present value is the reciprocal of the amount computed with the use of the first table.

Example (*c*). A building client expects to have to carry out certain alterations to a building in eight years' time and these are expected to cost £20 000. What sum should he invest now to provide sufficient funds at an operative interest rate of six per cent.

Estimated cost in eight years' time	£20 000
PV of £1 in eight years at six per cent	0.6274
Amount to be invested now	£12 548
Rounded off to	£12 550

Amount of £1 per annum Table

The purpose of this table is to determine the sum to which a series of deposits will accrue, if invested at the end of each year at a specific rate of compound interest. The table is based on the assumption that £1 will be invested *every year* for a given number of years at a certain rate of interest.

The table is constructed on the basis of the following formula

$$\text{Amount of £1 per annum after } n \text{ years} = \frac{(1+i)^n - 1}{i} = \frac{A-1}{i}$$

Where a person borrows a certain sum of money at the end of each year for a given number of years at a specified rate of interest, he may wish to know his total commitment in sums borrowed and accrued compound interest at the end of the period. Some examples will indicate the various uses to which this table can be put; part of an amount of £1 per annum table is illustrated in appendix 3.

Example (*a*). What sum will be obtained if £200 is invested every year for ten years at seven per cent.

Annual sum	£200
Amount of £1 pa for ten years at seven per cent	13.816
Accumulated outlay	£2763

Example (*b*). Owing to unsatisfactory trading conditions an industrial concern ceased production for four years. Throughout this period, repair and maintenance work was carried out and this cost £1500 per annum. Calculate the accumulated sum involved at the end of the four-year period, assuming a compound interest rate of six per cent.

Annual repairs and maintenance	£1500
Amount of £1 pa for four years at six per cent	4.375
Accumulated outlay	£6563

Annual Sinking Fund Table

This table gives the annual deposit which must be invested at the *end of each year* to provide a capital sum of £1 at the end of a certain number of years at a given rate of compound interest; an extract from this table is given in appendix 4. The present value of £1 table is based on the principle that if a sum of less than £1 is invested for a certain number of years at a given rate of interest it will accumulate to £1, whereas the annual sinking fund table is based on the assumption that if an equal amount is invested *every year* at a given rate of compound interest it will accumulate to £1. The table enables the sum to be invested *each* year to be calculated. The basis of the approach is as follows

$$\text{Annual sinking fund} = \frac{1}{\text{amount of £1 per annum}}$$
$$= \frac{i}{(1+i)^n - 1} = \frac{i}{A - 1}$$

This is the reciprocal of the amount of £1 per annum table, and its use is best illustrated by practical examples.

Example (*a*). A building client is expecting to have to replace various engineering services in a building in five years' time. He wishes to know what sum he should invest at the end of each year if the rate of interest obtainable is three per cent and the estimated cost of the replacements is £15 000.

Cost of replacements	£15 000
Sinking fund to replace £1 in five years at three per cent	0.1884
Amount of annual sinking fund	£2826

This total can be checked by using the amount of £1 per annum table.

Annual sinking fund payment	£2826
Amount of £1 pa in five years at three per cent	5.309
Capital sum to be replaced in five years	£15 003

NOTE: The small difference between this total and the cost of replacements stems from the restriction on the number of decimal places.

Example (*b*). A building client will need to replace a building at the end of twenty years, and the replacement cost is estimated at £60 000. Calculate the amount of the annual sinking fund assuming an interest rate of two-and-a-half per cent.

Cost of new building	£60 000
Sinking fund to replace £1 in twenty years at two-and-a-half per cent	0.03914
Amount of annual sinking fund	£2348.4

It is advisable to make a check as before

Annual sinking fund payment	£2348.4
Amount of £1 pa in twenty years at two-and-a-half per cent	25.545
Capital sum to be replaced in twenty years	£59 990

Present Value of £1 per annum or Years' Purchase Table

This table is used to convert a known annual sum which is received or required at the end of *each year* into an equivalent capital sum. It is an extremely useful table in costs in use calculations for obtaining the present lump sum value of annual payments for repairs, cleaning, lighting and heating a building throughout its effective life. Extracts from this table are given in appendix 5, and the basis of computation of the table follows.

$$\text{PV of £1 per annum} = \frac{(1+i)^n - 1}{i(1+i)^n} \text{ or } \frac{A-1}{i \times A}$$

which can be set down as

$$\frac{(1+i)^n}{i(1+i)^n} - \frac{1}{i(1+i)^n}$$

As n (the number of years) approaches perpetuity, the value of the second part of the formula becomes so small that it is really insignificant. Hence the present value of £1 per annum for perpetuity is

$$\frac{(1+i)^n}{i(1+i)^n} = \frac{1}{i}$$

An example will serve to illustrate the use of this table in cost planning.

Example. A building client wishes to compare the cost of purchasing one type of heating installation costing £2500, having a ten-year life, with another type having the same length of life but costing £1300 initially and £180 per annum for servicing and replacement of certain components. It is necessary to calculate the present value of the two installations for comparison purposes. Interest is to be taken at seven per cent.

Installation A			
Total cost over ten-year period			£2500
Installation B			
Initial cost		£1300	
Annual replacements and servicing	£180		
PV of £1 pa for ten years at seven per cent	7.024		
PV of annual costs		£1264	
Total present value			£2564

Installation A shows a small financial advantage over installation B when all costs are reduced to present value.

There are also dual-rate years' purchase tables whereby it is possible to use a lower rate of interest for the sinking fund than for the interest on capital. The annual sinking fund permits the capital to be replaced by the time the capital value of the investment has been dissipated. A normal tax-free sinking fund rate of interest is in the order of two to three per cent as it will be invested in a relatively risk-free asset such as gilt-edged securities. *Parry's Valuation Tables* also contain a series of years' purchase tables with rates of interest varying from four to twenty-five per cent, sinking funds ranging from one to ten per cent, and with the part of the income used to provide the annual sinking fund instalment subject to varying rates of income tax. The figures contained in the years' purchase table in appendix 5 contain no allowance for income tax, and are restricted to a single rate of interest. Hence they provide for a sinking fund to accumulate at the same rate of interest as that which is required on the invested capital, and ignore the effect of income tax on that part of the income used to provide the annual sinking fund instalment.

Costs in Use Calculations

A considerable number of worked examples involving the use of valuation tables are provided in chapter 11. In order to evaluate both initial and future costs, it is necessary to reduce all costs to present values or to annual equivalents and this requires a knowledge of valuation tables and their general application and use.

RENTAL VALUE

Rental value forms the basis for many valuation computations and, for this reason, its main characteristics and underlying influences are now examined.

Rent and its Relationship to Value

The rental value of a property is generally considered to be the value which the average tenant is prepared to pay for its occupation. With properties other than residential the value will be influenced by the profitability of the processes undertaken. Hence, in periods of prosperity rent levels can be expected to rise. Rent is, however, usually associated with leases operating for terms of several years so that there tends to be a timelag between changes in profit levels and the modification of rental values. The term *rack rent* is often used to describe the economic rent of a property or its true rental value, and is the rent which a property should command in the open market.

Factors influencing the Demand for Properties and Rental Value

(1) General prosperity of the country; in times of prosperity demand for properties and their rental values will rise, whereas in periods of recession the opposite will apply.

(2) Movement of population, where areas such as south-east England with increasing opportunities for employment, cause increased demand for properties with enhanced rental values.

(3) Improved transport facilities; for instance, the provision of new roads, underground lines, etc., open up new sites for development.

(4) Changes in character of demand, stemming from improved living and working standards, may create a demand for new buildings and result in a lowering of rental values of older properties.

(5) Rent for commercial, industrial and agricultural properties is a proportion of profit, and so the rent that occupiers are prepared to pay is influenced by anticipated profits.

Determinants of Rental Value

(1) The rent actually paid can form a basis for computation of rental value but it may be less than true rental value for one of three reasons, firstly, changes in rental values since commencement of lease; secondly, consideration may have been paid for lease by way of premium; or thirdly, a personal or business relationship exists between the lessor and lessee.

(2) Comparison of rents paid for similar properties in the district, and making allowance for old leases, special conditions of leases and variations in size, arrangement and condition of the properties. It is advisable to use common units for purpose of comparison, such as the hectare for agricultural land and the square metre of floor area for commercial properties, with 6 m zones for shops.

(3) Assessment on the basis of profits after computation of turnover and gross profit, less outgoings and interest on capital.

(4) Computation on the basis of a certain percentage of the capital outlay.

Rent in Relation to Market Value

A person who purchases property as an investment expects a reasonable annual return for the use of his capital. The annual return or net annual income will consist of the rent received, less outgoings in the form of repairs, management expenses and the like.

Example (a). If an investor expects an eight per cent return on his investment and the net income of the property is £100 pa, then he will be prepared to pay

$$£100 \times \frac{100}{8} = £1250$$

Hence

$$\text{Net income} \times \frac{100}{\text{Rate of interest}} = \text{Market value}$$

This process is described as 'capitalising' the net income and the multiplier is termed 'years' purchase' or YP. Thus net annual income × years' purchase = market value.

Example (b). A owns two sites, No. 15 and No. 185, at either end of a shopping street. On each site he has erected almost identical shops and No. 15 lets at £600 pa and No. 185 at £500 pa, both on full repairing leases. The cost of erection of each shop is £5000. Analyse the rents.

	No. 15	*No. 185*
Rent	£600	£500
less seven-and-a-half per cent return on outlay of £5000 (construction of shop)	375	375
Ground rent	£225	£125

This indicates that No. 15 is in a superior trading position to No. 185.

Example (c). X owns a plot of land which he recently purchased for £2000 and he intends to erect a house on the plot at a cost of £5000. Estimate the true rental value. Assuming a return of six per cent on the land,

$$\text{the rental value} = £2000 \times \frac{6}{100} = £120$$

A suitable return on the building would be seven per cent,

$$\text{and the rental value} = £5000 \times \frac{7}{100} = £350$$

Estimated true rental value £470

The main weakness in this calculation is that it is based on the landlord's expectations and not on what the market will necessarily be prepared to pay.

PREMIUMS

A premium is a sum of money paid by a lessee in consideration of a reduction in rent. The lessee purchases a profit rent and the landlord capitalises part of his future income. The primary advantages to the landlord are that he secures an immediate capital sum with restricted tax liability and the security of his income is increased. This may influence the landlord to grant a longer lease to the tenant and to give more readily any consents required under the lease.

Assessment of Premium

The lessee will not be prepared to pay, in combined premium and rent, a rental equivalent which exceeds the market value. He is entitled to expect a reduction in rent for the premium made up of both the interest foregone on the premium at the full leasehold rate, and a sum sufficient to recover the initial capital outlay by means of a sinking fund over the period of the lease. These two payments together constitute the annual equivalent of the premium.

Example. Y is taking a forty-year lease of a shop worth £1000 pa net and will pay a premium of £2000. What rent should he pay?

Full rental value	£1000
Reduction of rent on account of premium	
Lessee foregoes interest at, say, eight per cent on £2000 and will need to recover £2000 over forty years.	

The annual equivalent is

$$\frac{£2000}{\text{YP – forty years at seven per cent and two per cent (tax 40p)}} = \frac{£2000}{9.549} = 210$$

£790 pa

The landlord would stand to lose about £800 by this arrangement although it is likely that this loss would be more than offset by the tax relief on the premium, which would be available for investment over a forty-year period. The alternative valuations from the landlord's viewpoint follow.

No premium payable			
Rent		£1000	
YP – forty years at seven per cent		13.332	£13 332
Premium payable			
Premium		£2000	
Net income	£790		
YP – forty years at seven per cent	13.332	10 532	£12 532

SERVICE CHARGES

The landlords of blocks of flats and offices have to meet the cost of various outgoings from the gross income in rents obtained from the properties. The outgoings consist of rates (where not paid by tenants), repairs, maintenance of common parts of building and communal services, insurance and management. External repairs and maintenance of common parts of building could amount to ten per cent of gross rents, while internal repairs could account for a further eight to ten per cent. Typical annual service costs are: hot water and central heating – £15 per tap and £20 per radiator; lifts – small passenger-operated electric lift at £250 to £300; electric lighting – £4 per lighting point; porter, including uniform and cleaning materials – £700 to £1000, depending on whether porter is supplied with living accommodation; management – four to five per cent of total rents.

11 COSTS IN USE

WITH MANY PROJECTS cost planning cannot be really effective unless the total costs are considered, embracing both initial and future costs. This chapter examines the concept of costs in use, the various approaches, problems in application and its use in practical situations. A number of related issues such as discounting future payments, lives of buildings, and the relationship of design and maintenance are also considered.

CONCEPT OF COSTS IN USE

Nature of Costs in Use

The Institution of Civil Engineers[1] has emphasised the need to apply economic analysis to engineering projects in order to assess the *real* cost of using resources when establishing priorities between competing proposals. This hypothesis applies equally well to building projects, where the term real costs should embrace the running costs of maintaining and operating a building throughout its effective life. In some cases the appraisal should extend even further to include the benefits accruing to occupants from alternative designs.

The term costs in use is sometimes referred to as *ultimate cost* or *total cost*, a technique of cost prediction by which the initial constructional costs and the annual running costs of a building, or part of a building, can be reduced to a common measure. This is a single sum which is the annual cost in use or the present value of all costs over the life of the building. The technique is employed as a design tool for the comparison of the costs of different designs, materials and constructional techniques. It is a valuable guide to the designer in obtaining value for money for the building client. It can also be used by property managers or developers to compare costs against the value accruing from future rents.

Stone[2] has described how the costs-in-use approach enables the way in which a building functions to be expressed in terms of the costs of renewing and repairing the fabric and fittings, of heating, lighting and fire insurance, and of the labour needed to operate the building. These costs can be added to the amortised initial cost of the building to give the annual cost in use or ultimate cost. The use of this technique thus makes it possible to combine all the costs of the building and so enables the vast range of factors on which judgement is necessary to be reduced to a comparison of a single cost with the personal assessment of the value of the building. Stone[3] has also shown that the cost implications of building designs are often wider than the effect on the initial costs. For some types of buildings the equivalent of first costs is less than the running costs, and small changes in design have a much larger impact on running costs than on first costs.

Cullen[4] also asserts that the essential role of costs in use is to provide a rationale for choice in circumstances where there are alternative means for achieving a given object, and where these alternatives differ not only in their initial costs but also in their subsequent running costs. Cullen quotes a fundamental example of the thermal insulation of walls which is quite straightforward as there is no interference with the various functions of the wall or its appearance, and all that has to be considered is the initial cost and life of the insulation and the value of the expected heat savings. Unfortunately, few building design problems are as simple as this, and the alternatives usually exhibit differences in functional efficiency and aesthetic quality. The construction of factory walls in brick or asbestos cement sheeting is a typical illustration. Initial costs, maintenance costs, cost effects of different rates of heat loss, vulnerability to damage and appearance all have to be considered. Some may argue that aesthetic considerations are purely subjective and as such have no place in a financial appraisal. Yet there can often be financial implications; the improved appearance of a factory could result in more satisfied employees and the higher standing of the firm and its products.

Southwell[5] has described how total building cost refers to all costs and expenses throughout the life of a building, *irrespective of who pays them*, while Codgbrook[6] asserts that in any economic appraisal one should not ignore the inevitable future upkeep costs necessary for a building to perform its complete function. The costs of maintenance, heating, lighting, cleaning, etc. (that is running costs) must affect the true economic worth of a building in use.

Most design decisions affect running costs as well as first costs, and what appears to be a cheaper building may in the longterm be far more expensive than one with much higher initial costs. Some idea of the relationship between initial costs and running costs can be obtained from an examination of table 11.1, from which it will be

TABLE 11.1

BREAKDOWN OF COSTS IN USE FOR VARIOUS TYPES OF BUILDINGS

Type of annual cost	*Houses*	*High flats*	*Industrial buildings*	*Schools*	*Offices*
			Percentages		
By type of expenditure					
Maintenance and decoration	13	13	9	16	13
Fuel and attendance	24	24	50	18	29
Initial costs (amortised)					
(a) Building	48	57	39	50	48
(b) Land and development	15	6	2	16	10
Total costs in use	100	100	100	100	100
By component					
Land and development (amortised)	15	6	2	16	10
Fabric and fittings (excluding heating and lighting installations)	51	61	42	50	50
Heating (including installations)	30	30	46	25	23
Lighting (including installations)	4	3	10	9	17
Total costs in use	100	100	100	100	100

Source: P. A. Stone, *The Economics of Building Designs*[3]

seen that running costs often amount to about two-thirds of the annual equivalent of first costs, and for factory buildings can be as much as one-and-a-half times the equivalent. The proportions vary widely from one building to another.

Land costs are relatively low with industrial buildings which are erected on lower priced land with a higher utilisation factor, whereas houses and schools have high land costs being built to a low density on relatively highly priced land. Fabric and fittings represent the highest proportion of total cost with flats, and the least with industrial buildings. Heating costs of industrial buildings are high because of heavy heat losses and long periods of use, whereas offices and schools rank low with improved thermal insulation and reduced operational periods. Lighting of offices is an amazingly high cost item stemming from the high standards of illumination that are required. On maintenance and decoration schools rank high with extremely heavy wear-and-tear, while factories rank low on account of the low standards usually prevailing.

The relative importance of first and running costs respectively is influenced considerably by the financial interests of the building client. A developer building houses for sale will not usually consider running costs unless they affect the selling price. An industrialist will almost certainly be influenced by the greater tax savings obtainable for running costs as compared with first costs. An owner-occupier, on the other hand, will be concerned with the total effect of the design upon the costs of owning and operating the building. Stone[3] has also indicated that there are a number of effects of any particular design feature which cannot be costed directly, although their influence on costs may be considerable. Such factors as appearance, noise level and the level of illumination all have an influence on output, but it is difficult to evaluate their effect other than indirectly using a cost-benefit approach as described in chapter 14.

Volume and Impact of Building Maintenance Work

The significance of building maintenance work in the national economy is apparent from the statistics produced in table 11.2. By 1971 it was estimated that the annual building maintenance bill in Great Britain had reached £1800 million.

TABLE 11.2

APPROXIMATE ANNUAL EXPENDITURE ON MAINTENANCE WORK IN GREAT BRITAIN IN 1968

Construction maintenance including minor improvements and alterations.	£1190m
Estimated value of do-it-yourself construction maintenance, improvements and alterations.	£ 200m
Estimated value of construction maintenance carried out by directly employed labour forces within industry.	£ 200m
Estimated expenditure on general engineering maintenance of production plant and equipment by manufacturing industry.	£1100m
NOTE: New construction by contractors.	£3090m

Source: B. E. Drake, *The economics of maintenance*[7]

Speight[8] has described how a greater awareness has developed of the need to reduce the cost of maintaining buildings. The annual expenditure on construction maintenance is equivalent to the first cost of about 1000 teacher training colleges, and a saving of one per cent on maintenance would each year equal the capital cost of something like 2000 grammar school places. Hence it is vitally important that the probable maintenance and running costs of a building should be considered at the design stage, and due attention directed towards the maintenance implications of alternative designs. A reduction in initial constructional costs often leads to higher maintenance and running costs.

About forty per cent of the present building labour force is engaged on maintenance work covering about twenty-five per cent of the value of all building work. On present trends, and taking into account the fact that insufficient maintenance work is currently being undertaken, it seems feasible that the proportion of the building labour force employed on maintenance work could increase to sixty per cent by the year 2000. This imbalance must to some extent stem from the failure of earlier design teams to take full account of future costs.

The magnitude and seriousness of the problem resulted in a former Minister of Public Building and Works setting up a committee on building maintenance in 1965. The terms of reference of the committee were to keep the problem of maintenance under continuous review, concentrating particularly on the relationship between design and maintenance, the dissemination of information, and the establishment of priorities for research and development.[9] The Department of the Environment entered into arrangements with a number of universities and polytechnics in 1971 to undertake research projects into costs in use and performance standards, which is indicative of the interest and concern of both the Government and higher education establishments in this field. The building maintenance cost information service (BMCIS) commenced operation in 1971 with the support of the University of Bath, the Department of the Environment and the Royal Institution of Chartered Surveyors. This service has the stated objective of helping its subscribers to derive value for money from efficient property occupancy and economic building maintenance; further details about the service are given at the end of this chapter.

Maintenance is therefore an important part of construction but has tended to lag behind new constructional work as regards modernisation and mechanisation. While output per man in the construction industry as a whole rose steadily in the nineteen-sixties at between three and four per cent annually, there is little evidence of any significant increase in output per man on maintenance work. The labour-intensive nature of this section of work is largely inevitable, as many maintenance jobs can only be performed manually and demand is often dispersed and unco-ordinated. Building firms engaged on maintenance work are generally small in size and frequently under-capitalised. Small contractors have often been reluctant to invest in labour-saving small power tools, which could produce considerable reductions in the man-hours needed for many maintenance tasks.

Value for Money

Full consideration of maintenance aspects and possible future costs at the design stage is likely to result in the building client securing better value for his money. Maintenance work often falls into two categories – the necessary and the avoidable. Avoidable maintenance work may result from faulty design or poor workmanship. The Building Research Station[10] has investigated the maintenance aspects of local authority housing and found that the equivalent capital costs of maintenance compared with initial costs is relatively low, except for water services (eighty-four per cent) and external painting (100 per cent). Maintenance costs of water services could be reduced substantially by protecting the services against frost damage and electrolytic action and external painting costs could be reduced by the substitution of materials which do not require painting.

A large proportion of the annual charges on a building is attributable to heating and lighting, and these costs are influenced considerably by the planning and structure of the building. For example, the heat loss from a compact single-storey school with a storey height of about 2.40 m could be about thirty per cent less than from one with an irregular layout and a storey height of around 3.40 m. Furthermore, the initial costs of the building and the heating plant are likely to be about twenty per cent less.

The heating and ventilating costs of factories often amount to between one-third and two-thirds of the combined total of the running costs of the building and the annual equivalent of its initial costs, a proportion sufficiently large to justify close consideration. Roof design is an important factor and a flat roof is usually the cheapest solution from the heating viewpoint, resulting in reduced heat losses and less air to heat. The standard of insulation and amount of glazing must also be considered. It is the total costs that must be considered and the effect of each design decision should be studied with this in mind. For instance, the area of glazing affects structural, heating and lighting costs and it is advisable to adopt a form of cost analysis which shows the effect of each design on all three components. One approach is to convert the costs of each component into annual charges made up of annual equivalents of initial costs plus annual maintenance costs.

Difficulties in Assessing Costs in Use

There has been some reluctance on the part of many quantity surveyors to include running costs in their cost planning calculations because of the inherent difficulties in implementation. The major difficulties are now listed.

(1) The difficulty of accurately assessing the maintenance and running costs of different materials, processes and systems. There is a great scarcity of reliable historical cost data and predicting the lives of materials and components is often fraught with dangers. In these circumstances the quantity surveyor may be compelled to rely on his own knowledge of the material or component, or possibly on manufacturer's data in the case of relatively new products. Even the lives of commonly used materials like paint show surprising variations and are influenced by a whole range of factors, such as type of paint, number of coats, condition of base and extent of preparation, degree of exposure and atmospheric conditions.

(2) There are three types of payments – initial, annual and periodic, and these all have to be related to a common basis for comparison purposes. This requires a knowledge of discounted cash techniques, described later in this chapter.

(3) Income tax has a bearing on maintenance costs and needs consideration, as it can reduce the impact of maintenance costs. Taxation rates and allowances are subject to considerable variation over the life of the building.

(4) The selection of suitable interest rates for calculations involving periods of up to sixty years is extremely difficult.

(5) Inflationary tendencies may not affect all costs in a uniform manner, thus distorting significantly the results of costs in use calculations.

(6) Where projects are to be sold as an investment on completion, the building client may show little interest in securing savings in maintenance and running costs.

(7) Where the initial funds available to the building client are severely restricted, or his interest in the project is of quite short-term duration, it is of little consequence to him to be told that he can save large sums in the future by spending more on the initial construction.

COSTS IN USE TERMINOLOGY

It is considered advisable to define the costs in use terms currently in use, as problems have arisen in practice due to the varying interpretation of some of the terms. For instance some surveyors have understood the term costs in use to mean the cost of using the building, rather than the intended meaning of total costs or ultimate costs, made up of both initial constructional costs and all running costs, including maintenance costs. Southwell[11] has expressed the view that the construction field is particularly prone to loose interpretations, the term *maintenance* ranging from 'all construction work other than new capital works' to 'all work undertaken to keep a building in a satisfactory state of preservation'. Table 11.3 shows a breakdown of total costs with the object of isolating the significant aspects of futurc costs in the design process and thus assisting in the search for suitable expressions.

TABLE 11.3

BREAKDOWN OF TOTAL COSTS

Total costs (life span of building)		
Initial costs	*User costs*	
	Running costs	*Occupational charges*
Land Construction Professional fees	Maintenance Operating services (operating and cleaning) Fuel and power	Rates Insurance Modifications and alterations Estate control (management)

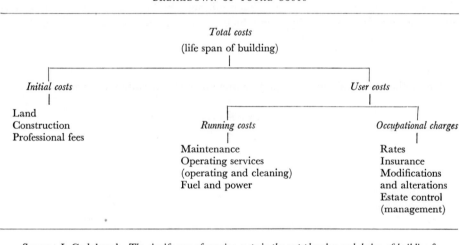

Source: J. Codgbrook, *The significance of running costs in the cost planning and design of buildings*[6]

Initial costs. The capital or initial expenditure on an asset when first provided.

User costs. These are synonymous with *future costs* and comprise both running costs and occupational charges.

Maintenance. This is defined in BS 3811 as 'work undertaken in order to keep or restore every facility, that is every part of a site, building and contents, to an acceptable standard'.

Operating services. These embrace cleaning, caretaking, operation of plant and equipment and other allied activities.

Fuel and power. These represent the energy costs for heating, lighting, air conditioning and the like.

Modifications and alterations. These are new works required to improve or adapt an asset.

THE TECHNOLOGY OF MAINTENANCE

White[12] sees maintenance as seeking to preserve a building in its initial state so that it continues to serve its purpose and is an essential component in the life cycle of a building. Yet there exists a general financial climate where minimum first costs are often the only consideration, risking future maintenance problems. Cost assessment formulae and yardsticks rarely take account of costs in use implications, and capital expenditure is all too frequently divorced from current spending. Benroy[13] asserts that a designer of buildings could be failing in his function as adviser if he does not understand the problems connected with the maintenance and running costs of buildings and fails to apply this knowledge at the design stage. He sees merit in the preparation of standard details and, where different qualities of material are involved, listing each with their effect on both capital and maintenance costs for the guidance of the client.

Robertson[14] believes that engineering or building work will be a discredit to the designer if its usefulness and convenience is permitted to fall below an acceptable standard. In his definition of acceptable standards he recognises three separate categories which serve to illustrate some of the complexities of maintenance.

(1) Functional performance, quality and reliability, which relate to user needs.

(2) Structural, electrical, fire and other safety aspects, for which maintenance personnel are generally responsible.

(3) The preservation of the asset and its amenities, in which the owner has the primary interest.

Expressed in another way the prime benefits of maintaining a building are to retain the value of the investment; to maintain the building in a condition in which it continues to satisfactorily fulfil its function; and to present a good appearance to the public.[15]

Types of Maintenance

Table 11.4 shows a method of classifying the main arrangements for maintenance work.

TABLE 11.4

TYPES OF MAINTENANCE

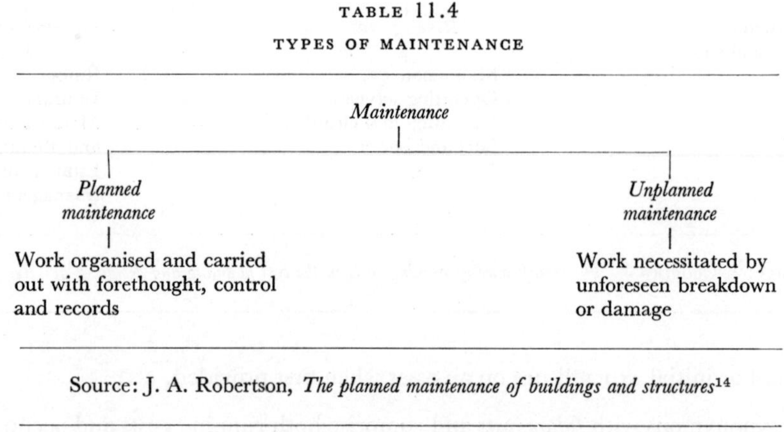
Maintenance
Planned maintenance
Unplanned maintenance
Work organised and carried out with forethought, control and records
Work necessitated by unforeseen breakdown or damage

Source: J. A. Robertson, *The planned maintenance of buildings and structures*[14]

Planned maintenance is also divided into three classes by BS 3811.

(1) Planned preventative running maintenance: work which can be done while the facility is in service.

(2) Planned preventative or corrective shut-down maintenance: work which can only be done when the facility is, or is taken, out of service.

(3) Planned corrective breakdown maintenance: work which is carried out after a failure, but for which advance provision has been made in the form of spares, materials, labour and equipment.

Chaplin[16] has described how

> 'the predominant characteristic of maintenance is the variety of factors that affect its incidence. These range from the initial design and cost involving the quality of materials and workmanship, the intensity of exposure, to the efficiency of the maintenance organisation. Their interaction directly affects the durability of the buildings and their components and the resultant maintenance work. The control of maintenance, if it is to be effective, should therefore commence at the time the building is designed and continue throughout its life.'

Origins of Maintenance Problems

Maintenance work is generated by a whole range of factors including weathering, corrosion, dirt, structural and thermal movement, wear, low initial expenditure, passage of time, incorrect specification, inferior design, poor detailing and damage by users. James[17] has also highlighted the maintenance problems stemming from the use of new materials and techniques. These require careful appraisal prior to use, but processes and products which satisfactorily withstand laboratory tests may not always be suitable in live situations. Nevertheless, the pace of technological development demands the use of such processes and products before their suitability over long periods has been adequately tested. The work of the Agrément Board will, however, help designers in their selection.

Some of the principal maintenance problems will now be considered on the basis of specific elements, components or services.

Concrete. Gray and Ransome[18] assert that good concrete with adequate cover to the reinforcement should be satisfactory for at least fifty years in the United Kingdom climate and is unlikely to be attacked by frost except in copings. However, sea water and sea spray can cause chemical attack on concrete as well as corrosion of reinforcement, and sulphate-bearing soils are a potential source of danger to concrete foundations. In these situations it is advisable to use special cements in well-designed and compacted concrete.

Brickwork. In this country frost is the chief cause of the decay of brickwork but this only occurs where bricks are frozen when saturated, principally in parapets, retaining walls and free-standing walls. Speight[8] has also drawn attention to the tendency to omit water shedding drips and projections, resulting in unsightly surface staining and maintenance problems.

Masonry. In the United Kingdom most sandstones are immune to frost attack and limestones are rarely affected except in exposed positions such as cornices, string courses and copings to parapets. The chief cause of decay is attack by atmospheric sulphur gases. Waud[19] has drawn attention to the need to avoid the use of metals, such as bronze, copper and aluminium in proximity to marble and polished granite because of the danger of permanent staining of the masonry.

Cladding panels. The use in modern building of prefabricated units of increasing size and complexity and framed structures with infill or cladding panels produces joints subject to relatively large movement. This has created several problems which are the subject of extensive research at the Building Research Station.[20] Information is required on the satisfactory design of joints, particularly between dissimilar components, tolerance of dimensions in manufacture and changing dimensions of joints with change in temperature and wetness. It is vital that the various sealants, mastics, putties and gaskets used for jointing shall be satisfactory under all conditions.[21]

Roofs. Speight[8] has described how frequently otherwise durable buildings have roofs of less permanence and that when preliminary estimates are reduced it is often roofs that suffer first. Flat roofs often produce serious maintenance problems due to lack of or haphazard falls, disregard of codes of practice, inadequate eaves, verge or fascia details, sharp granite chippings puncturing roofing felt with possibly sodden strawboard below supported on untreated and unventilated timber, inadequate thermal insulation, heavy condensation on underside of roof slabs, ill-conceived gutters, lack of walkways, etc. Walker[22] has given further emphasis to the high maintenance costs resulting from poor roof design and construction. His examples included moisture trapped in lightweight roof screeds between the asphalt and the structural roof, and asphalt melting a polystyrene underlay.

Timber. One of the main problems with timber is caused by shrinkage which occurs because the moisture content of the wood at the time of erection was too high and it has dried out subsequently. The resultant change in moisture content can cause distortion and deflection. Unprotected external timber looses its colour and the surface becomes rough and difficult to clean. Even western red cedar, a durable species, looses durability after prolonged leaching. Fungal attack can only take place if the wood attains a moisture content of twenty per cent or more and remains wet over an appreciable period of time. There is evidence to suggest that in houses over twenty years old the incidence of some measure of attack by insects or fungi or both is greater than sixty per cent, although in some cases the extent of infection is very restricted. Indications are that the total annual bill to the country in making good and eradicating troubles stemming from insect and fungal attack is probably between £25 million and £30 million.[23]

Gibson[23] postulates that many wooden casement joints are not strong enough to withstand the racking strains imposed by wind loadings on open windows, resulting in cracking of paint films at joints which permit the entry and trapping of moisture under the paint, and that there is often excessive shrinkage on glazing putty. He deprecates the common practice of storing timber for considerable periods on building sites in unsatisfactory situations prior to use, and the use of central heating for the rapid drying out of a building after the installation of interior joinery. Speight[8] has also commented on ill-fitting windows with frames of too light section.

Metals. In most environments steel corrodes at between 0.05 and 0.10 mm per year when freely exposed in air. In soils the corrosion rate is similar, but pitting often takes place and at localised points much higher corrosion rates may occur. Steel reinforcement in concrete should have a minimum cover of 50 mm. Copper requires little or no maintenance as it develops a green protective patina after a few years. Corrosion of aluminium and zinc roof coverings can take place around chimneys which emit soot particles. Unanodised aluminium and aluminium alloys are satisfactory provided the environment is not exceptionally corrosive.

The life of galvanised steel coldwater cisterns depends on the nature of the water and can vary from five to thirty years. In view of the difficulty of replacing them and the damage that can occur as a result of leakage, it is advisable to paint them with two coats of suitable bituminous paint before installation. In districts where the water is known to be aggressive towards galvanised steel cisterns it is desirable also to fit a sacrificial magnesium anode to provide cathodic protection and to encourage protective carbonate scale deposition on any exposed parts of the metal surface.[24]

Plastics. There is considerable confidence that suitably formulated phenolic resins, rigid polyvinylchloride (PVC), polymethyl methacrylate, glass fibre reinforced polyester, polyisobutylene and polychloroprene can withstand the effects of external weathering for periods of up to thirty years without maintenance.[18]

Painting. Paint is normally applied in thickness little more than 0.125 mm, and often in even thinner layers, yet a great deal is expected from the coating. The selection of the most economical time to repaint, the type of paint to use and the thickness to apply are not easy as most jobs have their own peculiar features. Bullett[25] believes that there is scope and need for careful observation and collection of data on large numbers of buildings so that evidence can be accumulated to form the basis of a rational study of the problem.

Gray and Ransom[18] consider that paint is the most vulnerable of building materials and its regular replacement constitutes the largest maintenance item resulting from weathering. Although the life of paint films has increased by twenty per cent over the last fifty years, there is considerable scope for improving painting maintenance by imposing standards of good practice. The Greater London Council, probably the leading authority in this field, has found that despite its known insistence on proper standards, seven per cent of unopened cans fail their standard and no less than one in three samples from painters' kettles on site are rejected, mainly because of unauthorised addition of thinners. The Council also found frequent omission of an undercoat from a four-coat paint scheme.[26] If so much unsatisfactory painting is found by an authority known to impose rigid standards one wonders whether a satisfactory job is ever obtained on the average uncontrolled building site. If the GLC methods of control could be imposed over the whole country the contribution to reducing building maintenance expenditure would probably be greater than any other single measure.

Services. A decision has to be made whether to expose, bury or duct services in a building. The first gives ease of maintenance and alteration but produces cleaning problems, the second seems to solve all problems until failure occurs or a major rearrangement becomes necessary, while the third is expensive, space-consuming and demands good access for inspection, day-to-day use of valves and switchgear and replacement or alteration.

Bennett[27] has described how mechanical plant produces two main problems: firstly, the extent to which a number of standard units may be used for a particular purpose, for example, the use of small fans for ventilation rather than one or more larger but less standard units; and secondly, the relationship between final cost and maintenance cost of individual items and manufacturers' standard goods.

Where appropriate, tenders for such items as lifts, escalators, boilers, etc. should include the necessary maintenance costs so that these may be evaluated at the time of ordering. This is particularly desirable for buildings where no engineering maintenance staff are available on the site.

Cleaning. Waud[19] has described how the most important and expensive single item relating to any building, after the rent and rates have been paid, is the cost of internal and external cleaning. In thirty years, or even less in some instances, the cleaning costs will have risen above the original cost of the building. It therefore follows that if buildings are designed with a view to reducing cleaning costs, the savings to the occupant can be considerable. Furthermore, it is possible for two similar buildings, of say 5000 sq m floor area each, to vary from £4500 to £10 000 per annum for internal and external cleaning.

The process of cleaning uses increasingly more sophisticated engineering equipment and is often based on complicated service schedules. The frequency of cleaning varies: floors are generally swept daily and polished weekly, windows washed monthly and flues swept every six months. Service schedules may also embrace painting for decoration and protection, cleaning of gutters and drains, and servicing lifts and central heating plant. The cost of maintaining floors varies from 18 p to 45 p per square metre (1971 prices).

Toilets require cleaning every day and so the surfaces should be specified for easy cleaning. Toilet pans should be placed on walls, toilet doors should not come down to floor level, wall tiling should extend to ceiling level, tops of skirtings and corners should be rounded, and sanitary disposal units should be provided in ladies toilets. Public entrance halls and common staircases should be surfaced with durable and easily cleansed materials. Ample storage space for cleaning equipment, water supply and sink should be provided on each floor. With tall buildings, the design should permit easy cleaning of all glazed areas.

Execution of Maintenance Work

Hill[28] has described comprehensively suitable arrangements for the execution of maintenance work. Full records should be kept of each property stating the location, age, condition, constructional details of each element, records of services, floor area and cubic content, accommodation provided, current use and any relevant planning decisions. Small organisations would probably use card records and large organisations will make use of computers. It is good policy to require contractors on new projects to supply maintenance manuals giving a physical record of each building, inspection and maintenance cycles for the various elements, list of specialist subcontractors and suppliers and information and instructions on maintenance for occupants.

Inspection cycles are a vital aspect of any maintenance system. Suppliers of services will normally prescribe inspection cycles for the plant they have provided and the fabric of a building should also be inspected at regular intervals, preferably related to the endurance period of a significant component or material. External painting is generally undertaken at five-year intervals and prior to this it is good policy to carry out a thorough inspection of the property to determine the extent of necessary repair work and its probable cost. Interim inspections should be carried out more frequently, possibly at twelve-monthly intervals to detect defects which would result in progressive deterioration if left unattended until the next cyclic inspection. It is advantageous to use standard report sheets to ensure uniformity of approach. Speight[34] sees the principal challenge in devising a maintenance programme as achieving the right balance between the cost of check inspections and the resultant benefits. Gutter clearance, drain rodding and checking flat roofs should always be programmed.

A yearly budget prevents adequate forward planning, and ideally the budget period should extend over a number of years, possibly matching the maintenance programme based on a five-year external painting cycle. This would provide a framework for an efficient maintenance system based on detailed cyclic inspections, accurate estimates and a firm long-term budget which would enable work to be properly planned and executed. Maintenance can be performed either by directly employed labour or by private contractors. A direct labour system permits full control of operatives but entails the provision of supporting facilities such as workshops, stores, transport, etc., and a high standard of supervision and control. Macey[29] has described the organisation for controlling the 250 000 houses owned by the Greater London Council, embracing ten district offices, mobile trade gangs with seven persons in each gang and covering 6000 to 8000 properties and painting gangs of twelve painters dealing with 2000 dwellings per gang.

There are various contractual arrangements which can be used for maintenance work undertaken by private contractors and these are fully described in *Contractual Policies and Techniques for Maintenance Work*.[30] The most common methods are:

(1) payment on the basis of time expended and materials used, plus agreed percentage additions for overheads and profit;
(2) payment on the basis of agreed measurements and an agreed schedule of prices; and
(3) a lump sum offer based on specification, bill of quantities and/or drawings.

A MPBW publication[31] reported that very few firms apply incentives to building maintenance and relatively few of these are successful. Nevertheless, it was claimed that incentive schemes could be applied to maintenance work for which an estimate had been prepared at fairly low cost and could result in increases in output ranging up to fifty per cent as well as providing a means of labour control and a check on profitability. Kenyon[32] has described how incentive bonus schemes are applied to direct labour housing maintenance work for Middlesbrough Corporation, with work values built up from time studies and making due allowance for attendance to personal needs, recovery from fatigue according to the arduousness of the work done (resting time) and travelling time. A daily allowance was also calculated to cover the drawing of stores, receiving instructions and obtaining tools. External painting work has been measured and evaluated, and the operation of financial incentives has resulted in a forty per cent increase in productivity and twelve per cent reduction in costs.

Stone[33] has analysed the probable effects of metrication on maintenance work and believes that the materials and components most likely to give rise to difficulties after all products have been metricated are wall, floor and ceiling tiles, wood floor blocks, partition units, standard doors and windows, standard joinery fittings, sanitary appliances in ranges, radiators, some engineering equipment and electrical fittings. For example, connectors will be needed to join imperial and metric pipes, tiles and paving slabs are slightly smaller in metricated sizes, a metric door is 12 mm narrower than the old 2 ft 6 in door, and metric standard joinery fittings are wider and longer but slightly lower. Whilst on the subject of metrication, readers may find the metric conversion table in appendix 6 useful.

CURRENT AND FUTURE PAYMENTS

One of the principal difficulties in making costs in use calculations is that every building project involves streams of payments over a long period of time (usually the life of the building). The payments are of three main types:

(1) present payments covering the building site, erection of the building and professional fees;
(2) annual payments relating to minor repairs, cleaning, heating, lighting, rents, etc.; and
(3) periodic payments such as external painting at five-year intervals and replacement of services or other parts of the building.

All these varying types of payments have to be converted to a common method of expression to permit a meaningful comparison to be made between alternative designs. The process is often referred to as discounting future costs and is based on the premise that if the money were not spent on the project in question it could be invested elsewhere and would be earning interest. £100 invested today at five per cent compound interest will accumulate to £162.88 after 10 years (see appendix 1 – amount of £1 table). In the reverse direction the present value of £1 table (appendix 2) shows that it will be necessary to invest £61.391 now at five per cent compound interest to accrue to £100 in ten years time. We are often concerned with annual payments throughout the life of a building, which is commonly taken as sixty years. The present value of £1 per annum table (appendix 5) shows that an expenditure of £1 per annum throughout the sixty-year period is equivalent to a single payment of £18.9292 today taking an interest rate of five per cent. Expressed in another way, if £18.9292 were invested today at five per cent compound interest it would provide sufficient funds to be able to pay out £1 per annum for each of the sixty years. Hence it is sometimes said that it is worth spending up to an extra £20 today on initial construction if this will reduce the expenditure on maintenance work by £1 per annum throughout the sixty-year life of the building.

There are two possible approaches in making costs in use calculations and both will be illustrated in the worked examples used later in the chapter.

The first one is to discount all future costs at an appropriate rate of interest, often taken at five or six per cent (long-term pure interest rate with no allowance for risk premium), and so to convert all payments to present value (PV) or present worth, using the valuation tables described in chapter 10.

The second one is to express all costs in the form of annual equivalents, taking into account the interest rate and annual sinking fund. A building client is entitled to interest on the capital he has invested in the project and requires a sinking fund to replace the capital when the life of the building has expired.

An example may serve to illustrate the discounting principle. A building is designed to last sixty years and can either be provided with a roof costing £500 which will last thirty years and then need replacing, or be covered with a roof costing £750 which will last the life of the building. It is necessary to determine which is the better proposition financially and to do this the payment in thirty years time has to be converted to its present value. The present value of a payment of £500 in thirty years time is found from the present value of £1 table (appendix 2) and is £500 × 0.23137 = £115.685, taking an interest rate of five per cent. The calculations can be summarised as follows:

	Cost	*Present value*
Roof A		
Initial construction	£ 500	£500
Replacement after thirty years	£ 500	£115.69
Total cost	£1000	£615.69
Roof B		
Initial construction	£ 750	£750

These calculations show roof A to be the better long-term proposition. It might be argued that is is over-simplified as, for instance, it takes no account of the cost of demolishing the old roof or any temporary work that may be necessary to protect the occupants and contents of the building in the case of roof A.

The selection of a suitable discounting interest rate is also extremely difficult. Interest rates vary considerably over time and we are concerned with exceptionally long periods. Wilson[35] points out that there are three ways of assessing the interest rate to be used, as shown below.

(1) The social time preference rate. This is a positive rate of interest which expresses the value persons place on having assets now, rather than at some time in the future and, adopting the kind of life tables used by insurance companies, a social time preference rate could be as low as two per cent.

(2) The rate of interest at which the Government lends and borrows, and is roughly the risk free rate of interest, possibly about four to six per cent.

(3) The opportunity cost rate of interest. This is the rate of interest which could operate if the project being evaluated were not carried out, and so freed the capital for an alternative opportunity, possibly in the range of seven to twelve per cent.

Building clients must either borrow money to finance the project or sacrifice an alternative use for their money. A realistic rate of interest is therefore either the market rate for money borrowed on the security of the building, or the average return which the building client can secure for money invested in his own business. Stone[3] asserts that these rates will be considerably lower than the rate of interest often assumed when predicting the return on a business investment. The actual rate of interest to be used for cost prediction purposes will depend on such factors as the financial standing of the client and on predictions of the long-term movement of rates of interest. The worked examples that follow will be based on a six per cent rate of interest as it seems likely that we are moving into a period of permanently high interest rates.

MAINTENANCE AND RUNNING COSTS

In 1970, The Royal Institute of British Architects drew attention to the need to balance capital costs against subsequent maintenance and running costs. Their report described how economies in finishes today will undoubtedly lead to inflated maintenance costs in future and that skimping external works or attempting to increase densities at the expense of environmental facilities, such as play areas, may well reduce the acceptable life of a housing estate. These aspects were further highlighted at a conference on cost control of hospitals[37] where it was shown that the costs of servicing a hospital may be six times greater than the building costs.

Housing

In the year 1966/67 the total maintenance costs on Greater London Council houses were £8½ million, which amounted to £38 per dwelling per annum. The most expensive maintenance items were internal decorations – £2.5 million, external decorations – £1.1 million, structural repairs – £0.6 million, structural fixings – £0.8 million, and sanitary services – £0.9 million. Table 11.5 shows the relationship of the major items of maintenance costs to their initial capital costs.

A survey of maintenance work on local authority houses revealed the distribution of costs shown in table 11.6. The greatest economies in maintenance can be obtained by concentrating on the most costly items, that is decorations and plumbing, by reducing the amount of external paintwork, and designing better plumbing systems well-protected from frost, aggressive water and electrolytic action. Another analysis of housing maintenance

TABLE 11.5

LOCAL AUTHORITY TRADITIONAL HOUSING:
CAPITALISED MAINTENANCE COSTS AS A PERCENTAGE OF INITIAL COSTS

Item	*Percentage*
Water services	84
Sanitary appliances	22
Heating, cooking and lighting	33
Internal structure and finishes	7
Main structure	5
External services and site works	20
External painting	100

Source: P. A. Stone, *Design economics – building costs today and tomorrow*[2]

TABLE 11.6

SOURCES AND CAUSES OF TYPICAL LOCAL AUTHORITY HOUSING MAINTENANCE COSTS

Item	*Percentage of total maintenance costs*	*Main causes*
Structural and cladding repairs	10	Roofs, windows and external doors
External redecoration	25	Protection of wood and metal-work
Internal repairs and renovations	25	Redecorations and minor repairs
Services, installations and sanitation	40	Ball valves, tanks, cylinders, burst and blocked pipes

work carried out by a large local authority revealed that 12.5 per cent of the work resulted from faulty materials and workmanship and twenty per cent from design or specification faults. Furthermore, the cost of maintaining pre-war houses was roughly double that of post-war houses, and that of pre-1930 houses could be two to three times as high as that of post-war houses. In general, maintenance expenditure increases with the age of the house at a rate of about £0.40 to £0.50 per annum.[38]

Comparison of Different Buildings

The relative importance of the maintenance expenditure incurred on different parts of buildings does however vary considerably with the type of building as shown in table 11.7. Roof repairs account for between one-quarter and one-half of the expenditure on *structure*; much of that on *partitions* is spent on doors; and heating services account for between one-quarter and one-half of the *services* costs. The higher proportion of the cost on services in factories and hospitals is mainly due to the greater complexity of service arrangements in these buildings.

The annual equivalent costs in use for a wide range of buildings are given in table 11.8 to show the variations in unit costs and between the three cost heads of construction, maintenance, and fuel and attendance.

The floor area per unit varies considerably between the different buildings as does also the constructional

TABLE 11.7

DISTRIBUTION OF MAINTENANCE EXPENDITURE BETWEEN DIFFERENT BUILDINGS

Building element	*Percentage of annual maintenance expenditure*				
	Houses	*Factories*	*Schools*	*Hospitals*	*Average*
Structure	15	28	12	10	17
Partitions	14	–	4	6	11
Decorations	36	29	52	29	31
Fittings	–	–	3	7	5
Services	30	43	25	41	31
Other	5	–	4	7	5
Total	100	100	100	100	100

Source: *Ministry of Public Building and Works, Conference on maintenance of buildings*[39]

TABLE 11.8

ANNUAL EQUIVALENT COSTS IN USE FOR TYPICAL BUILDINGS EACH PROVIDING 100 UNITS OF ACCOMMODATION

Type of building	*Offices*	*High flats*	*Houses*	*Light industrial buildings*	*Hospitals*	*Secondary schools*
Type of unit	*Work spaces* £	*Bed spaces* £	*Bed spaces* £	*Work spaces* £	*Bed spaces* £	*Pupil spaces* £
Construction	2640	4750	2900	4490	29 600	1800
Maintenance	700	1000	800	1250	8800	600
Fuel and attendance	1650	2000	1450	2250	7000	700
Total	4990	7750	5150	7990	45 400	3100

Source: B. E. Drake, *The economics of maintenance*[7]

costs per square metre of floor area. The ratio of running costs to contruction costs ranges from about eighty per cent with houses and light industrial buildings to ninety per cent with offices.

The annual cost of decorating a typical factory of 9300 m² floor area is likely to be around £1700 (1971 prices), maintenance of the fabric about £1900 to £2000, cleaning about £650, upkeep of heating installations, £800, and other services and fittings, £1600.[40]

Schools

A study of the annual maintenance costs of schools showed that prewar schools (average age thirty years) cost double those of post-war schools (average age five years).[41] Table 11.9 shows the distribution of school maintenance costs over the main elements.

Offices

The annual running costs of a conventional four-storey office building per square metre of floor area in 1971 were likely to be in the following order

Maintenance	£1.15
Operating services	1.60
Fuel and power	1.15
Total	£3.90

In studies of Crown office buildings[42] total costs in use comprise, at an interest rate of ten per cent, approximately two-thirds capital or initial cost and one-third running and maintenance costs; the one-third sector consists of three approximately equal parts: fuel, electricity and gas; cleaning; and repairs, redecorations and minor new works. The proportion of initial cost to the remainder is influenced considerably by the rate of interest used to amortise the initial cost for conversion to annual equivalent values. High maintenance costs can arise from low first costs coupled with inadequate specifications, or through complex services and elaborate finishes. It is interesting to note that in the study of Crown buildings the costs were based on internal redecoration at eight-year intervals and washing down at the third and sixth years; external redecoration at six-year intervals; and

TABLE 11.9

DISTRIBUTION OF SCHOOL MAINTENANCE COSTS
(average costs/93 sq m pa – 1965 prices)

Element	*Pre-war schools: age thirty years*		*Post-war schools: age five years*	
	Primary £	*Secondary* £	*Primary* £	*Secondary* £
Main structure	8.81	5.89	4.76	1.57
Internal construction	1.27	1.37	1.00	1.00
Finishes and fittings	5.51	2.26	1.34	3.50
Plumbing	5.34	2.16	1.39	0.84
Mechanical services	8.77	5.28	3.77	2.33
Electrical services	0.48	3.07	0.47	1.19

Main structure: Walls, roofs, chimneys, rainwater goods, windows and external doors, including glazing.
Internal construction: Floors, stairs, partitions, internal doors and screens, including glazing.
Finishes and fittings: Wall, floor and ceiling finishes, ironmongery, shelves, built-in furniture, and miscellaneous carpentry.
Plumbing: Pipes, taps, ball valves, tanks and cisterns, sanitary appliances.
Mechanical services: Boilers, flues, steam and hot water distribution, air-conditioning, ventilation, gas installations.
Electrical services: Electrical wiring, switch and control gear, appliances and fittings.

Source: M. A. Clapp and B. D. Cullen, *The maintenance and running costs of school building*[41]

replacement of bitumen felt roofing after twenty years, boilers after twenty-five, internal pipework, thirty, storage tanks, twenty-five, electrical installation wiring, fifteen, distribution switchgear, twenty-five and passenger lifts, twenty years. Whilst for houses, Knight[43] suggested renewal of twenty-five per cent of gutters after forty years, repointing fifty per cent of brickwork after forty years, renewing external doors after thirty years, renewal of ironmongery after twenty to forty years, renewal of cupboards and sanitary appliances after twenty years and hot and cold water services and electrical wiring every thirty years. The average annual charges for services maintenance in 1971 were £1.0/m² for heated offices and £1.60/m² for air-conditioned offices.[52]

Hospitals

An analysis of the 1961 hospital costing returns[44] showed a wide variation between different hospitals. The average maintenance cost for all hospitals was £13 per 30 m³ while the middle two-thirds of hospitals had maintenance costs ranging between £8.70 and £17.40 per 30 m³. Similarly, the average power, lighting and heating cost per 30 m³ was £12.80, with a two-thirds range from £9.40 to £16.20.[45]

Elemental Costs

Roofs. Built-up felt roofing at an initial cost of about £1.25/m² requires more maintenance than asphalt at about £2.00/m² and may require replacement at twenty-year intervals. Zinc sheeting may have a limited life in industrial areas.

Walls. Brickwork requires repointing at thirty to forty-year intervals. Stonework may give rise to much greater maintenance problems through its laminated structure and consequent frost damage and bad weathering. Large areas of rendering will always prove costly in maintenance.

Windows and curtain walling. Cheap timber windows are likely to prove troublesome particularly when they open up at joints. Painted hot dip galvanised steel windows are cheaper in first cost than aluminium but require periodic repainting and over the life of the building, aluminium may be competitive.

External decoration. The cost of external painting can be reduced by using materials which do not require decorating, such as precast concrete, asbestos cement and plastic rainwater goods, and self-coloured cladding panels. Large surfaces of painted wall will require redecoration at about five-year intervals and will thus be a costly maintenance item. On tall buildings roof anchors should be provided to carry painters' cradles and reduce painting costs.

Internal decoration. Emulsion paint costing about £0.25/m^2 is more hardwearing than washable distemper at about £0.19/m^2 and it allows the wall to dry out without damaging the paint film. Wall tiling, from £3.00/m^2 and terrazzo at about £4.00/m^2 are much more expensive in initial cost but require little maintenance.

Services. Service pipes should be carefully located to give protection from frost and yet be accessible for inspection and repair. Pitch-fibre drainpipes compare very favourably with glazed vitrified clay, particularly where long lengths are involved or ground conditions are bad.

Refuse disposal. Taking the annual overall cost of supplying, replacing, emptying and disposing of the contents of dustbins to flats at a base cost of 100, then the relative overall costs of other alternatives are communal containers, 110, chutes, 146 and Garchey system, 354.[46]

THE LIFE OF BUILDINGS AND COMPONENTS

Buildings

Costs in use comparisons are concerned with buildings and their component parts, and the longest period over which comparisons need be made should not exceed the expected life of the building. This emphasises the need to be able to predict with reasonable accuracy how long buildings should be expected to last. One approach would be to determine the lives of a representative sample of our present stock of buildings and to assume that the average life for each class of building represents half its effective life, but there are a number of weaknesses inherent in this hypothesis. Cullen[4] has described how in work undertaken at the Building Research Station it has been generally assumed that the life expectancy of a new building is in the order of sixty years, although they often last longer, but as the annual equivalent of a capital sum is virtually constant after sixty years there is little point in assuming a longer life.

Stone[47] asserts that the period taken should be realistic, and will normally be the period over which the building is expected to earn an income or provide a service, for it is during this period that the costs will be recovered. Switzer[48] distinguishes between *structural life* and *economic life* of buildings. Structural or physical life is the period expiring when it ceases to be an economical proposition to maintain the building, whilst economic life is concerned with earning power and is the period up to replacement to increase the income or reward. The structural lives of buildings can be very long indeed as the continued existence of many factories constructed during the Industrial Revolution shows. Stone[3] estimated the average life of houses in 1960 at fifty years and a quarter of a sample of 200 factories in the Midlands investigated in 1956 were over fifty-seven years old. Switzer[48] does, however, emphasise that buildings are more usually demolished because of changes in demand rather than through becoming worn out. Optimum life is therefore determined primarily by the earning power of the building, and only secondarily by the structural stability. Changing social and economic conditions can have a considerable influence on the life of a building which can become ill-suited to present day needs and its demise may also be accelerated by the increasing ratio of land to building costs. Switzer[48] suggests that wherever possible the aim should be to extend the economic life of a building by making the structure adaptable and by careful management and control of the surroundings. Stone[3] has also described how the actual physical life of a building is frequently much greater than its economic life but that it is often demolished before the physical life has expired in order to permit a more profitable use of the site, or because it is found cheaper to clear and rebuild rather than to adapt the building to a change in requirements.

Southwell[5] continues the same theme.

> 'Recent studies have shown how little significance the physical life of buildings has in our social economy. Long before a building would have reached an age when physical collapse was a possibility, it has been pulled down and replaced by a new one, or it has been converted to another use, so that little of the original structure is left. The obsolescence of buildings arises from a number of causes, which are nearly all (because buildings are essentially marketable assets) due to economic factors. It is therefore customary to refer to the period from their construction to their obsolescence, as being their economic life, but it should be borne in mind that economic factors may not wholly determine this.'

Components

The next consideration is the life of the materials and components used in the building. For example, we may need to compare the costs in use of brick cladding with asbestos cement sheeting, or paints as a wall finish with a plastic sheet finish, or a variety of materials for use in rainwater goods or as flat roof finishings. In each case we need to know the life of the materials, so that it will be possible to compute how many times they will need to be replaced during the life of the building, and we would also need to know what maintenance treatment will be required and how often. But as Cullen[4] has observed, there is little reliable information on the life of materials or on maintenance frequencies.

Stone[47] has postulated that the lives of building materials and components can be determined on the basis of observed probability of failure, but this data is rarely available and it is frequently necessary to predict the life on the basis of a knowledge of the age of early failures. Such information is often incomplete since it does not record the successes or the numbers at risk, and hence suggests a higher rate of failure than actually occurs. Stone[47] believes that the best procedure would be to prepare a tabulated list of the estimated lives of as many materials and components as possible, with the list being continually updated both as regards new materials and estimated lives. The author submits that other factors such as degree of exposure, amount of wear, amount of atmospheric pollution, and similar matters would also need recording and considering. Doubtless, the new Building Maintenance Cost Information Service, currently operated by the Department of the Environment and the Royal Institution of Chartered Surveyors, will provide valuable information on lives of building materials and components in varying situations and on their maintenance involvement.

PRACTICAL COSTS IN USE EXAMPLES

A number of worked examples follow to show the application of costs in use techniques to design problems involving complete buildings, components and services.

Alternative Building Designs

The first two examples are designed to show the PV and annual equivalent approaches to cost in use calculations.

Example (a). To find the present value of the running costs of a building with a life of sixty years, given that annual cleaning costs are £400, annual decorations, £150 and annual repairs, £100, external painting, £1000 every five years, and a new roof will be required every thirty years at £10 000. Interest is to be taken at six per cent.

It is feasible to add together the three annual costs of cleaning, decorations and repairs, although it might possibly be argued that no decorations will be needed in the last year of the building's life. This is a little problematical as the decorations could be undertaken at the beginning of each year and the building might secure a reprieve.

PV *Cleaning, decorations and repairs* £650 × 16.1614	£10316.60
(PV of £1 pa for sixty years at six per cent)	
External painting £1000 × 2.83661	
(PV of £1 at five year intervals at six per cent)*	2836.61
Roof replacement £10 000 × 0.17411	1741.10
(PV of £1 in thirty years at six per cent)	
PV of running costs	£14894.31

* PV of £1 in	five years	at six per cent	0.74725
,, ,,	ten years	,,	0.55839
,, ,,	fifteen years	,,	0.41726
,, ,,	twenty years	,,	0.31180
,, ,,	twenty-five years	,,	0.23299
,, ,,	thirty years	,,	0.17411
,, ,,	thirty-five years	,,	0.13010
,, ,,	forty years	,,	0.09722
,, ,,	forty-five years	,,	0.07265
,, ,,	fifty years	,,	0.05428
,, ,,	fifty-five years	,,	0.04056
			£2.83661

Example (b). To find the annual equivalent cost over the life of the building, with an initial constructional cost of £100 000, annual costs of £4000 for cleaning and minor repairs, quinquennial repairs of £10 000 and replacement costs of £20 000 every twenty years. The life of the building is to be taken as sixty years and interest at six per cent (ASF at three per cent).

Annual equivalent			
Building	£100 000 × 0.06613		£ 6613.00
Interest	0.06		
ASF for sixty years at three per cent	0.00613		
(annual sinking fund)			
	0.06613		
Cleaning and minor repairs			£ 4000.00
Larger repairs	£10 000 × 2.83661	£28 366.10	
(see computation in previous example)			
Replacements	£20 000 × 0.40902	8180.40	
(0.31180 + 0.09722)			
		36 546.50	
To convert PV to annual equivalent			
multiply by interest + ASF		0.6613	
			2424.13
Annual equivalent of costs in use			£13037.13

Example (*c*). Compare the costs in use of the following alternative building schemes.

Scheme A. Total cost of building is £50 000, including architect's and surveyor's fees on a site costing £10 000. Annual running costs are estimated at £1500. Certain services and finishings will require replacing at a cost of £6000 every twenty years. Other services have an estimated working life of thirty years and a replacement cost of £8000.

Scheme B. Total cost of building is £65 000, including architect's and surveyor's fees on a site costing £10 000. Annual running costs are estimated at £1200. Certain services and finishings will require replacing at a cost of £4000 every twenty years. Other services have an estimated working life of thirty years and a replacement cost of £5000.

In both cases the estimated life of the building is sixty years. Take an interest rate of six per cent and an annual sinking fund of three per cent.

NOTE: No annual sinking fund need be applied to the site as it will still be available at the expiration of the life of the building.

Scheme A			
Cost of *site*		£10 000	
Annual equivalent in perpetuity at six per cent		0.06	
			£600.00
Cost of building		£50 000.00	
First replacement cost in twenty years	£6000		
PV of £1 in twenty years at six per cent	0.31180		
		1870.80	
Second replacement cost in forty years	£6000		
PV of £1 in forty years at six per cent	0.09722		
		583.32	
Replacement cost in thirty years	£8000		
PV of £1 in thirty years at six per cent	0.17411		
		£1392.88	
PV of building and replacement costs		£53847.00	
Annual equivalent over sixty years			
Interest at six per cent	0.06		
ASF to replace £1 in sixty years at three per cent	0.00613		
		0.06613	
			3560.90
Annual running costs			1500.00
Costs in use of Scheme A			£5660.90

Scheme B			
Cost of *site*			£600.00
Cost of *building*		£65 000.00	
First replacement cost in twenty years	£4000		
PV of £1 in twenty years at six per cent	0.31180		
		1247.20	
Second replacement cost in forty years	£4000		
PV of £1 in forty years at six per cent	0.09722		
		388.88	
Replacement cost in thirty years	£5000		
PV of £1 in thirty years at six per cent	0.17411		
		870.55	
PV of building and replacement costs		£67506.63	
Annual equivalent over sixty years		0.06613	
			4464.21
Annual running costs			1200.00
Costs in use of Scheme B			£6264.21

Scheme A is financially more favourable than scheme B, as the considerably lower initial and replacement costs in A are not offset entirely by the reduced running costs in B, after discounting future costs.

Example (*d*). A building which is to be demolished in twenty-five years time requires repainting now and will also require repainting every five years until demolition. The cost of each repainting is estimated at £300. In ten years time £2000 is to be spent on alterations, and £150 will be spent at the end of each year on sundry repairs. What sum must be set aside now to cover the cost of all the work, assuming that the rate of interest obtainable on investment is six per cent, and ignoring the effect of taxation?

Cost of *painting*		£300	
Present repainting	1.00000		
PV of £1 in five years at six per cent	0.74725		
,, ,, ten years ,,	0.55839		
,, ,, fifteen years ,,	0.41726		
,, ,, twenty years ,,	0.31180		
		3.03470	
			910.41
Cost of *alterations*		£2000	
PV of £1 in ten years at six per cent		0.55839	
			1116.78
Cost of *sundry repairs*		£150	
PV of £1 pa for twenty-four years at six cent		12.5503	
			1882.55
Sum to be set aside			£3909.74

It might be prudent to raise this to £4000 to meet some of the possible future increased costs.

Example (*e*). A temporary building is to be replaced in fifteen years time by a new building which it is estimated will then cost £60 000. What sum must be set aside at the end of each year, if the interest rate on investment (after deducting for tax) is three per cent, to accumulate to the building cost figure in fifteen years?

Cost of new building in fifteen years' time	£60 000
Sinking fund to provide £1 in fifteen years at three per cent	0.05376
Sum to be set aside each year	£3225.60

Heating and Other Services

The examples that follow show how the heating system with the lowest initial cost can involve the heaviest long-term expenditure.

TABLE 11.10

COMPARATIVE HEATING COSTS

		School 1		School 2
Capital cost of heating system		£13 450		£ 8680
Annual running cost	£710		£1630	
Capitalised annual running cost		13 455		30 890
Total capitalised cost (PV)		£26 905		£39 570

Source: *Ministry of Public Building and Works*

In this comparison a rate of interest of five per cent and a building life of sixty years has been used. Under these circumstances the installation of school 2 could be fifty per cent more expensive than school 1. The break-even point for this set of cost figures operates at an interest rate of approximately twenty per cent. A worked example will help to emphasise the need to consider running costs when comparing alternative heating schemes.

Example (*a*). The following heating schemes and costs have been submitted in connection with a proposal for a new three-storey office block. The building client requires the total costs to be assessed for the two alternatives and a recommendation to be made.

The following are central heating proposals to maintain an even temperature of 17°C.

(1) Electric storage heaters: initial cost of £1600 and estimated annual running costs of £800.
(2) Oil-fired ducted hot air: initial cost of £3000 and estimated annual running costs of £500.

(1) *Electric storage heaters*		
Initial cost	£1600	
Annual equivalent over sixty years at six per cent and three per cent	0.06613	
		£105.81
Annual running costs		800.00
Total costs in use with electric storage heaters		£905.81

(2) *Oil-fired ducted hot air*		
Initial cost	£3000	
Annual equivalent over sixty year at six per cent and three per cent	0.06613	
		£198.39
Annual running costs		500.00
Total costs in use with oilfired ducted hot air		£698.39

The oil-fired ducted hot air has a twenty-three per cent cost advantage over electric storage heaters when both initial and capital costs are considered. The calculations do not however include any allowance for replacement of heating equipment during the life of the building but their inclusion in this instance would not change the order of preference. Variations in the future prices of oil and electricity can also change their relative positions. The oil-fired ducted hot air system is recommended on the grounds of lower total costs and reduced loss of usable space.

Stone[36] has formulated a method of tabulating present and future costs of alternative designs, systems and components in a way which is most meaningful and easily understood. His approach is illustrated in table 11.11.

Example (*b*). A choice is to be made between solid fuel and underfloor electric heating for a new block of flats. The cost particulars relating to installation and maintenance follow.

Solid fuel

Initial costs	£175 per flat	
Running costs	annual flue cleaning	£ 0.70
	every ten years – boiler descaling, etc.	£ 16
	every twenty years – replacement of boiler	£ 34

Electric underfloor heating

Initial costs	£100 per flat	
Running costs	every ten years replace thermostat	£ 8
	every fifteen years replace panel fire	£ 14
	every thirty years renew cables	£110

The calculations in table 11.11 are based on a building life of sixty years and an interest rate of six per cent. The total costs shown do not give the complete picture as they do not include running costs. If running costs were taken into account it is likely that electric underfloor heating would lose its cost advantage.

TABLE 11.11

COMPARISON OF COSTS OF INSTALLING AND MAINTAINING HEATING SYSTEMS

Heating system	*Costs*	*Frequency*	*Factor*		*PV*
Solid fuel	£175.00	initial	1.000		£175.00
	0.70	yearly	16.161	(PV of £1 pa for sixty years)	11.31
	16.00	ten years	1.196	(PV 10 + 20 + 30 + 40 + 50)	19.14
	34.00	twenty years	0.409	(PV 20 + 40)	13.91
				Total costs	£219.36
Electric underfloor	£100.00	initial	1.000		100.00
	8.00	ten years	1.196	(PV 10 + 20 + 30 + 40 + 50)	9.57
	14.00	fifteen years	0.664	(PV 15 + 30 + 45)	9.30
	110.00	thirty years	0.174	(PV 30)	19.14
				Total costs	£138.01

Example (*c*). This example relates to glazing and illustrates the need to consider the cost effects of glazing on heating, including changes in running costs.

	Initial Cost of Different Glazing Systems	
	Single glazing	*Double glazing*
Glazing	£ 750	£3500
Heating installation	9500	8250
Total initial costs	£10 250	£11 750

The double glazing involves an extra initial cost of £1500 (£11 750 – £10 250), but offset against this will be savings in the running costs of the heating installation, which are estimated at £150 per annum. Taking a six per cent rate of interest and a sixty years' life of building, this annual sum is equivalent to £2424 (£150 × 16.1614). The double glazing thus has an equivalent first cost advantage of £924 (£2424 – £1500). This calculation is somewhat oversimplified as it ignores the effects of the differing maintenance costs of the various heating and glazing systems, tax relief on repairs and other related matters.

External Works

Questions often arise in practice as to the comparative long-term costs of grassed areas as against various forms of paving. The following example will serve to illustrate the approach.

Example An architect has requested advice on the comparative total costs for an area of 400 sq m which is either to be paved with 50 mm precast concrete paving slabs or to be finished with 150 mm of vegetable soil sown with grass. Advise the architect as to the most economical proposition taking a period of sixty years and six per cent interest.

Paving slabs	£/m^2
Initial cost:	
75 mm bed of ashes	0.25
50 mm slabs on mortar bed and grouting	1.40
Cost/sq m	£1.65

Cost of 400 sq m = 400 × £1.65 = £660
Annual equivalent over sixty years at six per cent and three per cent
= £660 × 0.06613 = £43.65

Maintenance. Average of two days attendance per annum of one craftsman and one mate plus allowance for materials (replacement of cracked and broken slabs).

Sixteen hours at £2.00	£32.00
Materials	8.00
Cost/sq m	£40.00

Vegetable soil sown with grass	£/m²
Initial cost:	
Wheel, spread and level soil from soil heap	0.20
Sow grass seed and rake soil	0.06
Grass seed	0.04
Cost/sq m	£0.30

Cost of 400 sq m = 400 × 0.30 = £120
Annual equivalent over sixty years
at six per cent and three per cent
= £120 × 0.06613 = £7.94

Maintenance. Grass will require cutting fairly frequently and receive top dressing every year to keep it in good order. Allow ten man-days for grass cutting each year.

	£/400 m²
Eighty hours at £0.95	76
Hire of machine, fuel, sharpening cutters, etc.	6
Top dressing and fertiliser	12
Cost/400 sq m	£94

Summary of costs.

	Paving slabs	*Grass*
Initial costs	£43.65	£ 7.94
Maintenance	40.00	94.00
Total annual costs	£83.65	£101.94

Hence, in the longterm, paving slabs are more economical than grass even though they have a much higher initial first cost. It might be possible to equate the two by cutting the grass less frequently if a lower standard of maintenance were acceptable.

Components

Selection of components frequently involves a comparison of total costs as the initial costs of the various alternatives under consideration will not always indicate the best solution. Several worked examples follow to show the approach.

Example (a). This example is a comparison of asbestos cement, vitreous enamel and painted cast-iron rainwater goods, adopting a building life of eighty years and interest rate of six per cent. It is assumed that dismantling and erection costs of replacements will be ten per cent more than that of initial provision. The method used for tabulating the cost information for ease of appreciation follows that used by Stone.[36]

Asbestos cement	expenditure	£150	£165	£165
	years	0	30	60
Vitreous enamel	expenditure	£115	£247.50	
	years	0	60	
Cast-iron	expenditure	£210	£20	£231
	years	0	at five year-intervals (painting costs)	60

Material	*Installation costs*	*Renewal costs*	*Total c/u*
Asbestos cement	£150	£165 × 0.2044 (PV 30 + 60) = £33.73	£183.73
Vitreous enamel	£225	£247.50 × 0.0303 (PV 60) = £7.50	£232.50
Cast-iron	£210	£231 × 0.0303 (PV 60) = £7.00 Painting: £20 × 2.9191 (PV 5 + 10 + 15 + 20 + 25 + 30 + 35 + 40 + 45 + 50 + 55 + 60 + 65 + 70 + 75) = £58.38	£275.38

Asbestos cement gutters and downpipes show a longterm cost advantage, but it must be emphasised that factors other than cost may determine the final decision. In this case it might be felt advisable to select a more expensive material than asbestos cement for improved appearance.

Example (*b*). A cost comparison is made of copper, zinc, nuralite and asbestos-based felt as covering to a roof with an area of 200 sq m, taking a sixty years' building life and six per cent rate of interest.

		Costs in use	
Copper			
Initial cost: 200 m² at £6.15		£1230	£1230
No replacement costs			
Zinc			
Initial cost: 200 m² at £4.42		884	
Replacement in thirty years	£884		
Plus additional dismantling and fixing costs (ten per cent)	84		
	£968		
PV of £1 in thirty years at six per cent	0.1741		
		169	
			£1053

Nuralite

Initial cost: 200 m² at £2.00			£400	
Replacement every twenty years		£400		
Plus additional dismantling and fixing costs (ten per cent)		40		
		£440		
PV of £1 in twenty years at six per cent	0.3118			
PV of £1 in forty years at six per cent	0.0972			
		0.4090		
			180	
				£580

Asbestos-based felt

Initial cost: 200 m² at £1.80			£360	
Replacement every fifteen years		£360		
Plus additional dismantling and fixing costs (ten per cent)		36		
		£396		
PV of £1 in fifteen years at six per cent	0.4173			
PV of £1 in thirty years at six per cent	0.1741			
PV of £1 in forty-five years at six per cent	0.0727			
		0.6641		
			263	
				£623

On costs alone, nuralite shows a definite advantage but other factors would also need consideration, such as appearance and the nuisance value of roof replacement. It would also probably be advisable to allow a ten per cent margin either way on each of the total cost figures to make allowance for prediction errors; these will be considered later in the chapter.

Example (*c*). Advice is required on which of the following alternatives for windows is the most economical in total costs.

(1) Softwood windows costing initially £1800 fixed complete and requiring repainting every fifth year at a cost of £150.

(2) Hardwood windows costing intially £2400 fixed complete and requiring treatment every tenth year at a cost of £60.

(3) Anodised aluminium windows at a cost of £4200 fixed complete and not requiring any maintenance.

Assume a sixty years' life of building and a six per cent rate of interest.

			Total costs
(1) *Softwood windows*			
Initial cost		£1800	
Repainting every five years			
£150 × 2.83661		426	
(PV 5 + 10 + 15 + 20 + 25 + 30 + 35 + 40 + 45 + 50 + 55)		——	£2226
(2) *Hardwood windows*			
Initial cost		£2400	
Treatment every ten years			
PV of £1 in ten years at six per cent	0.55839		
" " twenty years "	0.31180		
" " thirty years "	0.17411		
" " forty years "	0.09722		
" " fifty years "	0.05428		
£60 ×	1.19580 =	72	
		——	£2472
(3) *Anodised aluminium windows*			
Initial cost			£4200
No maintenance			

Softwood windows are the most economical proposition in this case but other factors, such as appearance and aspirations of the building client may dictate a choice different from that of the lowest cost solution.

An interesting comparative study of different forms of cladding for a factory of 1860 m² is given in table 11.12, based on an investigation by Stone[2] in 1960; the costs have been suitably updated to 1971 figures. The costs in use are based on the factory being heated to a temperature of 18°C for a normal working week over a heating season of about thirty weeks.

If the brickwork were to be faced, the cost of this form of cladding would be increased by about £65.

Davey[51] investigated the maintenance to capital cost relationship of a large multistorey office building with a total floor area of about 23 000 m² erected in Bristol in 1964. He was restricted to a four-year occupancy period (1965–8), but within these limitations the figures are, nevertheless, quite useful and a summary of them follows.

Element	*Brief constructional details*	*Average annual maintenance to capital cost (percentage)*
Roof	Concrete slab and asphalt	0.100
Wall cladding	Precast concrete panels	0.026
Wall glazing	6 mm plate and 4 mm sheet glass in bronze frames (cleaning costs)	0.761
Internal partitioning	100 mm block and demountable partitions of aluminium frames with vinyl-faced chipboard	0.637
Decorations and finishes	Floor finishes mainly vinyl tiles; ceilings primarily of stove enamelled suspended modular ceiling panels and acoustic tiles; decorations mainly emulsion paint to plaster	1.254
Lifts	Five high speed lifts and one service lift	1.185
Air-conditioning	Oil-fired boilers for heat load of 1 900 000 W	0.542

TABLE 11.12

COMPARISON OF COST OF FACTORY WALLS
(annual equivalent costs)

Component	*Asbestos cement sheets*	*Asbestos cement sheets lined with plasterboard*	*Asbestos cement sheets lined with insulated plasterboard*	*225 mm cavity brick panels*
Initial costs				
Structure	£ 57	£ 57	£ 57	£ 42
Cladding and sheeting rails	87	81	81	138
Lining	–	50	53	–
Glazing	28	28	28	28
Doors	34	34	34	34
Extra over for facing bricks	–	–	–	12
	206	250	253	254
Maintenance costs				
Repairs and replacements to asbestos cement sheeting and repainting sheeting rails	£ 22	£ 12	£ 12	£ –
Repainting frame	10	10	10	4
Repair plasterboard	–	10	10	–
Decorate plasterboard	–	53	53	–
Repoint and repair brickwork	–	–	–	3
Repaint internal brickwork	–	–	–	50
Other maintenance	33	33	33	33
	65	118	118	90
Annual costs of heating				
Walls	£495	£150	£108	£108
Doors and windows	78	78	78	78
Total costs	£844	£596	£557	£530

PREDICTION ERRORS

Very real problems arise in attempting to predict building prices or costs; either of initial constructional work or of future maintenance and running costs. As described in chapters 7 and 8, there is a marked variability in the prices quoted for the same building work, making the cost assessment of future projects extremely difficult. Again, the cost estimates are influenced by the predictions made for such factors as rates of discount, durability of materials and components, maintenance and operating costs, future relative prices, taxation and the expected life of the building.[47] For instance, we have seen that interest rates are likely to fluctuate within a range of five to eight per cent or even greater, and that this can have a significant effect on predicted costs. Table 11.13 shows how prediction errors in the lives of buildings where the lives are relatively short will have an appreciable effect on annual costs.

Prediction errors may also arise due to changes in requirements during the life of the building, often stemming from changes in fashion or taste. Technological changes may outdate present methods and materials before they are worn out, and the likely future impact of environmental changes generally is difficult to predict. With all these sources of possible predictive errors it is advisable to state costs in use sums in rounded figures to avoid implying a degree of accuracy which cannot possibly be achieved. Ranges of figures are often more meaningful.

TABLE 11.13

EFFECT OF ERRORS IN PREDICTING LIFE OF BUILDINGS

Life *x* (*years*)	*Annual equivalent per £100 of first cost with interest at five per cent* £	*Percentage errors when life taken as x* *instead of* 40	*instead of* 60
20	8.02	+38	+52
30	6.51	+12	+23
40	5.83	0	+10
50	5.48	− 1	+ 4
60	5.28	− 9	0
70	5.17	−11	− 2
80	5.10	−13	− 3

Source: P. A. Stone, *The economics of building designs*[3]

Quantity surveyors would be well advised not to press unduly design solutions which show only marginal benefits on costs in use calculations, bearing in mind all the prediction problems outlined previously and the fact that other design considerations such as increased amenity could conceivably outweigh a small cost advantage. A quantity surveyor has to be realistic as well as technically sound in his advice to building clients.

EFFECT OF TAXATION AND INSURANCE

Stone[47] believes that the incidence of taxation can have a considerable effect on the design economics of buildings. For instance, with industrial buildings some relief can be obtained on the initial cost, through depreciation allowances, investment, initial and cash allowances, their actual form and impact varying from time to time. Amounts spent on maintenance and repairs, heating and lighting, and other running expenses, are classified as business expenses and are deductible from profits in the case of all types of buildings. The exact incidence of taxation varies with the circumstances of the taxpayer. Public authorities do not normally pay tax and so are not affected. Stone[47] postulates that current regulations and levels of taxation tend to favour alternatives with low construction costs and high running costs, since a typical industrial concern might obtain about £45 tax relief for each £100 spent on running costs but only about £25 on each £100 of initial costs. The total costs of buildings can thus be influenced considerably by the form of taxation. For example, an added value tax would eliminate the tax advantage previously accruing to running costs.

The Government Committee on Building Maintenance set up a working party 'to consider whether there is evidence that tax considerations affect decisions on the balance of capital expenditure and subsequent maintenance expenditure on buildings, and if so, to examine the practical effects', and its report was published in 1969.[49] The Income Tax Act 1952 admitted for deduction against liability for income tax and, more recently, corporation tax, the whole of maintenance expenditure on buildings, including running costs, but exclusive of any element of enlargement or betterment. Items of capital expenditure are excluded from deduction. However, in 1945, a depreciation allowance was introduced on all industrial buildings and a system of initial allowances operated which accelerated the process of depreciation for tax purposes. These allowances have varied from time to time and currently stand at four per cent annual depreciation on cost and fifteen per cent initial allowance, which together assume an effective asset life of twenty-two years. There are no allowances on commercial or residential property.

In certain circumstances industrial premises receive even more favourable treatment, for instance, plant and machinery in a building are eligible for depreciation, generally on the basis of thirty per cent initial allowance plus annual depreciation on the written-down value of the asset at the rate: fifteen per cent for plant with a life of more than eighteen years, twenty per cent for fourteen to eighteen years life and twenty-five per cent for less than fourteen years. Even the building itself is eligible to be treated as plant for tax purposes if it is integral with the plant it houses, as with a petrochemical installation.

From 1954–66 investment allowances, allowable against tax, were given on industrial buildings as well as on plant and machinery. When the Industrial Development Act 1966 replaced the investment allowances by cash grants, the grants were confined to plant and machinery used in industry. In 1967 and 1968 the grants amounted to twenty-five per cent of cost, or forty-five per cent in development areas, and in 1969 the corresponding rates were twenty and forty per cent. Plant which is eligible for a grant does not attract initial allowance, and depreciation is allowable on the cost less grant. In addition, the Local Employment Act 1963 provided for the payment of a grant towards the cost of buildings erected in development areas which will enable new jobs to be created. The normal rate of grant is thirty-five per cent.

There is thus a wide variation of fiscal relief against building expenditure, ranging from the total absence of relief against investment in commercial or residential property, through the general run of investment in industrial property, to the favoured case of a building treated as plant for tax purposes and situated in a development area. The case has often been argued that whilst maintenance expenditure is wholly allowable against liability to tax, and capital expenditure, subject to the incidence of grants and allowances, is not allowable, then a given volume of maintenance work must be less expensive to the property owner than a corresponding volume of new construction; hence building expenditure is liable to bias against new construction in favour of maintenance, even when maintenance would otherwise be uneconomic. If this is so, the demand for maintenance is increased at the expense of demand for new construction, which would put the same volume of physical resources to more productive and less labour-intensive use. Maintenance-saving investment is also stifled; the use of buildings is prolonged beyond their natural life, existing buildings are put to uneconomic uses, and the quality of the environment deteriorates.

The MPBW working party[49] asserts that there can be alternative explanations. First of all, despite every fiscal inducement, it remains genuinely uneconomic to retire apparently obsolete buildings; fiscal policy has yet to realise the full development potential of land; and finally there may be an innate resistance among businessmen to investment in construction unless necessary, or for prestige.

It might be argued that if the belief exists, rightly or wrongly, there is at least an unconscious tendency for investment decisions to be biased in favour of higher maintenance and lower capital costs. The working party[49] believes that fiscal considerations have but a marginal influence on investment decisions in new building, and that the harmful effects of fiscal discrimination between new construction and maintenance may thus be less prevalent than is supposed. Furthermore, Drake[50] postulates with some justification that the argument described earlier is conceptually in error since money saved by building more cheaply initially would be invested elsewhere to produce at least an equal return and consequentially equal tax liability. Table 11.14 attempts to show the possible effects of tax allowances and grants on the costs of different types of building.

Stone[3] has described how the design and layout of buildings may also influence rating valuations and premiums payable for fire insurance. For industrial buildings, floor space is rated according to the level of amenities provided; thus upper floors, and areas which are unheated, have low storey heights or can only carry low loads will be assessed at lower rates. However, an attempt to reduce rateable value by lowering standards may adversely affect efficiency and flexibility. Fire insurance premiums are related to the degree of fire risk and reductions in premiums may be made for design features which are likely to reduce fire spread, such as the use of noninflammable materials or those which resist the spread of fire, and the provision of fire-fighting equipment like sprinklers. It may not pay to instal sprinklers or automatic fire alarms where the annual equivalent cost of provision and maintenance is greater than the reductions in fire insurance premiums.

MAINTENANCE COST RECORDS

Form of Records

Southwell[11] has suggested that maintenance cost records may be kept to fulfil three separate functions.

(1) Budgetary control – to produce the annual or other periodic sum which needs to be set aside to provide for maintenance and operating services.

TABLE 11.14

THE EFFECT OF ALLOWANCES AND GRANTS ON CAPITAL AND MAINTENANCE COSTS IN DIFFERENT TYPES OF BUILDING

Capital expenditure

The incidence of taxation varies dramatically with the type of building. The following figures illustrate the actual cost in terms of net present value (NPV) to a profit-earning company of investment in three types of new construction. The capital cost in each case is assumed to be £100 000.

	Cost to company (net present value basis)
Commercial building No tax relief	£100 000
Industrial building Initial allowance fifteen per cent Annual allowance four per cent straight line	£ 77 730
Building structure treated as plant for tax purposes in a development area Investment grant forty per cent Annual allowance fifteen per cent reducing value	£ 47 109

Maintenance expenditure

Maintenance expenditure is generally allowable as a revenue expense in calculating taxable profits; the table below lists the NPV of an annual expenditure of £1000 on maintenance both before and after tax relief is taken into account (columns 2 and 3); the NPV after tax relief is shown as a percentage of the NPV cost to the company shown above (columns 4, 5 and 6).

(1)	(2)	(3)	(4)	(5)	(6)
			After tax NPV *as a percentage of* NPV *cost of:*		
Period (years)	NPV *before tax*	NPV *after tax*	*Commercial building*	*Industrial building*	*Plant structure*
10	£ 6418	£3878	3.9%	5.0%	8.2%
20	9129	5516	5.5%	7.1%	11.7%
30	10 273	6207	6.2%	8.0%	13.2%
40	10 757	6500	6.5%	8.4%	13.8%
50	10 961	6623	6.6%	8.5%	14.1%

Calculations are based on following assumptions

Discounting rate	nine per cent
Tax delay	one-and-a-half years
Investment grant delay	one year
Date of study	1 June 1969
Life of plant structure	twenty-two years

Source: *Ministry of Public Building and Works, Implications of taxation provisions for building maintenance*[49]

(2) Management control – to permit the day-to-day control over maintenance expenditure.

(3) Design-cost control – to contain full information concerning causes of failures, types of failure, design faults and similar particulars. Records of roof repairs are of little use unless, for instance, they show the type of tile, quality of tile, method of laying, angle of pitch, degree of exposure, etc.

Chaplin[16] has suggested a hierarchical system of grouping or classification of items for recording maintenance work and costs. A primary grouping would be of value to organisations who did not wish to prepare very detailed records, and would also form the base for more detailed recording. Chaplin's primary grouping was:

(1) external decoration;
(2) internal decoration;

(3) main structure;
(4) internal construction;
(5) finishes and fittings;
(6) plumbing and sanitary services;
(7) mechanical services, including heating, ventilation and gas installations;
(8) electrical services and kitchen equipment;
(9) external and civil engineering works; and
(10) miscellaneous and ancillary works.

A secondary grouping of plumbing and sanitary services could be hot water and heating services and a tertiary grouping of the latter could embrace pipes of different materials, and valves and tanks and cylinders of different materials.

Building Maintenance Cost Information Service

This service commenced the distribution of building maintenance cost analyses in 1971 and as the service develops it should provide valuable basic information and will indicate clearly the areas where greatest economies can be made. Each analysis gives details of the type, age, accommodation, dimensions, construction and maintenance organisation relating to the particular building. The analysis also breaks down the building into elements – improvements and adaptions; decoration; fabric; services; cleaning; utilities; administrative costs; overheads; and external works; and these are further subdivided into subelements. Against each subelement is recorded the total cost in a stated financial year and this is converted to the cost per 100 m^2 of floor area as a unit of comparison.

The following cost ranges from a limited number of analyses may be of interest. All figures are 1969 annual costs per 100 m^2 of floor area.

Improvements and adaptions: £10 to £570 (hospitals having the costliest improvements)
External decorations: £35 to £120 (factories having the highest decoration costs)
Internal decorations: £13 to £60
Fabric: £0–£30 (factories being the highest)
Heating and ventilation: £1 to £67 (libraries being the most expensive)
Lifts: £3 to £7
Cleaning: £27 to £310 (libraries having the highest cleaning costs)
Utilities: £127 to £270 (hospitals being the most costly)
Administrative charges: £25 to £287
Property insurance: £3 to £162
Rates: £31 to £158

These annual cost figures from a very restricted sample show extremely wide variations and in some cases, such as improvements and decorations which do not occur every year, the costs must be used with discretion. The overall annual cost ranges were hospitals: £475 to £1042; libraries: £650 to £747; factories: £416 to £816; and an office block: £813 per 100 m^2 of floor area (1969 costs).

In addition, the Building Maintenance Cost Information Service (BMCIS) issue excellent design/performance data sheets which give illustrated details of faults in constructional work, the way in which they have been remedied and suggestions as to improved design. They cover a wide range of topics from the need for chair rails on lightweight precast demountable partitions and substitution of painted plywood or glass fibre panels for obscure glass cladding where impact damage may occur, to excessive thermal movement on dark-coloured surfaces exposed to considerable sunlight and use of external plywood in preference to boarding for wide roof fascias. The need for a combination of a ventilated cavity and vapour barrier in flat roof construction, structural screeds of sufficient strength to withstand point loads and reinforcement in concrete thresholds is also reiterated.

REFERENCES

1. INSTITUTION OF CIVIL ENGINEERS. An introduction to engineering economics (1969)
2. P. A. STONE. Design economics – building costs today and tomorrow. *The Chartered Surveyor* (January 1960)
3. P. A. STONE. The economics of building designs. *Journal of the Royal Statistical Society. Series A (General)*, **123.3** (1960)
4. B. D. CULLEN. Costs in use. *The Quantity Surveyor* **23.4** (1967)
5. J. SOUTHWELL. Total building cost appraisal. Research and Information Group of Quantity Surveyors' Committee, Royal Institution of Chartered Surveyors (1967)
6. J. CODGBROOK. The significance of running costs in the cost planning and design of buildings. *Nottingham Regional Conference no. 3, Trent Polytechnic*. Royal Institution of Chartered Surveyors (1967)
7. B. E. DRAKE. The economics of maintenance. *The Quantity Surveyor*, **26.1** (1969)
8. B. A. SPEIGHT. Maintenance in relation to design. *The Chartered Surveyor* (October 1968)
9. J. E. CAPITO. The work of the MPBW committee on building maintenance. *The Quantity Surveyor*, **25.6** (1969)
10. BUILDING RESEARCH STATION. *Building Research Station Digest 109*. Building economics: cost planning. HMSO (1958)
11. J. SOUTHWELL. Methodology in the use of maintenance costs for systematic design. *Nottingham Regional Conference no. 3, Trent Polytechnic*. Royal Institution of Chartered Surveyors (1967)
12. W. WHITE. Design and maintenance. *Building Maintenance* (May 1971)
13. E. BENROY. An architect's view of maintenance related to design. *The Architect and Surveyor* (January/February 1971)
14. J. A. ROBERTSON. The planned maintenance of buildings and structures. *Paper 7184S, Proceedings of the Institution of Civil Engineers* (1969)
15. R. ANDERSON. Good maintenance pays off – the benefits in efficiency, productivity and morale. *Proceedings of Profitable Building Maintenance Conference, London*. Ministry of Public Building and Works (1967)
16. M. F. CHAPLIN. Analysis of maintenance. *Proceedings of Maintenance of Buildings Conference, London*. Ministry of Public Building and Works (1965)
17. D. B. JAMES. Maintenance technology. *The Quantity Surveyor*, **26.2** (1969)
18. V. R. GRAY and W. H. RANSOM. The weathering of building materials. *Proceedings of Conference on Technology of Building Maintenance, Bath University*. Ministry of Public Building and Works (1968)
19. A. G. WAUD. Cleaning costs – they start on the drawing board. *Proceedings of Conference on Profitable Building Maintenance, London*. Ministry of Public Building and Works (1967)
20. MINISTRY OF PUBLIC BUILDING AND WORKS. The relationship between design and building maintenance (1970)
21. H. GRAHAM, R. T. KELLY and H. G. STANTIALL. Some considerations in the choice, use and maintenance of building components. *Proceedings of Seminar on Maintenance by Design, London*. Ministry of Public Building and Works (1969)
22. M. P. WALKER. The building as an investment – some guidance on maintenance policy. *Proceedings of Profitable Building Maintenance Conference, London*. Ministry of Public Building and Works (1967)
23. E. J. GIBSON. Timber technology and maintenance. *Proceedings of Technology of Building Maintenance Conference, Bath University*. Ministry of Public Building and Works (1968)
24. K. A. CHANDLER. Protection and maintenance of metals in buildings. *Proceedings of Technology of Building Maintenance Conference, Bath University*. Ministry of Public Building and Works (1968)
25. T. R. BULLETT. Paint in buildings. *Proceedings of Technology of Building Maintenance Conference, Bath University*. Ministry of Public Building and Works (1968)
26. GREATER LONDON COUNCIL. Development and materials bulletin 8 (second series) (1967)
27. P. BENNETT. How to brief your architect. *Proceedings of Profitable Building Maintenance Conference, London*. Ministry of Public Building and Works (1967)
28. J. F. HILL. A review of maintenance practice in property management. *The Quantity Surveyor*, **26.3** (1969)
29. J. P. MACEY. Maintaining 250 000 houses. *Journal of the Institution of Municipal Engineers* (April 1968)
30. MINISTRY OF PUBLIC BUILDING AND WORKS. Contractual policies and techniques for maintenance work (1969)
31. MINISTRY OF PUBLIC BUILDING AND WORKS. Incentive schemes applied to building maintenance by small firms (1969)
32. J. A. KENYON. Work study and its application to housing maintenance in Middlesbrough. *Journal of the Institution of Municipal Engineers* (April 1968)
33. L. J. F. STONE. An analysis of the metrication of maintenance. *Building Maintenance* (April 1971)
34. B. A. SPEIGHT. Formulating maintenance policy. *Proceedings of Second National Building Maintenance Conference, London*. Ministry of Public Building and Works (1969)
35. A. G. WILSON. The technique of cost-benefit analysis. *Cost-benefit analysis*. Institute of Municipal Treasurers and Accountants (1968)
36. P. A. STONE. *Building Design Evaluation: Costs in Use*. Spon (1967)
37. SOUTH-EAST METROPOLITAN REGIONAL HOSPITAL BOARD. *Conference on the Future of Cost Control in the Health Service, Eastbourne* (1970)
38. I. H. SEELEY. *Municipal Engineering Practice*. Macmillan (1967)
39. T. W. PARKER. Building research and maintenance. *Proceedings of Conference on the Maintenance of Buildings, London* (1965)
40. P. A. STONE. The economics of factory buildings. Factory building studies no. 12. HMSO (1962)
41. M. A. CLAPP and B. D. CULLEN. The maintenance and running costs of school building. *Building Research Station Current Paper*, **72/68** (1968)
42. MINISTRY OF PUBLIC BUILDING AND WORKS. The relationship of capital, maintenance and running costs: A case study of two Crown office buildings (1970)
43. H. KNIGHT. Capital cost and cost in use. *The Chartered Surveyor* (February 1971)
44. MINISTRY OF HEALTH. Hospital costing returns. HMSO

45. B. D. CULLEN and I. M. JEFFERY. Design paper 65: Running costs of hospita lbuildings. Building Research Station (1967)
46. BUILDING RESEARCH STATION. *Building Research Station Digest 40 (Second Series)*. Refuse disposal in blocks of flats (1963)
47. P. A. STONE. The application of design evaluation techniques. *Building* (20 March 1970)
48. J. F. Q. SWITZER. The economic life of buildings. *Proceedings of* RICS *Regional Conference no. 3, Trent Polytechnic, Nottingham*. Royal Institution of Chartered Surveyors (1967)
49. MINISTRY OF PUBLIC BUILDING AND WORKS. Implications of taxation provisions for building maintenance, (1969)
50. B. E. DRAKE. Do buildings need to last? *Proceedings of Maintenance by Design Seminar, London*. Ministry of Building and Public Works (1969)
51. C. R. DAVEY. Maintenance Study: Robinson building, Bristol. *Building* (24 September 1971)
52. BUILDING RESEARCH STATION. Operating costs of services in office buildings. Building Research] Station Digest 138 (February 1972)

12 LAND USE AND VALUE DETERMINANTS

THIS CHAPTER INVESTIGATES the factors which influence land use patterns and land values. It is also concerned with the whole spectrum of matters which bear upon the development of land, such as site characteristics, planning and other statutory controls, encumbrances and easements.

CHANGING LAND USE REQUIREMENTS

Land Use Patterns

Clawson, Held and Stoddard[1] have judged our present day basic concern to be the conflict between the demands of an expanding economy and a fixed area of land. Wibberley[2] has shown how Britain has a high man/land ratio and a pattern of increasing old and new uses. The area of England and Wales is approximately fifteen million hectares; as against a 1971 population of about 48.59 million this gives a man/land ratio of about 320 persons per km^2. In the year 2000 with a population probably around sixty-six million, this ratio will have changed to over 430 persons per km^2. Comparable present day man/land ratios for some other countries are 350 persons/km^2 for the Netherlands, 250 for Japan and twenty for the United States.

Land is unique in that each parcel or plot has a specific location with its own particular geography. Height above sea level, slope, latitude and longitude, soil and subsoil, rainfall, sunshine, temperature, wind exposure, drainage and distance from other places – all vary from one plot to another. Some of these variables can be partially controlled by the use of other resources – capital and labour for instance, but there is no homogeneity or easy interchangeability, each site has its own peculiar characteristics.[2] The changing pattern of urban growth has been well portrayed by a number of writer. Jones[3] refers to the general broad pattern of growth, decay, rebuilding, central business and shopping core, and outer residential suburb.

On the other hand, Chapin and Weiss[4] foresee that in a twenty-five to forty-year period, scientific advances and technological change may profoundly affect patterns of urban development; and the shifting character of life styles and consumer tastes in large segments of an increasingly affluent society may further complicate the problem of prediction. These writers describe how land development is the consequence of many decisions and implementing actions of both a public and private nature. They illustrate by giving examples of how priming actions often trigger secondary actions which taken together produce the total pattern of land development. Thus an industrial or commercial location decision may set in motion a whole chain of other decisions and actions, for example location decision of households, firms, institutions, etc. Alternatively, a highway location decision, or a decision on building a new school, or a combined series of decisions of this nature can serve to prime such secondary decisions and actions. A priming action has two main characteristics – structuring effect on the distribution of land development and the timing effect in fixing the sequence of development.

Best and Coppock[5] have traced the changing pattern of urban development in this country from the compact towns of the Industrial Revolution to the more open development of interwar years assisted by the increased mobility of the urban population through improved public transport services. These developments now require extending to include the more compact and higher density residential layouts of the nineteen-sixties. Ciriacy-Wantrup[6] has investigated the changes in competition for land arising from urbanisation and industrialisation, and has shown how at the margin of urban-industrial development, agriculture is quickly priced out of the land market. Another important implication which follows from the increasing competition for land for urban uses is the irreversibility of the results, so different from merely replacing one agricultural use by another. Chisholm[7] asserts that the central problems in the economics of land use are those of location and of competition between alternative users and uses to command each particular site, although he omits to include the influence of legislative measures.

Urban Land Use

Guttenberg,[8] in his investigation of land use classification methods in the United States, has adopted a threefold approach to urban development.

(1) The user's interest: quality (age, deterioration, basic utilities, modern amenities).

(2) The investor's interest: economic durability (physical condition, adaptability of design, public services, environment, title, management, etc.).

(3) The public interest: social impact (legal aspects, sanitary condition, aesthetic attributes, extent of development).

Best[9] has shown that housing is easily the greatest urban land use, and in the New Towns of England and Wales it accounts for just over one-half of the total urban area. This is a higher percentage than in the older towns where the areas used for residential purposes generally amount to about forty-three to forty-five per cent, as indicated in table 12.1.

TABLE 12.1

PROPORTIONATE COMPOSITION OF URBAN AREAS

	Housing per cent	*Industry* per cent	*Open space* per cent	*Education* per cent	*Residual uses* per cent
New towns (London region)	50.5	9.3	17.4	8.9	13.9
New towns (Provincial)	50.2	7.7	25.3	7.4	9.4
New towns (England and Wales)	50.4	8.9	19.3	8.6	12.8
New towns (Scotland)	38.8	14.1	24.2	7.7	15.2
New towns (Great Britain)	48.9	9.6	19.9	8.4	13.2
County boroughs (79)	43.9	9.4	19.6	6.1	21.0
Large town map areas (186 over 10 000 population)	44.7	8.9	18.8	6.5	21.1

Source: R. H. Best, *Land for New Towns*[9]

Furthermore, following the recommendations of the former Ministry of Housing and Local Government,[10] the more recent New Town proposals show a noticeable reduction in space standards. This is particularly so with housing, where the newly planned residential areas have an average net residential density of about eight hectares per 1000 population, which is under half the land provision recommended by the New Towns Committee.[11]

Best[12] has calculated the area of agricultural land in this country transferred to urban and other uses, from which the figures in table 12.2 have been computed and converted to hectares.

These figures show a continuous reduction in the area of land devoted to agricultural use but at a much reduced rate compared with the prewar trend. During the war land was acquired by service departments at a rate exceeding 40 000 hectares per annum, and in the postwar years one-half of this has been returned to agricultural use. The annual figures produced by Best for the postwar years show an intimate association between economic trends and transfers of agricultural land to urban use. The 'stop-go' oscillation in economic controls has been reflected by a marked fluctuation in urban demands on land from year to year.

TABLE 12.2

ANNUAL TRANSFERS OF AGRICULTURAL LAND TO URBAN AND OTHER USES IN ENGLAND AND WALES

Transfers to: *period*	*Building and general constructional development* *ha*	*Sports grounds* *ha*	*Total urban area* *ha*	*Service departments and miscellaneous* *ha*	*Allotments, woodlands and forestry development* *ha*
1927–1934	15 300	3600	18 900	500	not applicable
1934–1939	20 000	4300	24 300	6000	not applicable
1945–1950	13 200	3600	16 800	+15 000	7900
1950–1960	13 600	1300	14 900	+ 2100	8700
1960–1965	14 200	1250	15 450	+ 1000	6500

Plus rates indicate net gains to agricultural area; ha = hectare

Lean and Goodall[13] describe how land use within an urban area can often be conveniently subdivided into four separate and distinct districts or zones.

(1) Central business district which is the optimum location for many economic activities, such as shops, offices, theatres and hotels, having maximum accessibility.
(2) Zone of transition surrounding the central business district where older buildings are being replaced.
(3) Suburban areas which are developed for residential purposes at moderate densities on cheaper land.
(4) Rural-urban fringe accommodating commuters who wish to live in rural surroundings.

Impact of Changing Use Pattern on Agriculture

Agriculture loses land in two main directions – some of its poorest land to afforestation at about 8000 hectares per annum, and some of its better land to urban uses at about 14 000 to 16 000 hectares per year. The latter rate of transfer is directly influenced by the density of the urban development.

Wibberley[2] estimates that by the year 2000 about seventy-six per cent of the total land surface in this country may be available for agriculture, and that British agriculture will tend to adjust itself to the area of land left for its use. Increases in population and rising real incomes create demand pressures for more land at present in agricultural use, but improvements in the physical efficiency of food production economise on land area. Again if imports of food products are increased, home land can be saved for other uses. If pressures lead to a decrease in food imports, more home land will be pressed into more intensive service. Edwards[14] has also referred to the possible development of synthetic foodstuffs.

Jones[15] holds the view that the present average increase in agricultural productivity of 1.3 per cent per annum will more than offset the land likely to be needed for urban uses. This view is supported by Wibberley[16] and Edwards[14] provided that the choice of new urban sites is made with sensible discrimination and, in particular, that development is not concentrated to an unnecessary extent on our most highly productive farmland.

LAND USE PLANNING

Objectives

A universal interest has developed in land use planning and the determinants of land use, and there is an increasing awareness of the advisability of forward planning. The author once wrote that town and country planning, or physical planning as it is now more commonly described, is necessary to ensure that all development is co-ordinated with an eye to the future, and carried out in such a way to assist in producing a community environment that will advance human welfare in health, well-being and safety.[17] Many defects are apt to arise from unplanned development, such as waste, congestion, disharmony, undesirable mixture of incompatible uses, lack of social services and unnecessary loss of good agricultural land. Chapin[18] also sees the need for the city planner to view land use in the context of 'health, safety and general welfare' and generally to ensure that development is in the public interest. Hoyes[19] has expressed the view that the broad objective of planning is to ensure that land is put to the best use from the point of view of the community, and to secure a proper balance between competing demands for land. Furthermore, the regulation and control of the use, in urban areas, of land resources within proprietary land units is necessary, in order to prevent the repetition of the undesirable mixture of land uses which has emerged over time in towns and cities.[20]

The World Health Organisation[21] has shown the importance of thinking of the metropolitan area as a coherent whole and to recognise the interplay of social, political and economic factors, which must be taken into consideration. This organisation defined planning objectives as 'a model of an intended future situation' and 'a programme of action and predetermined co-ordination', illustrating the dynamic nature of the process and of the need to improve human conditions.

Arvill[22] has established certain principles in land use planning – the unity of the environment (fusion of town and country), comprehensiveness (controlling many activities, often conflicting) and quality of the environment. He also contended that planning must accept that there is always a limit on the resources available and so it must actively encourage management and development – public and private – to accept and work for strategic goals. Various writers have emphasised the need for land use plans to take full account of social and economic factors, with the general objective of maximising social net benefit or public interest.[6,23,24,25] Planning the countryside of Britain is largely a matter of conserving natural resources. The basic objective is to make full use of our minerals, soil, water and wildlife, while ensuring that we do not allow exploitation to endanger their future supply. Planning our towns, on the other hand is fundamentally concerned with the development of land resources with buildings, roads and other urban services, to accommodate changing demands.[26]

Until the early nineteen-sixties it was generally considered that land use planning objectives could be achieved by a control mechanism which operated at local planning authority level (county councils and county borough councils), but since then it has become accepted that many of the activities we want to influence take place over an area larger than that administered by a single authority. Hence regional and subregional planning has been undertaken and there is a pressing need for a larger degree of national planning.

Willhelm[27] has described how in land use planning we control land use, and that in controlling land use we are controlling people. Hence it is vitally important that the public are permitted to participate actively in the planning process – the subject investigated by the Skeffington Committee. The use of a series of seminars in the first stages of preparation of the plan for Milton Keynes New Town is a practical illustration of its application.[28]

Finally, it is interesting to examine the geographical and historical reasons for the differences between British and American patterns of urbanisation. The town planning movement in Britain had its roots in a reaction against urban growth as represented by the nineteenth-century industrial town, and it sought an orderly countryside and contained towns which would not spread across the countryside unchecked. In the United States a different attitude is adopted towards land use. The vastness of the North American continent and the tendency of its inhabitants to move on periodically from one location to the next has resulted in successive exploitation and abandonment. This has given rise to the expression 'God's own junkyard', describing many Americans' attitude to land – a resource that can be used, even squandered, with little thought for the future.[29]

Economic Aspects of Land Use Planning

The economic theory which had competitively established price as its keystone, concluded that since consumer desire and producer capacity tended towards a balance at any moment in time, the most efficient allocation of resources was ever present through market forces. In certain spheres, especially in the use of land, this conclusion is suspect. The analysis is too narrow as it ignores a whole of side effect. Furthermore, many forms of land development are relatively inflexible, as for example land developed with houses is not readily returned to agriculture, and there is no opportunity to correct the adverse effects of bad decisions.[19]

Lean and Goodall[13] postulate that, subject to certain provisos, if town planning leads to higher land values than would exist without it, then a better or more efficient use of resources has been achieved, based on rent theory. This is a dubious hypothesis, as the planner may artificially reduce the supply of land for a particular kind of use and thus force up its price – putting the owner in a monopolistic situation. To ensure the best use of resources, the economic consequences of alternative courses of action should be considered, although non-economic implications frequently deserve attention.[31] Difficulties arise in attempting to place monetary values on nonrevenue–producing public sector development, and the valuation of the revenue-producing investment can only reflect from empirical data the imperfections of market values which are of interest in plan evaluation, and they form an inadequate basis for the latter function.[32]

Furthermore, the fragmentation of the ownership of land may prevent it being used for the most efficient purpose. Some public urban land uses such as roads, parks, schools and sewage disposal works will be nonrevenue-producing, but the profitable uses often depend on the nonrevenue uses. Urban planning can, by control and reorganisation of both types of use, lead to a more efficient use of urban resources. If urban planning increases the accessibility within an urban area, then this again is likely to increase efficiency and land values. Hence the economic approach to the study of urban land use patterns is an important one. The patterns that emerge are the sum total of a large number of individual appreciations of locational values. Successful planning must be based on an intimate knowledge of the economic forces at work within cities. Through planning we aim to produce a new geography, a better distribution of activity and land use related to contemporary social and economic needs, yet fully aware that needs change more quickly than physical forms.[33]

Mention should perhaps be made of the Land Commission Act 1967 which established a Land Commission with wide powers of acquisition, disposal and management of land, and who were empowered to collect a levy on betterment (forty per cent of development value), although it was subsequently abolished in 1970. The main aims of the Act, as stated by the Minister, were:

(1) to secure that a substantial part of the development value created by the community returns to the community;

(2) to secure that the burden of the cost of land for essential purposes is reduced; and

(3) to secure that the right land is available at the right time for the implementation of national, regional and local plans.

In practice, the price of building land rose appreciably and there was an acute shortage of building land in various parts of the country. Betterment levy introduced in 1967 and abolished in 1970 was a modified successor to the 100 per cent development charge introduced by the Town and Country Planning Act 1947 and abolished in 1953.

LAND VALUES

Land Value Determinants

While various users are in competition for sites, the sites vary considerably in their suitability for different purposes. The attributes of sites can be divided into three main groups: physical, locational, and legal consents as to use. The prices of site are very much influenced by the use to which they can be put. For example, in 1962, Peters[34] gave the average price of vacant farms as £333 per hectare. At the same time, Stone[35] calculated the average price for residential sites at £12 400 per hectare.

Wendt[36] has stated a general model for determination of land values

$$\text{land value} = \frac{(\text{aggregate gross revenues}) - (\text{total expected costs})}{\text{capitalisation rate}}$$

Revenues are influenced by the investors' expectation of the size of the market, income spent for various urban services, urban area's competitive pull, supply of competitive urban land and prospective investment in public improvements. Expected costs are the sum of local property taxes, operating costs, interest on capital and depreciation allowances. The capitalisation rate is affected by interest rates, allowances for anticipated risk and expectations concerning capital gains.

According to land economics theory, these factors are taken into account in the property market. Users of land bid for sites in accordance with what will maximise their profits and minimise their costs. Land users in retail business and services tend to bid for space at the highest prices, and land best suited for these activities shows the highest value.[18]

Like most other goods, the value of land is influenced considerably by the interaction of supply and demand. The supply of land is fixed although its use may change. Similarly with landed properties (buildings) the supply, particularly in the short-term, is relatively inelastic, although the supply of a particular type of building may on occasions be increased fairly quickly by conversion. In the long-term, where there is an obvious need for a certain class of property, it is possible to acquire land and erect fresh stocks of the particular building.

In practice, the market prices of the developed real properties determine the land values (the residual figure in the developer's budget, described in chapter 13, which the developer can afford to pay for the site). Land values are influenced by a variety of factors, such as accessibility and compatibility. If there is a shortage of land available for a certain use, prices will tend to rise until further land is transferred to this use, unless land use planning frustrates it.

Probably the most important factors influencing land values are:

(1) supply and demand – a limited supply of building land, or fierce demand for it will force up the price;

(2) the permitted use to which it can be put under planning regulations, of which central area uses such as shops, offices and theatres, are the most valuable;

(3) location, highly priced land often being the most accessible;

(4) physical characteristics which affect the cost of development and suitability for a given purpose, industrial areas for example needing extensive flat sites;

(5) availability of public service, such as roads, sewers, water mains and electricity cables;

(6) form of title (freehold or leasehold) and any restrictive convenants or other encumbrances which will affect its use; and

(7) general nature of the surrounding development and whether compatible.

Values are best established by reference to the price paid for comparable properties in the market.[37] In practice, there are few really comparable sites or buildings and there is all too fiequently little information available on property transactions. Generally, auction prices are accepted as the truest indicators of market prices.[38]

Effect of Town Planning on Land Values

According to the Uthwatt Report[39] the effect of town planning is to shift land values but not to destroy them. Planning schemes frequently cause a redistribution of values, where the permitted land uses differ from the existing patterns of land use. If, for example, a town map limited the supply of land for residential purposes and permitted high densities on the outskirts and lower densities in inner districts, then the land on the outskirts would in all probability have a higher value than that nearer the centre; a shift of values has therefore occurred. Where the shift in land values, which is caused by a redistribution of the profit uses of land, settles, will partially depend on the planned redistribution of the nonprofit uses, such as open spaces.[13]

Farmland near towns often has a potential development value, described by the Uthwatt Committee as 'floating value' which increases as the likelihood of development becomes more certain.[40] This enhanced value was the betterment of which the Land Commission required a forty per cent share under the Land Commission Act 1967 and which was abolished in 1970. The aim was to collect some share of increases in value representing the community's actions, while leaving some share to the benefit of the private individual, thus allowing the market to continue to operate.[41] It did, however, result in a reduction in the amount of land available for development.

Successful planning should seek to understand the economic and social forces which shape our environment and assist in the allocation of land uses to meet those needs in a manner beneficial to the whole community. This involves ensuring an adequate supply of land to meet various anticipated demands.[42] In the latter half of 1971 the Federation of Registered House Builders considered the shortage of residential land at that time to be an artificial scarcity resulting in unjustifiably high land prices. The Federation thought that the only remedy was for local planning authorities to release more land on the market. Wates[43] also believed that the high cost of housing land was the direct and unavoidable result of restrictive planning procedures, and that were land released in sufficient quantities, its price would drop substantially.

Pattern of Land Values

Commercial and similar uses are located in city centres, as they are able to pay the high land prices and secure the benefits of maximum accessibility and convenience. Hence rents serve to act as sorters and arrangers of land use patterns, and planning control alone does not decide land use. It has been suggested that the outgrowth of this market process of competitive bidding for sites among the potential users of land is an orderly pattern of land use specially organised to perform most efficiently the economic functions that characterise urban life.

Generally commercial and industrial uses can attract land away from residential uses. Competition between firms to be in the desired positions will force the land values above those of the surrounding land used for residential purposes. If all the land in a given part of the town is used for complementary purposes this is likely to enhance the land values, whereas if they are incompatible, this may lower the land values. For example, if there is a residential district well-served by schools, open spaces, transport, etc., persons will wish to live there and both property and land values will be higher than if it lacked these facilities.[30] Developments in transport systems may also lead to changes in urban land values. For example, the extension of a bus route, the building of a railway station or the carrying-out of major road improvements, may cause changes in land values in adjacent areas.

Table 12.3 shows the rapid increase in land prices in a London residential suburb between 1950 and 1960, well in excess of the general inflationary trends.

Over the period 1960–2, the median price for residential sites was about £12 400 per hectare, the corresponding price for industrial sites was £20 000 per hectare and for commercial sites £62 000 per hectare. Maximum prices recorded over this period were £124 000 per hectare for residential sites, nearly £250 000 per hectare for industrial sites and almost £1 700 000 per hectare for commercial sites.[35]

TABLE 12.3

INDEX OF LAND AND RETAIL PRICES, 1913–1960

Year	1938 = 100 *Land prices*	*Retail prices*
1913	60	64
1919	67	139
1920	69	159
1925	75	112
1930	81	101
1935	87	92
1938	100	100
1939	102	103
1946	150	154
1950	175	185
1953	238	216
1955	307	242
1958	640	272
1959	765	273
1960	875	274

Source: J. S. Andrews, *MSc thesis, University of London*[44]

The greater the permitted residential density, the higher the price per hectare which can be obtained. For example, an increase in permitted densities from twenty-four to thirty-six dwellings per hectare would, in 1962, tend to reduce site costs per dwelling by only £50 to £100 but would raise the land price per hectare by £5000 to £7500.[38] Denman[45] has shown that a developer will pay more per hectare for the marginal piece of land needed to complete a development site than he will for the site as a whole.

Further substantial increases in residential land prices occurred between 1965 and 1969. Median London prices rose from £50 000 to £93 000 per hectare, although in the East Midlands the rise in prices was not quite so spectacular – £12 400 to £17 500 per hectare.[46] Part of this substantial increase in land prices could conceivably stem from the payment of betterment levy to the Land Commission (since abolished) and which could produce distorted land values.[47] Thorncroft,[37] in his survey of world property prices over the two decades since the last world war, shows that the trend in property values has been a persistent rise.

Hillmore[48] asserts that average building land prices doubled between 1963 and 1970 and are still escalating. But land prices vary widely in different parts of the country. It is cheapest in parts of eastern and northern England and dearest within a 50 km radius of London. In consequence, even supposing a builder obtains permission to build thirty houses per hectare on land costing £50 000/hectare, it will cost him at least £2000 a plot, after roads, drainage and other services have been provided, but before a brick is laid on the site. Allan[49] has described how land hunger is a brake on the housing programme and a blight on the building industry. Where land is available for development, competition forces up the price which, coupled with rapidly increasing building prices, may cause projected building to fail to meet the criteria for investment return or rents to reach levels which occupants cannot or are very reluctant to afford.

Finally, the benefits of land value maps as a bridge between valuers and planners has been clearly demonstrated by Anstey.[30] The planner needs to consider land values, how they may change, and how far present values should be maintained in the interests of urban efficiency.

FACTORS INFLUENCING DEVELOPMENT

There is a wide range of factors influencing the development of a building site, from the physical characteristics of the site itself to legal restrictions, planning controls and building regulations.

Land Ownership

Legally persons hold interests in land but do not own it, as all land is held from the Crown on tenure. Nevertheless a freeholder with a freehold interest in the land is the absolute owner of the property in perpetuity, and he can do as he likes with it provided he does not contravene the law of the land or interfere with the rights of others.

A freeholder can create a lesser interest out of his absolute one, such as a leasehold interest where the leaseholder will hold the land for a limited period and subject to the payment of rent. The landlord or lessor retains an interest in the land, known as a reversion, and is entitled to receive the agreed rent. The tenant or lessee has exclusive possession of the property for the period of the lease, provided he pays the rent and observes the covenants or conditions attached to the lease. As a general rule the lessee can assign or sell his interest to another person or grant a sublease for a shorter term than his own lease. The Leasehold Reform Act 1967 permits certain tenants to purchase the freehold of the property which they occupy. With a building lease the lessee pays a ground rent for the land and undertakes to erect and maintain suitable buildings on the land. For the period of the lease, often ninety-nine years, the lessee receives an income from his possession of the buildings and land but on the termination of the lease, both buildings and land revert to the freeholder.

Site Characteristics

Each site has its own characteristics, which have an important influence on its suitability for development for a particular purpose. The main characteristics are now considered.

Soil conditions. The subsoil should have a reasonable load-bearing capacity, as poor soils create foundation problems and increase constructional costs.

Water table. It is desirable that the site should be well above the water table and free from the possibility of flooding. Working in wet conditions is difficult and a permanently wet site can give rise to unhealthy conditions for occupants, and deterioration of the buildings.

Contours. A reasonably level site will reduce constructional costs, particularly where the buildings cover large areas as with factories. Steeply sloping sites require stepped foundations and extensive earthworks, and may involve special land drainage installations as well as being inconvenient to users of the site.

Obstructions. These may take various forms and all involve additional expenditure in site clearance work. Humps and hollows require levelling or filling, inconveniently sited buildings or trees need demolishing or felling, ponds require filling and possibly involve some drainage work, and ditches may need piping.

Services. The availability of essential services, such as sewers, water mains, electricity cables and possibly gas mains, of adequate capacity, is an important consideration. Private sewage works, water supply installations or electricity generating plant are expensive in both provision and operation. Culverts or sewers crossing a site may also prove costly in realignment and it may be difficult to secure adequate falls on a longer route.

Access. Satisfactory access to the site must be available and some types of development will require good access roads leading to the site.

Aspect. Ideally the site should be on a gentle slope facing in a southerly direction to secure maximum sunlight and protection from the cold northerly and easterly winds.

The site requirements for different uses vary substantially but in nearly all cases the sites have to be accessible to users, and the use of the site needs to be compatible with the uses of adjoining sites. A consideration of the requirements of two widely differing uses will serve to illustrate these points.

School Site Requirements

(1) Sited centrally in relation to the catchment area from which pupils will be drawn.

(2) Away from main roads and resultant noise and vibration but readily accessible to network of distribution roads, and also bus routes where possible.

(3) Sufficient area to accommodate buildings and playing fields with ample space for probable future extensions.

(4) Preferably of regular shape and to even and gentle falls, of porous subsoil and well above water table.

(5) Freedom from major obstructions or adverse restrictive convenants, etc.

(6) Adequate underground services.

(7) Compatible with adjoining uses.

(8) Not excessively expensive.

(9) Planning consent forthcoming.

Factory Site Requirements

(1) Good access by road and in some cases access by water and/or rail is beneficial.

(2) Suitably located in relation to raw materials, markets and workers.

(3) Adequate underground services.

(4) Adequate area for factory and possible future extensions together with ancillary uses such as car parking, and of regular shape.

(5) Reasonably level site with subsoil of suitable load-bearing capacity above water table.

(6) Freedom from major obstructions or adverse restrictive covenants, etc.

(7) Compatible with adjoining uses.

(8) Not excessively expensive.

(9) Planning consent forthcoming.

Planning Controls

Planning controls stem from the operation of the Town and Country Planning Acts of 1962 and 1968, and are administered principally by the county councils and county borough councils, termed local planning authorities, although these powers may be transferred to district councils in 1974. In the case of Greater London there is a division of responsibility between the borough councils and the Greater London Council. Local planning authorities were required under the 1947 and 1962 Acts to prepare development plans for approval by the appropriate minister, indicating the manner in which the land should be used and the stages by which the development should be carried out. The 1968 Act prescribed a new planning system to be introduced progressively whereby structure plans were to be prepared first, subject to confirmation by the minister, followed by detailed local plans where no confirmatory action was required. Planning proposals will have direct relevance to deliberations at the brief stage as not only will they indicate the uses to which the land can be put and permitted densities, but will also show the manner in which adjoining areas are to be developed and the location and extent of major public works, such as new roads and improvements of existing roads. A road widening may sterilise part of the site, whereas a new road may give improved means of access.

Planning permission is required for most forms of development and this is obtained by making application to the local planning authority in the prescribed manner. The local planning authority can give unconditional permission, permission subject to conditions or refuse permission altogether and there is a right of appeal from this decision to the minister. Applications can be in outline, giving brief particulars of the development in order to secure permission in principle possibly before the land is purchased, or detailed. The development has to commence within five years of the first approval.

Development is defined in the Town and Country Planning Act 1962 as 'the carrying out of building, engineering, mining or other operations in, on, over or under land, or the making of any material change in the use of any building or land'. The Town and Country Planning (Use Classes) Order 1963 prescribed a number of classes and where a change of use keeps within the same use class it does not constitute development and does not therefore require planning permission. A change from one use class to another, such as from an office to a shop, will require permission as the change is then material. The conversion of a single dwelling into two or more flats also constitutes development. Certain types of development are, however, expressly excluded from planning control and are deemed to be *permitted development*, and some typical examples follow.

(1) Maintenance, improvement and alteration to the inside of a building provided it does not materially affect the external appearance.
(2) Repair or replacement of underground services.
(3) Certain highway works.
(4) Enlargement of buildings up to certain limits in volume and with some other restrictions.
(5) Work on gates, fences and boundary walls up to certain specified heights.

Planning proposals in the form of development plans, town maps, and local plans will normally indicate the uses to which the land may be put and the permitted maximum density to which it can be developed. In the case of residential development the unit of density will normally be the number of persons or habitable rooms permitted per hectare of site. With nonresidential buildings the unit of density is often the floor space index (FSI) or plot ratio. The floor space index of a building equals

$$\frac{\text{total amount of floor space of the building (on all floors)}}{\text{total area of building site} + \text{half width of adjoining streets.}}$$

Whereas plot ratio is the ratio of floor area of the building to plot area, excluding any reference to adjoining roads. The floor space index has the additional merit of making allowance for the benefits of light and air accruing to a property from adjoining streets. Typical FSI guide figures for different classes of building are: shops – 1.5; offices – 2.0; wholesale warehouses – 2.25; and light industry – 1.5. The rigid application of low floor space indices or plot ratios on central area sites can result in development being uneconomical.

Buildings of Special Interest

The Department of the Environment has compiled a list of buildings of special architectural or historic interest, and these buildings cannot be demolished, extended or altered in such a way as to seriously affect their character without obtaining a *listed building consent* from the local planning authority. The only exceptions are where the works are urgently needed for the safety or health of persons, preservation of the building or the safety of neighbouring property. Breach of these requirements or of the conditions prescribed in a consent constitutes a criminal offence. There is provision for *enforcement notices*, requiring the owner or occupier to restore the property to its former state, and for *purchase notices* whereby the owner requires the local planning authority to purchase his interest in the property because of the reduction in value resulting from the building being listed. It will be appreciated that a listed building can constitute a major obstacle in a development scheme.

Tree Preservation Orders

Under the Town and Country Planning Act, a local planning authority is empowered to make a tree preservation order for the preservation of trees or woodlands in order to preserve amenity. Such an order prohibits the felling, topping, lopping or destruction of specified trees, groups of trees or woodland without consent, and may also require the replanting of woodlands felled during forestry operations. Compensation is payable when consent is withheld and orders cannot apply to trees which are dead, dying or dangerous. Once again, tree preservation orders can operate to the disadvantage of a developer in that they may restrict the form and extent of the development.

Other Statutory Requirements

Industrial development certificates. It is often necessary to obtain an industrial development certificate (IDC) before an industrial building can be erected, re-erected or extended, or a nonindustrial building converted to industrial use. Some small industrial building projects are exempt from these requirements; the limits of floor area vary in different parts of the country and are subject to exemption orders made under the Control of Offices and the Industrial Development Acts 1965 and 1966, but are often around 465 m^2. The certificates are issued by the Department of Trade and Industry and will only be issued if the Department considers that the proposed development is consistent with the proper and balanced distribution of industry. Where a certificate is refused by the Department of Trade and Industry there is no right of appeal and the local planning authority is unable to grant planning permission for the project.

Office development permits. Under the provisions of the Control of Offices and Industrial Development Act 1965 it is frequently necessary to obtain an office development permit (ODP) before applying for planning permission to erect or extend office buildings or to convert other buildings into offices. Districts where permits are required include the London area, south-east England and much of the Midlands. The situation varies from time to time and exemption limit orders prescribe the latest controls and their area of operation. The limit of size beyond which a permit is necessary also varies from 93 m^2 in some areas to 279 m^2 in others. Office development permits are issued by the Department of Trade and Industry.

Highways requirements. By virtue of the Highways Act 1959, many local authorities, particularly highway authorities, operate new street by-laws which regulate the widths, levels and form of construction of new streets. A developer is normally required to submit plans and other details of any new streets for approval. Where the local authority believes that the new street is likely to become a main thoroughfare they may require the street to be of greater width than would normally be necessary, and if they require the street to be widened more than 6 m above the normal width, the developer must be reimbursed the extra cost. The developer is also entitled to compensation if he is required to amend his street layout to provide improved junctions with existing streets. The by-laws prescribe minimum widths of carriageway according to the function and length of the street and whether it is open at one or both ends, and also the number and minimum widths of footpaths.

Typical street by-law requirements are as follows.

(1) Principal street in residential neighbourhood: minimum carriageway width of 6.70 m and two footways each with minimum width of 2.75 m or 3.66 m for shopping frontages.
(2) Normal residential estate road: minimum carriageway width of 4.90 m and footway to each developed frontage with a minimum width of 1.80 m.
(3) Residential cul-de-sac not exceeding 45 m in length or residential street bordering on to open space: minimum carriageway width of 4.0 m and single footway with minimum width of 1.80 m.
(4) Principal shopping street in main business area: minimum carriageway width of 13.40 m and two footways each with a minimum width of 4.60 m.

Highway authorities, being local authorities with highway responsibilities, often prescribe *improvement lines* alongside highways to reserve sufficient land for future road improvements. The developer is very restricted in what he is able to build between the improvement line and the street. In other cases the highway authority may prescribe a *building line* on either side of a street and it frequently consists of a line joining the fronts of existing buildings. With minor exceptions, no building work is permitted in advance of a building line. The practical effect of both improvement lines and building lines on development schemes at the design stage is to sterilise strips of land adjacent to highways. Where a developer wishes to close or divert a public right of way which is crossing his site and preventing the use of the land to best advantage, he may apply to the local authority for an order extinguishing or diverting the path. Such an order requires approval by the local planning authority and confirmation by the minister. Where a road is involved it will be necessary for the highway authority to apply to the magistrates or obtain an order from the minister. The ministry involved is the Department of the Environment.

Building requirements. Local authorities are responsible for ensuring that all building work is carried out to certain minimum standards of construction. For most of England and Wales, the requirements are laid down in the Building Regulations, formulated under powers derived from the Public Health Act 1961, whilst the Greater London area is subject to a different set of regulations administered by the Greater London Council. Even comparatively minor alterations and improvements require consent under Building Regulations, and the Regulations cover such as aspects as fitness of materials; preparation of site and resistance to moisture; structural stability; structural fire precautions; thermal insulation; sound insulation; stairways; refuse disposal; open space, ventilation and room heights; chimneys, flue pipes, hearths and fireplaces; drainage; sanitary conveniences; and wells, tanks and cisterns. The quantity surveyor needs to be familiar with Building Regulation requirements and their likely impact on alternative design proposals.

Plans of proposed building work have to be submitted to the appropriate local authority, as distinct from the local planning authority, for approval under Building Regulations. After approval the contractor is required to submit notices to the local authority at various stages of construction, in order that the appropriate officer to the authority (building inspector, building control officer or building surveyor) can inspect the work and ensure that it complies with the approved plans and the Building Regulations. Where contraventions occur the authority can take enforcement action.

Certain classes of building are subject to additional controls. For instance, the Factories Act 1961 prescribes certain constructional and operational requirements for factory buildings, which must affect the layout of a new factory and also its constructional and operational costs. The requirements include a minimum working area of just over 10 m^3 per employee; a normal minimum temperature of 16°C; minimum lighting and ventilation standards; minimum provision of sanitary appliances, hot and cold water supply, drinking water and storage accommodation for employees' clothes; adequate fire-fighting and warning systems and means of escape in case of fire; and suitable protective devices to lifts, hoists and openings. Other statutory provisions affecting factories include the Clean Air Acts, 1956 and 1968, and the Thermal Insulation (Industrial Buildings) Act 1957. The Clean Air Acts aim at reducing atmospheric pollution and are likely to necessitate taller factory chimneys with consequently higher costs, whilst the Thermal Insulation Act prescribes certain minimum thermal insulation requirements for factory roofs which have a significant effect on both constructional and operational costs.

New offices and shops have to comply with the Offices, Shops and Railway Premises Act 1963, which prescribes detailed requirements on lines similar to those contained in the Factories Act for industrial buildings. The minimum working space requirements are 3.72 m^2 per employee.

There are a number of statutes which prescribe minimum provision in relation to means of escape in case of fire and fire-fighting appliances. These include the Cinematograph Acts 1909 and 1952 for cinemas, the Licensing Act 1964 for certain clubs, the Housing Act 1969 for certain residential properties occupied by more than one family and the Public Health Act 1936 for theatres, some restaurants, churches and places of public assembly. In addition, proposals for hotels and restaurants are examined by the public health inspector to the local authority under the Food Hygiene Regulations, and proposals for licensed premises need the approval of the licensing justices. Special precautions have to be taken in the construction of structures to be used for the storage of petrol and in garages to house more than twenty cars, by virtue of the Petroleum (Consolidation) Act 1928, whilst insurance companies usually prescribe requirements for the storage of large quantities of oil. The need to obtain a variety of consents from different authorities can result in a lengthening of the design period, greater probability of amendments to design and the possibility of increased costs.

ENCUMBRANCES AND EASEMENTS

On occasions land is subject to restrictions of one kind or another which adversely affect the development of the land or its enjoyment by owners and occupiers. Such restrictions can make development more costly and even sterilise parts of the site. Hence they need careful investigation in any feasibility study. The two main types of restriction are restrictive covenants and easements, but these should be considered against the background of natural rights enjoyed by occupiers of land.

Natural Rights

At common law an occupier of land has certain natural rights which impose obligations on neighbours, although some of these rights can be suspended by the granting of easements. Typical examples include the right of an occupier of land to have the support of the soil in its natural state from the land of the neighbour, although the neighbour is not made responsible for the support of a building on this land. There is also a natural right to water flowing in a defined course as a natural stream over the occupier's land, subject to the rights of other owners along its banks; these are termed *riparian owners*. In addition, water or air passing from land in one ownership to that in another should not be polluted, as an occupier of land has a natural right to pure water and air.

Restrictive Covenants

The nature and objectives of restrictive covenants vary widely; they may have been imposed on the land in previous conveyances or the freeholder of a large parcel of land may himself impose restrictions when disposing of smaller plots in order to secure satisfactory development of the whole area. Restrictive covenants impose conditions which govern the use of the land; sometimes they are so onerous as to restrict the use of the land considerably, whilst in other cases the effect is marginal.

Covenants can be of two kinds: positive and negative. A positive covenant imposes an obligation on the purchaser to do something, such as erect a building. Restrictive covenants are negative in character in that they restrict the purchaser in the manner in which he can use the land. For instance a restrictive covenant may specify that the density of residential development shall not exceed ten houses per hectare and also lay down a minimum floor area for new houses and prescribe conditions as to materials to be used or other matters affecting the external elevations, or the submission to and approval of house plans by the vendor. Restrictive covenants created since 1925 on a freehold must be registered as land charges if they are to be enforceable against all purchasers, by virtue of the Land Charges Act 1925. With building schemes where the purchaser of each plot enters into covenants with the vendor, each purchaser or his assignees can normally sue or be sued by every other purchaser for breach of covenants, provided there is a common vendor; the covenants apply to all plots and are for the benefit of all purchasers who have bought their plots on this understanding.

Where restrictive covenants are reasonable in their requirements and are framed to secure the orderly development of the area, they assist in maintaining values in that area, but where the character of the district has so changed that the restrictions are completely outdated and no longer relevant, then they are likely to retard normal development with consequent loss of value. It is accordingly essential to investigate any restrictive covenants attaching to land before its purchase is completed and to take account of the cost effect of their operation in any feasibility study.

In some areas much-needed development is thwarted by unreasonable and unacceptable restrictions imposed by restrictive covenants. For these reasons the Law of Property Act 1925, as subsequently amended by the Lands Tribunal Act 1949, the Landlord and Tenant Act 1954 and the Law of Property Act 1969, enables a person to apply to the Lands Tribunal for the modification or removal of a restriction if the changes in character of the property or the neighbourhood or other material circumstances make the restriction obsolete, or such as to impede the reasonable use of the land and do not secure any practical benefit to other persons or is contrary to the public interest, and that money will be adequate compensation for the modification or discharge. These Acts do not apply to restrictions imposed on land given gratuitously or for a nominal consideration for public purposes, nor to leaseholds of less than forty years or for recent ones of over forty years. A feasibility study must take account of any compensation payable for the modification or discharge of restrictive covenants, as it can be quite costly.

Easements

The ownership of some properties may give the owner the right to enter another property (the *servient tenement*) and take something from it, other than water, such as sand, gravel, timber or even fish. This right is termed a *profit à pendre*. A more common right attaching to the ownership of a property is an *easement*, which the owner of the *dominant tenement* secures over another property (the servient tenement). The tenements must be in different ownerships, the right must be capable of being granted, the servient owner must not be involved in any expenditure in complying with the easement and it must not involve the removal of anything other than water from the servient tenement. A legal easement is made by a grant from the owner of the servient land or by prescription (long use of the privilege by the dominant owner under certain conditions), and is binding on all persons who occupy the servient tenement.

The more common easements relating to building development are as below.

(1) Right of light: the right of light to a building becomes legally protected after it has been enjoyed for a period of twenty years. A right of light prevents the owner or occupier of land from erecting buildings on it which will obstruct the light passing on to the other person's land.

(2) Right of support: the ownership of land carries with it the right to support from the adjoining land but not for the support of any buildings subsequently erected on it.

(3) Right of way: a private right of way may be presumed if it has been enjoyed for twenty years and becomes absolute if enjoyed for a period of forty years. Public rights of way are not easements.

(4) Right of drainage or pipe easements: the right to drain across the property of another. An easement which permits the dominant owner to lay pipes across the land of the servient owner, whether for purposes of drainage or water supply, may sterilise the strip of land through which it passes and provision will have to be made for access to and inspection of the pipes.

Land subject to easements is normally reduced in value as restrictions are placed upon the owner's full use and enjoyment of the site. As with encumbrances it is essential that their effects should be taken into account in assessing the value of the land and in any feasibility studies.

MATTERS DETERMINING LAND USE AND VALUE

In concluding this chapter it might be helpful to the reader to summarise the main categories of factors which determine land use and, in consequence, land values.

(1) Location: primarily from standpoint of accessibility and compatibility with adjoining uses.

(2) Economics: with particular emphasis on supply and demand, and through this the satisfying of a need or want.

(3) Topography: with the main emphasis on the physical characteristics of sites acting as sorters of suitability for specific uses.

(4) Tenure: freehold or leasehold and if leasehold the term and conditions of the lease.

(5) Servitudes: existence or otherwise of easements such as rights of way, light and drainage, and/or restrictive covenants.

(6) Legislation: such as Town Planning Acts and through them the operation of planning schemes, Building Regulations and other statutory controls.

REFERENCES

1. M. CLAWSON, R. B. HELD and C. H. STODDARD. *Land for the Future.* John Hopkins Press, Baltimore (1962)
2. G. P. WIBBERLEY. Land scarcity in Britain. *Journal of the Royal Town Planning Institute,* **53.4** (1967)
3. E. JONES. *The City in Geography.* London School of Economics and Political Science (1962)
4. F. S. CHAPIN and S. F. WEISS. *Factors Influencing Land Development.* Institute for Research in Social Science, University of North Carolina (1962)
5. R. H. BEST and J. T. COPPOCK. *The Changing Use of Land in Britain.* Faber and Faber (1962)
6. S. V. CIRIACY-WANTRUP. The 'new' competition for land and some implications for public policy. *Natural Resources Journal,* **4.2** (1964)
7. M. CHISHOLM. *Rural Settlement and Land Use.* Hutchinson (1966)
8. A. Z. GUTTENBERG. New directions in land use classification. *American Society of Planning Officials,* Chicago. (1965)
9. R. H. BEST. Land for New Towns. Town and Country Planning Association (1966)
10. MINISTRY OF HOUSING AND LOCAL GOVERNMENT. Planning Bulletin no. 2: Residential areas – higher densities. HMSO (1962)
11. NEW TOWNS COMMITTEE REPORTS, Cmd. 6759, 6794 and 6876. HMSO (1945–46)
12. R. H. BEST. Extent of urban growth and agricultural displacement in postwar Britain. *Urban Studies,* **5.1** (1968)
13. W. LEAN and B. GOODALL. *Aspects of Land Economics.* Estates Gazette (1966)
14. A. EDWARDS. Land requirements for United Kingdom agriculture by the year 2000. *Town and Country Planning* (March 1969)
15. E. JONES. Resources and environmental restraints. *Urban Studies,* **6.3** (1969)
16. G. P. WIBBERLEY. Pressures on Britain's land resources. Tenth Heath Memorial Lecture, University of Nottingham, School of Agriculture (1965)
17. I. H. SEELEY. *Municipal Engineering Practice.* Macmillan (1967)
18. F. S. CHAPIN. *Urban Land Use Planning.* University of Illinois Press (1965)
19. T. HOYES. The evaluation of alternatives in land use planning. *Proceedings of the Nottingham Symposium on Subregional Studies.* Regional Studies Association (1968)
20. D. R. DENMAN and T. HOYES. Land systems and urban development. International Federation of Surveyors (1968)
21. WORLD HEALTH ORGANISATION. Technical report No. 297. Environmental health aspects of metropolitan planning and development. World Health Organisation, Geneva (1965)
22. R. ARVILL. Planning and the future. Forward in Europe, Strasbourg (1968)
23. R. W. KIEFER. Land evaluation for land use planning. *Building Science,* **1**. Pergamon Press (1965)
24. F. J. MCCULLOCH. *Land Use in an Urban Environment – the Social and Economic Determinants of Land Use.* Liverpool University Press (1965)
25. K. CLEMENS. Planning policies and organisation. International Federation of Surveyors (1968)
26. R. J. S. HOOKWAY and J. A. HARTLEY. Resource planning. Countryside Commission (1969)
27. S. M. WILLHELM. *Urban Zoning and Land Use Theory.* Free Press of Glencoe, New York (1962)
28. R. G. BELLCHAMBERS. An exercise in participation. *Town and Country Planning* (September 1969)
29. P. COWAN. Developing patterns of urbanisation. *Urban Studies,* **6.3** (1969)
30. B. ANSTEY. A study of certain changes in land values in the London area in the period 1950–1964. *Land Values.* Sweet and Maxwell (1965)
31. W. LEAN. *Economics of Land Use Planning: Urban and Regional.* Estates Gazette (1969)
32. N. LICHFIELD. Evaluation methodology of urban and regional plans: A review. *Regional Studies,* **4.2** (1970)
33. TOWN PLANNING INSTITUTE. Research for urban planning. Report F: Geographical factors in the location and size of settlement, (1963)
34. G. H. PETERS. Farm sales prices in 1962. *Estates Gazette,* **185** (1963)

35. P. A. STONE. The price of sites for residential building. *The Property Developer* (1964)
36. P. F. WENDT. Theory of urban land values. *Land Economics* (August 1957)
37. M. E. T. THORNCROFT. Appraisal technique and practice. International Federation of Surveyors (1968)
38. P. A. STONE. The price of building sites in Britain. *Land Values*. Sweet and Maxwell (1965)
39. UTHWATT COMMITTEE. Expert committee on compensation and betterment. Cmnd. 6386 (1942)
40. N. LICHFIELD. *Economics of Planned Development*. Estates Gazette (1966)
41. P. HALL. The land values problem and its solution. *Land Values*. Sweet and Maxwell (1965)
42. P. H. CLARKE. Site value rating and the recovery of betterment. *Land Values*. Sweet and Maxwell (1965)
43. R. W. WATES. Homes for the nineteen-eighties. *The Chartered Surveyor* (April 1969)
44. J. S. ANDREWS. A study of the development of land values in a north London dormitory area, relating the changes to demand and supply factors, with particular reference to planning legislation and practice in the post-1945 period. Unpublished MSc. (Est. Man.) dissertation, University of London (1961)
45. D. R. DENMAN. Land in the market: Hobart Paper No. 30. Institute of Economic Affairs (1964)
46. J. MCAUSLAN. Price movements for residential land: 1965–1969. *The Chartered Surveyor*, **102.3** (1969)
47. W. BRITTON. Control through fiscal policy. Paper 2: Public control of land use. College of Estate Management, (1966)
48. P. HILLMORE. One damned plot after another. *The Guardian* (2 June 1972)
49. T. ALLAN. Building into space. *The Guardian* (12 August 1971)

13 ECONOMICS OF BUILDING DEVELOPMENT

THIS CHAPTER IS concerned with the basic criteria for development undertaken in both the public and private sectors, problems of land acquisition, financial considerations and sources of finance. An investigation is made of matters contained in a developer's budget and its practical application to various types of development projects.

THE ESSENCE OF DEVELOPMENT

Cannell[1] asserts that every development whether it be for a public authority, industrialist or private investor, has a 'market value' – a potential worth or earning power. Even civic buildings, hospitals, churches and polytechnics have an assessable value to the community – a cost above which it is not reasonable or feasible to build. Within certain limits of aesthetics, function and performance, the most economic development is that which shows the greatest return to the community for the minimum capital invested. This does not imply that the cheapest is the best; often the opposite is the case.

The art of phasing development to give an early return which can be used to pay for the less remunerative items is one of the objectives of a skilful developer. In this connection, an amalgamation of public and private agencies in development can be of great benefit to the community. It is also essential to integrate the planning of large scale redevelopment. For instance, the retention of an existing road layout even if only to provide a pedestrian precinct or parking facilities can economise in new construction, and the potential earning power of existing facilities should not be overlooked. Time is also important in the planning process; for maximum economy the time between capital expenditure and completion of a project should be kept to a minimum.

The private developer or industrialist will require a financial appraisal or feasibility study to determine the likely capital expenditure and probable revenue in order to arrive at the anticipated return on the money invested. Whether the project is to be financed by public or private funds, it is important to know the cost implications at the outset in order to be able to appraise the viability of the scheme. It is necessary for the developer to know the nature and extent of the proposed development, its cost and the time required to complete it. Indeed, the whole development process is becoming more sophisticated. Schemes need to be appraised from every aspect – aesthetic, fiscal and social. The long and frequently frustrating negotiations to assemble sites, obtain planning permission, barter with local authorities and secure finance, demand a truly professional approach. As to the scale of development, about £2000 million is being spent annually by property companies.[2]

Hence a developer usually wishes to know whether the investment of capital in a project will be justified by the return which he can expect to receive. It is therefore necessary to assess as accurately as possible the value of all the expected returns and benefits and to compare them with the estimated costs. One problem is to express all the benefits in monetary terms, as there may well be some indirect and intangible benefits, such as more contented employees or greater prestige value, which flow from a project. The quantity surveyor is frequently called upon to make cost comparisons of different design proposals with varying capital, maintenance and running costs. The task of the quantity surveyor is to inform the developer which is the most economical scheme after taking all these costs into account.

To this end a working party of the Royal Institution of Chartered Surveyors[3] strongly advocated increased co-operation between surveyors in general practice and in quantity surveying. The report of the working party recommended that general surveyors should have a general knowledge of building costs in addition to a detailed knowledge of values, whilst quantity surveyors ought to have a general knowledge of values as well as a detailed knowledge of building costs. There is an evident need for the quantity surveyor to be aware that the general practice surveyor, in making his general financial appraisal, has to consider the capital cost of the works, land purchase, compensation for extinguishment of leases, bridging finance, long-term finance, rental and capital values, profitability, and maintenance and other outgoings. The key factor in any successful project is generally believed to be the triangle of valuation surveyor, quantity surveyor and architect. The general practice surveyor will be primarily concerned with the broad economics of development, the architect with the design of the project and the quantity surveyor with the interaction of these two aspects of development. Some of the more important matters to be resolved by the development team include:

(1) ensuring development to maximum plot ratio;

(2) planning the most economical use of available floor space;

(3) implications of different methods of placing the building contract;

(4) the speed of construction balanced against financial considerations, such as the cost of bridging finance and loss of rent or interest, etc.;

(5) the effect of incurring extra capital costs including expensive finishings, balanced against additional net rental value (if any); and

(6) the effect of incurring extra capital costs balanced against a consequent reduction in future maintenance costs, including depreciation allowances.

DEVELOPMENT PROPERTIES

The term *development properties* refers to properties in which a developer invests capital to secure a greater and more profitable return than that previously received. Typical examples are the development of agricultural or accommodation land for residential purposes and the redevelopment of central area sites which are under-developed or occupied by obsolescent buildings. One method of assessing the value of sites available for development is the *residual method*, as illustrated in the following example.

Value of site when developed for the most profitable permitted use (gross development value)	£250 000
less	
Estimated cost of development and allowance for risk and developer's profit	180 000
Estimated value of existing property	£ 70 000

Central area premises will command much higher development values than premises on the outskirts of a town, as they will produce much higher rents. The most profitable permitted use will depend upon the policy of the local planning authority and the nature of the planning proposals for the area. The advice of a valuer is vital to secure this objective. The location will also influence the decision as to the type of design and standard of finish, as a central location will justify a higher standard of design and finish than a suburban site.

BUDGETING FOR PUBLIC AND PRIVATE DEVELOPMENT

Public Finance

Central government funds are obtained from two main sources: internal and external. Internal sources are mainly confined to nationalised industries where internal cash flows provide a source of finance for the industry's investment; external sources of finance are taxes and borrowing.

In the case of local authorities, money may be obtained by grant from central government funds or may be raised by the local authority's own borrowing or taxation (rates). In general, local authorities meet current expenses out of the rate fund which is produced by local taxation and may be augmented by the profits of successful local authority enterprises. Local authorities borrow money for capital investment by issuing securities in the ordinary market and also borrow from the Public Works Loan Board (a government agency), in addition to receiving money from central government funds and accepting money on deposit at interest.[4] In 1970 the sources of local government current finance were forty-five per cent grants from central government; thirty-three per cent rates; one per cent trading surplus; nineteen per cent rent and two-per cent interest; while capital finance comprised twenty-eight per cent current surplus; nine per cent grants from central government; thirty-eight per cent borrowing from central government and twenty-eight per cent loans and receipts from other sources.[28]

Developer's Return

The majority of building clients are seeking a financial return on the capital they invest in building projects. Nevertheless, there are some projects which are of a semi-commercial nature like local authority housing, and others which are non-commercial such as schools and churches. With the latter category of project the economic return approach will be difficult to apply and local authority housing falls between the other two classes of project with its complex political and social implications. From the quantity surveyor's point of view, it is important that the developer, in whatever category he may fall, shall prescribe the upper limit of his budget at the outset, in order that the quantity surveyor may formulate a cost plan with the object of securing value for the money expended and a realistic and advantageous distribution of costs throughout the various parts of the building. The general procedure and problems associated with budgeting for each category of development will now be considered.

Budgeting for Commercial Properties

Allan[5] has described how the major financial houses have come to recognise that property, particularly commercial and industrial property, has a good investment value and compares favourably with other investment opportunities. Sources of finance for commercial development projects are examined later in the chapter. Investors often require, as part of the consideration for lending, a share in the profits. This has resulted in the investor showing a very real concern in the situation, design and profitability of the building. Projects must be well-conceived and efficiently executed; built to designs which are attractive, satisfy the local planning authorities and meet the needs of occupants, as well as being profitable. In large-scale developments, land assembly may have taken years involving heavy expenditure with no immediate return and, in these circumstances, the developer's judgement can no longer be merely intuitive. It must be backed by expert opinion to ensure that the investment is a secure one and will produce a return which compares well with other available investments. Allan[5] has described how assessment is not just a matter of building and land costs matched with a rent income. It concerns people and their working and shopping habits, communications and trading trends and forecasts of growth.

Modern office and shop developments have always been in good demand for investment purposes and, of recent years, industrial and warehouse property has become more acceptable, although requiring a higher return.[6] The one noticeable omission in the investment field is that of rented housing. Rent control, doubtless introduced with the best of motives and retained for short-term political expediency, has forced landlords to subsidise their tenants. However, a Government White Paper in 1971[7] contained proposals for the establishment of fair rents and rent rebates or allowances to tenants of unfurnished properties who cannot afford to pay the full rents. The implementation of the provisions could alter the situation significantly by reducing the shortfall in rented living accommodation, encouraging the improved maintenance of existing houses, and raising rents to a proper market level so that funds can once more be set aside for housing on a profitable basis.

Financial institutions like to maintain a balance in their property portfolio between shop, office and industrial premises, but whatever type of property is considered, certain factors have to be taken into account. These factors include the return on capital invested both now and in the future, the financial security afforded by the tenant, the quality of the location and its future prospects, terms of lease and liability for insurance and repairs.[6]

Relf[8] described how in 1970 the property investment and finance market was becoming much more selective as to the quality of property and the need to obtain a balanced distribution as between different types of property.

Offices

A common type of commercial development project is a block of offices, where the developer needs to be assured of a reasonable excess of income over expenditure. He will be laying out a substantial capital sum now in anticipation of a larger return later. The capital outlay will be for site and buildings and the return is often in the form of rents. It would be quite absurd to spend a large sum of money on a site for a proposed office block unless and until an expert financial appraisal has been carried out indicating the overall supply/demand situation and rent levels for offices in the area. This appraisal would involve consideration of many related aspects, such as growth of commerce in the area, demand for all types of office space, site characteristics, transport facilities, public utility services and associated services, such as banks and markets.

The rent level is influenced considerably by the quality of the accommodation, so the standard of quality must be determined at an early stage. The amount of space to be provided will depend on a number of factors, including estimated demand, site area, planning restrictions and rights of adjoining owners. The site value can be determined by the residual method, whereby building costs and developer's profit are deducted from the gross development value, and this approach will be examined in some detail later in the chapter. The quantity surveyor will prepare estimates of building costs based on the architect's preliminary designs and the standard of building needed to attract the rents set by the valuer. It is vital that the quantity surveyor's estimate is realistic, otherwise the developer may have difficulty in obtaining his desired profit margin.

Shops

The complexities surrounding schemes of large scale commercial redevelopment may not be generally appreciated. Developers must ascertain the requirements of the large space users and in many cases a large measure of preletting is necessary so that the development is planned around them. Furthermore, the scheme is likely to take several years to complete. Erdman[9] has described how even after securing anchor covenants, because of ever-rising costs and retailer's lack of capital for expensive shopfitting, it takes time and knowhow to let all the remaining shops, with a certain loss of interest on capital in the meantime. More existing shopping streets will be pedestrianised and the ultimate pattern of shopping provision is likely to be the fully covered air-conditioned shopping centre, involving developers in management and promotion activities.

Apart from yield rate, inflation has caused developers and their advisers to consider future growth and the need for periodic rent reviews. Unlike offices, shop tenants of good standing have to invest capital on shop fitting and resist earlier rent reviews than seven years, as it takes time for a new branch to achieve full earning capacity. The large space users whose covenants are sought after often hold out for a fourteen-year review period, particularly if their existing branch already forms part of the development site and rehousing is necessary. This sometimes causes difficulties in obtaining finance, and Erdman[9] advocates a basic minimum rent plus a percentage rent based on turnover to eliminate rent reviews, but it is not likely to be favoured in this country by shop tenants. Where developers collaborate with local authorities in partnership arrangements for the redevelopment of central areas, it is essential that the developers should receive an adequate share of the return, commensurate with the work, expertise and risks involved.

Factories

Annual investment in industrial buildings is around £175 million and is now only second in size to the housing programme. Mergers and modernisation seem likely to increase this figure. Much of the present factory and warehouse provision is outdated and ill-suited for today's needs, whilst some of the newly erected factories are sited in unsatisfactory situations, sometimes as a result of government policies and local authority action. New industrial estates are not confined to new and expanding towns. Industrial developments in expanding towns and the problems of industrialists have been described elsewhere by the author.[10]

Private Housing

The majority of owner-occupiers finance their house purchase by borrowing from building societies. Attenburrow[11] asserts that while most housebuilders can arrange ninety or ninety-five per cent mortgages over a twenty or twenty-five year term on their new houses, they need to be aware of the financial standing and circumstances of the kind of purchaser they aim to attract to arrive at a viable price range; for the maximum advance a building society will make is usually limited to three times a man's basic income. For this reason most speculative builders concentrate on low cost houses. The private housing developer has no captive market, in the form of a housing list, on which to draw but has to sell each and every one of his houses to an individual purchaser. The percentage of owner-occupied houses in Great Britain rose from 29.5 in 1950 to 50.1 in 1970.[29]

In the nineteen-sixties developers found it necessary to offer a range of house types on any one development. The range is one of price rather than plan or elevational difference within the same price range. The progression is usually in steps of around £150 to £200, starting with a two-bedroom flat or semidetached bungalow, through various sizes of semidetached houses and ending with a large three or four-bedroom detached house. Attenburrow[11] has shown how the economical viability of any development assumes first priority in the design process. The developer has the choice between restricting his activities to well-tried and popular forms of traditional housing in desirable areas, and thus satisfying a demand which can be measured by instinct, experience or crude forms of market analysis, or pioneering a new form of development with all the accompanying risks. Most developers steer a middle course and as described by a North Midlands builder: 'It's a compromise in all directions. Every one of our clients would like a 'one-off' house but he can't afford it. Every builder would like to make his estate more attractive, but economics stop him.' There is always a conflict between the economies to be gained from long production runs and the desire to bring in new designs to stimulate the market. The design must also suit the requirements, tastes and pockets of purchasers whose ideas do not often keep pace with architectural and planning opinion. There is, for instance, considerable sales resistance to terraced houses, most purchasers considering that they should be rented and not bought.

Nicholson[12] believes that the main problems of the smaller house building firms are fivefold.

(1) The full extent of competition is often unknown in exact terms.
(2) Customers' exact requirements are unknown.
(3) House designs could be simplified, and they are often unnecessarily expensive.
(4) Marketing is usually carried out by untrained and inexperienced persons.
(5) Builders provide a poor aftersales service and do not offer any formal maintenance contracts.

Improvement of Older Houses

The Housing Act 1969 provided three types of grant payable to owners of properties which are in need of modernisation but have an effective life of at least fifteen years. It was estimated that there were four-and-a-half million homes in this country in that category. The more generous provisions in the 1969 Act aimed at providing better living conditions for a large number of people, improving the general environment and, at the same time, providing work for the building industry. The houses concerned are generally well-situated for workplaces, transport, public utilities and general amenities; the occupiers generally do not wish to move to new residential districts and the disruption of redevelopment is avoided. That the proposals have met with some success is shown by the increase in number of grants, for instance 40 800 grants were paid in Great Britain in the first quarter of 1971 (see also chapter 5). The view has however been expressed that the grants are being used principally by owner-occupiers, housing associations and local authorities, while private rented accommodation has hardly been touched. The three types of grant follow with the 1971 values.

Standard grants. These cover half the cost of works required for the provision of standard amenities up to the limits indicated: fixed bath or shower (£45); hot and cold water supply at fixed bath or shower (£67.50); wash-hand basin (£15); hot and cold water supply at wash-hand basin (£30); sink (£22.50); hot and cold water supply at sink (£45); water closet (£75). Half the cost of additional features up to a maximum of £375 is also payable to cover piped water supply, septic tank installation, adaption or conversion of buildings to accommodate a shower and like items.

Improvement grants. These are discretionary awards which are not restricted to standard amenities and cover half the cost of the approved works which must not be less than £100. The grant is limited to £1200 for dwellings provided by conversion of a building on three or more storeys or £1000 for other dwellings.

Special grants are also discretionary and cover works required for the provision of standard amenities in houses in multiple occupation. The amount payable for each amenity is the same as for standard grants.

In addition the government announced in 1971 a scheme to operate for a two-year period to cover higher grants for the improvement of older houses in development and intermediate areas, estimated to cost £46 million.

Budgeting for Local Authority Houses

A local authority has to balance income and expenditure relating to its housing provision but can do this in a number of different ways. The local authority has to determine the effect of each new housing scheme on its housing revenue account.

The *revenue* side of the account will contain some or all of the following items:

(1) rents for dwellings received from tenants of houses;
(2) amounts paid in housing subsidies by the government;
(3) receipts from sales of houses, etc.; and
(4) contributions by the local authority out of the general rate fund.

The *expenditure* side of the account will largely be made up of:

(1) loan charges on money borrowed to finance the housing scheme (cost of site, demolition, clearance, compensation for disturbed interests, buildings, siteworks, roads, sewers and other services, and professional fees);
(2) expenditure on housing management and administration;
(3) expenditure on repairs and maintenance, including renewals; and
(4) any other costs associated with housing, such as taxes and other charges.

The rents obtained from local authority houses are insufficient to make them self-supporting, let alone produce a profit. It is generally accepted that it is the local authority's function to provide a service and not to make a profit and this concept was clearly demonstrated in a television play *Cathy come home* which attracted considerable attention. Rent policy is left to each local authority to determine and the biggest single problem stems from the substantial difference between the amount paid by tenants and the economic rent required to balance the housing revenue account.

Local authorities are, however, usually concerned with the tenant's ability to pay. For many years it has been national housing policy for local authorities to operate differential rent schemes under which tenants pay an economic rent, less any rebates to which they may be entitled according to their financial position. Even with differential rent schemes, the majority of local authorities find it necessary to make some contribution from the rates to the housing revenue account. The government particularly favours the provision of houses for elderly persons and to replace slum clearance properties.

Most local authorities have steadily increased the rents of their older houses, largely because of their inability to charge true economic rents on all new houses. It is reasonable to suppose that the time is approaching when the rent burden on older houses will have reached its limit. As local authorities have developed up to their boundaries, the pressures for urban land have resulted in multistorey developments, with consequently higher costs and social problems.[13] Luder[14] has drawn attention to the exceptionally heavy financial burden flowing from the sixty-year loan period for houses, coupled with high maintenance and management costs. This has resulted in local rates being called upon to provide an ever larger additional subsidy, to such a degree that local authority houses in some urban areas are being supported from local rates to the extent of £5 or more per week. The greater the proportion of council house tenants in a local authority area, the fewer remaining ratepayers there are to provide the financial support.

Exchequer subsidies have been paid to local authorities as a result of Housing Acts, and in some cases extra subsidies were paid for expensive sites, excessive foundation costs and houses for special needs. In 1971, the government published a white paper, *Fair deal for housing*[7] aimed at reforming the housing finance structure. The government proposals had three basic objectives: a decent home for every family at a price within its means; a fairer choice between owning and renting a house; and fairness between one citizen and another in giving and receiving help towards housing costs. Achievement of these objectives has been thwarted in the public sector in the past, largely because of the existence of indiscriminate subsidies used regardless of the needs of tenants. Housing subsidies from taxpayers and ratepayers cost about £220 million in 1970–1.

The government proposed to phase out indiscriminate housing subsidies, and tenants who can afford it will move towards a fair rent based on the size, quality, location and state of repair of their dwelling. It was proposed that a rent rebate or allowance would be given to all tenants who cannot afford the full rent, and it was anticipated that, in many cases, tenants would pay considerably less than the fair rent for their home. The move to fair rents would be carefully staged so that the average increase in any one twelve-month period will not exceed 50p per week. Once fair rents have been determined for council dwellings they will be redetermined at three-yearly intervals. Housing subsidies will be concentrated on authorities with the worst housing problems and, in particular, there will be an entirely new subsidy for slum clearance to meet seventy-five per cent of the loss arising out of slum removal operations. The introduction of fair rents would increase the demand for houses for purchase.

Loan charges are usually paid over a sixty-year period, although the lives of the dwellings will probably be much greater. The general order of cost of local authority dwellings is determined by design standards and cost yardsticks emanating from the Department of the Environment. Local authorities are thus subject to some control on designing their houses and, in consequence, are also restricted to the extent to which they can reduce building costs.

The difference between amortised annual capital costs plus expenditure on housing management and maintenance on the one hand, and revenue from rents and subsidies on the other, must be made up by a contribution from the general rate fund. This becomes ever more distasteful as rates are continually being increased and other services are cut back due to lack of resources. New housing projects have to be viewed against this background and the rent and rate resources constitute a built-in upper cost limit. Rising building costs have caused some local authorities to suspend house building operations rather than increase house rents or rates any further, and there is little room for reducing constructional costs.

The financial effect of a new housing project is influenced by the local authority's past history of housing work. Where a local authority has provided large numbers of houses over a long period, then the effect of any one new scheme will probably be no more than marginal. If substantial numbers of houses were built prewar at very low cost, these will assist in subsidising the rents of the newer and more expensive dwellings, provided the tolerable rent limit for the older houses has not been passed. Each local authority's resources, commitments and problems in the housing field are quite different and the solutions reached will depend on the interplay of a number of factors.

The theory of a housing surplus forecast as long ago as 1960 by Professor J. Parry Lewis is beginning to take shape in some areas of the country. It may be that the total housing programme may eventually settle down at around 300 000 dwellings per annum, with consequent limitations on local authority building programmes. Furthermore, it has been suggested that there are about 400 000 council tenants, aged under forty-five, whose incomes are sufficient to enable them to obtain mortgages on homes of their own. The Co-operative Party believes that local authorities should consider transferring some of their existing housing estates to the joint ownership of the tenants in a co-operative scheme, and has drawn upon Oslo as an example where the bulk of the municipal housing has been transferred to a number of co-operative societies.[15]

The operation of cost planning is of considerable value in the public housing field, as the quantity surveyor can determine whether the architect's preliminary design is within the cost limits prescribed by the local authority after considering its housing revenue account. It is also important that an economical housing layout should be produced, consistent with achieving satisfactory standards of amenity, convenience and safety.

Budgeting for Non-commercial Projects

Probably the largest single group of non-commercial buildings are schools. It is not possible to produce a balance sheet of revenue and expenditure as there is no income received from local authority schools. It is therefore necessary to use some other criteria to establish cost standards and to ensure that local authorities receive value for money. The Ministry of Education, since superseded by the Department of Education and Science, established a set of cost standards for different types of school and a method of cost planning which is generally known as the elemental system and is described in chapter 7.

A similar problem arises with the building of churches. The church authorities have some idea of the size and quality of building required but the architect is rarely given a cost target. Yet the church authorities often experience considerable difficulty in raising sufficient funds and may have to resort to whist drives, garden fêtes and even direct appeals to meet the heavy cost of building new churches and of maintaining and operating existing ones. It seems evident that all church projects should be carefully cost-planned and all necessary steps taken to ensure that the client obtains the best possible value for his money.

It is unsatisfactory to leave the matter entirely to the architect without any brief as to costs. Full consideration ought to be given to the actual cost of past projects analysed on the basis of common units of measure, such as places in schools and seats in churches, as well as costing on the basis of per square metre of floor area. Extensive cost records and analyses of past projects must be maintained, to enable realistic yardsticks or metresticks to be established from known costs set against known standards of construction and quality. On occasions the building client or developer establishes the cost target by reference to known costs of projects undertaken on his behalf in the past. Nevertheless adjustments to costs may be needed to allow for differences in building design and site and market conditions, apart from the updating aspect. In addition it should always be appreciated that cost is only one of the factors to be considered in the design of a building project, and that other factors such as appearance, quality and utility must also receive adequate consideration.

LAND ACQUISITION PROBLEMS

There are three principal methods of site acquisition available to the private developer: by private treaty or negotiation, by public or private auction, and by tender. The private treaty method is favoured by the majority of developers. Next in preference come public auctions and very much lower on the list the tender system of acquisition. Public authorities also have powers of compulsory purchase.

Private developers have often suffered from a shortage of suitable land and this has contributed to the slowing down of the housing programme. In late 1970 the Secretary of State for the Environment called on planning authorities to make generous and immediate releases of land, except where this would conflict with green belt policy. He also urged wider consultations between local authorities and the building industry, particularly in areas where builders were experiencing difficulties in developing land because of the inadequacy of main services. The Minister urged local authorities to reassess the land available in relation to housing needs to ensure that there was sufficient land to meet the demand for the next five years, coupled with continuous monitoring of the balance between house building rates and the granting of planning permission. Local authorities were requested to sell land for private development wherever there was a surplus above the authorities' foreseeable requirements. This government action coincided with the abolition of the Land Commission which in its short life failed either to stabilise land prices or to contribute significantly to the release of land for development in areas of acute shortage. Developers have found that their only course of action to increase the amount of land available was through direct pressure on local planning authorities through the time-consuming and costly appeals procedure. In 1972 the Government set up a fund to finance land purchase by local authorities for private development.

FINANCIAL CONSIDERATIONS

Investment and Real Property

The essential nature of an investment is the release of a capital sum in return for income to be received over a period of time. Hence every purchase of real property is an investment. The income in this case is either a cash payment or savings in rent which would otherwise be payable. Purchasers of real property can be broadly classified into two categories: those who purchase as an investment and those who purchase for occupation and use. In both cases it is advisable for the would-be purchasers of real property to consider the available alternatives of other investments and renting of property. For investment purposes the rate of interest is the cost of holding money, or the reward for parting with it.[16] The government operates a monetary policy through its control over interest rates and credit to influence the level of economic activity. When the bank rate is increased, the Bank of England almost invariably uses open market operations to restrict the creation of credit by the banks.

Lean and Goodall[16] have described how the investment aspect of real property is based on its durability and use over time. The owner can transfer the right of use to another person in return for a periodic payment termed rent. The owner is treating the transaction as an investment; he could relinquish ownership in return for a capital sum but prefers to retain it and receive an income over time.

Needs of the Developer

Where a developer intends to build and then sell the completed development, he will only require finance for a limited period. If, on the other hand, he wishes to retain the building as a permanent investment, then he will need to raise two types of finance: short-term finance to purchase land and pay the contractor, and long-term finance which can be raised either by selling an interest in the development or by borrowing against the security of the completed building.[17]

Short-term money is relatively expensive to borrow because there is limited security in the land and the building under construction, and the interest rate will also be influenced by the financial status of the borrower. It is advantageous to the developer to keep his short-term borrowing to a minimum because of the high interest rates. In planning a development project it is important to programme expenditure so that capital remains unproductive for as short a period as possible. For instance, by phasing housebuilding, sales of the first houses completed could be used to finance construction in later phases.

Needs of the Lender

The lender of short-term finance will first consider the security of the capital which he lends and the prospects of receiving interest until the debt is repaid. He will also have regard to the risk in securing repayment of his money at the end of the period of loan. Both capital and income are at an appreciably high rate of risk. A well-established property company should not experience any great difficulty in raising short-term finance, as interest payments can be covered from income accruing from other property. The capital advanced is additionally secured by the value of the uncharged equity of the company. Small housebuilders with limited financial resources are normally able to borrow about seventy to eighty per cent or more of the value of the land and buildings.

Long-term finance is secured on completed buildings, and so the covenant of the borrower is of less importance. The margin between interest payable on capital borrowed and income arising from property will indicate the degree of security of income and consequently of capital. The borrower of money for property development is competing in the general market for his finance, so that interest rates will tend to follow the general market trend. Short-term borrowing rates will be subject to more frequent fluctuations as they will be related to the current bank rate. Property as a long-term investment is likely to remain attractive since it provides, if selected with care, security of capital.[17]

Relationship of Building Costs and Valuations

At an early stage in a development project it is necessary to decide what to build and to forecast the likely financial consequences. A valuer can advise on the optimum form of development and its ultimate value, followed by the quality of construction and finish best suited to meet the needs of the prospective market. The position of the building may be of paramount importance, whilst special architectural features are unlikely to justify higher rents in the eyes of tenants. Nevertheless, there are many developers who take a pride in their projects and are willing to incorporate attractive but possibly costly features which will improve the urban scene but not necessarily increase rental values.[18]

If sites are scarce and competition is keen, the developer will need to offer a fair price to acquire the site. The value of the site is generally computed from a residual calculation based on estimated development costs and rental values coupled with the estimated capital value of the completed job. The main headings of expenditure are usually:

(1) estimated building cost;
(2) fees payable to architect, quantity surveyor and consultants;
(3) legal charges for letting of completed building;
(4) agent's commission on letting the accommodation;
(5) advertising costs;
(6) contingencies to cover increased building costs and unforeseen problems; and
(7) interest on finance to cover cost of money invested in the building during the construction period.

After assessing all these costs and deducting the value of the property on completion, the difference will represent the value of the site and the risk allowance on the venture. The developer is then able to make his bid for the land, which may be by way of cash, ground rent, or part ground rent and part premium.[18]

In the housing field, the average price of new houses mortgaged to the Nationwide Building Society in the first half of 1971 exceeded £5500, varying from £7431 in London and the south-east to £4560 in the north-east. The average estimated value of the building plots on which these houses stand was approximately £1250, representing 22.5 per cent of the purchase price. Site values ranged from 29.7 per cent of total cost in London and the south-east to 11.5 per cent in Scotland.[19] The average price for a plot of land for private house building in England and Wales doubled between 1963 and 1970, with the average 1970 price per hectare of land in London approaching £80 000, nearly five times the national average.

Possible Causes of Loss to Developer

Developers may suffer loss for a number of reasons, and some of the main ones are:

(1) payment of exorbitant price for the land;

(2) unexpected capital expenditure stemming from such matters as problems with underground services, extra cost of work needed to satisfy building regulations or town planning requirements, or to comply with easements or restrictive covenants;

(3) unattractive layouts or provision of dwellings of types for which demand is limited, resulting in selling problems; and

(4) organisational weaknesses, such as inadequate or ineffective advertising, poor supervision and execution of work in the wrong sequence.

Impact of Development on the Quantity Surveyor

James[20] has drawn attention to the need for an understanding by the quantity surveyor of the relationships between anticipated rents obtainable from commercial investment in building projects and the building cost limits set by these rents. For this it is necessary to know which parts of the building will be used exclusively by tenants and which parts will bc in common use. Electrical fittings, telephones and, quite often, partitions are not part of the building client's expenditure. The client will wish to maximise the proportion of floor area available for use by individual tenants. The arrangement of stanchions to project outside the building can help.

James[20] has also described how the quantity surveyor needs to have a background understanding of the cost of raising finance for a building project – interest rates, amortisation periods for repayment of building loans, and their relationship to the income accruing from the finished building, whether by way of rents, manufactured goods or even school fees. In the case of public buildings such as schools and houses with prescribed cost limits, the quantity surveyor must ensure that he is adequately equipped with up-to-date knowledge and effective techniques to advise the client and the architect on how to build within these limits and yet achieve the best possible standards.

Quantity surveyors may be called upon to give advice to building clients on approximate cost ranges of building development before any other professional advisers have been appointed. This requires an ability to translate the client's requirements in layman's terms, such as 'a meeting hall to seat 200 people' into floor area needs and other technical requirements of the building. For instance, with the meeting hall the quantity surveyor should be able to assess the floor area and requirements for cloakrooms, toilets, fire exits, services, car parking and many other matters. It will also involve an assessment of all the professional fees, possibly including those of lawyers and estate agents, and even the cost of furniture, fixtures and fittings.

Indeed the quantity surveyor may be required to give cost advice on a whole range of development problems, such as:

(1) cost comparisons of alterations and additions to existing buildings, or of demolishing and rebuilding;

(2) cost comparisons between using a building client's direct labour force (probably his existing maintenance force) for all or part of a project, or of using independent contractors;

(3) advice on the taxation aspects of capital investment in building projects;

(4) cost comparisons between different industrialised building systems or between these and traditional buildings to perform the same function; and

(5) advice on the relationships between capital costs and subsequent running and maintenance costs of various elements of the building.

SOURCES OF FINANCE

Capital for Property Development

The sources of capital will be influenced by whether money is required short-term or long-term. In general a developer prefers to have as few variable elements in his calculations as possible and is accordingly inclined to pay a slightly higher overall rate of interest if he can be assured that the rate will remain fixed for the period of the loan. Merchant banks are prepared to advance money at fixed rates of interest for periods of up to two years, and will often finance property development on favourable terms where an equity interest in the development company is obtainable in addition to the short-term loan. In such cases it is often possible to arrange a revolving credit of short-term money so that as one project is completed and the money repaid, it is immediately available for another. Merchant banks are unlikely to lend money for longer terms than two years and hence cannot normally assist with the financing of major development projects, such as large office blocks and shopping centres. Some money is also available from private sources but not in very large sums. Long-term finance is normally provided by institutions such as insurance companies, pensions funds and investment trusts. Money from insurance companies comes mainly from life funds, which are invested in order to pay the contributor either a fixed sum in the future or alternatively an income for life. Public officers, school teachers and employees of nationalised industries are also provided with pension schemes.[17]

A property company can raise capital by selling shares whereby the purchaser is entitled to receive a share of the profits as and when distributed. There is legislation which imposes controls on the raising of money from the general public. Profits fluctuate according to demand for the development, the competence of the company administration and the effectiveness of its financing policy. Equity capital is used initially for the purchase of sites and for financing building contracts, to the extent that this money cannot be raised by normal loans in the money market. The long-term capital requirements of a development company are provided either by borrowing against the security of the completed building or by disposing of an interest in it for cash in return for the payment of an amount to the purchaser.[17]

Short-term Finance

Short-term finance operates for periods ranging from a few days up to a maximum of about two years. The principal sources of finance are the commercial or merchant banks, hire purchase companies and trade credit. Probably the most important source is borrowing from a bank on an overdraft; a facility on which interest is charged at a rate which is usually one or two per cent above bank rate (the rate at which the Bank of England will lend money for short periods to accredited customers on first-class security). British banks generally prefer to restrict their lending to advances to meet short-term requirements, such as arise from seasonal fluctuations in trade or the need for a bridging operation during the establishment of an asset pending the acquisition of finance of a more permanent nature.[4] In 1969, of the total bank advances in Britain of £10 570 million, £384 million was advanced to builders (about three per cent of the total). An analysis of bank advances in August 1971 showed 5.4 per cent to construction.[28]

Medium-term Finance

The most difficult financing problems are concerned with obtaining finance for periods ranging from about three to eight years. These loans are required to finance major expansions, machinery, plant and equipment. Ward[21] asserts that the major expansion of a company involves risks when the expansion is neither sufficiently large or sound to justify the issue of equities or debentures, nor small enough to be covered by short-term funding. The Industrial and Commercial Finance Corporation (ICFC) and the Finance Corporation for Industry (FCI) were formed with the main objective of investing in firms for periods of about five to seven years. This usually takes the form of holding shares in the company and foregoing annual dividends for the prospect of long-term growth. When the firm is firmly established, the Corporation in question sells the shares and reinvests elsewhere. The size of loans varies substantially between about £5000 and £300 000.[21]

Long-term Finance

Long-term finance for periods of about ten years or more may be required to finance new buildings, equipment and permanent expansion of company assets. Traditionally, long-term finance was provided by mortgage. The borrower had the advantage of receiving money at a fixed rate of interest over a relatively long period and did not have to repay it until the end of the term. The main disadvantage to the lender was that the money was lent for a considerable period of time often at a fixed rate of interest and the effect of inflation was to depreciate the value of his capital when repaid. Nevertheless, the security was reasonably good as the money advanced represented less than the value of the property and the mortgagee had substantial rights of enforcing payment of the interest and repayment of the debt. It has proved a useful form of investment for Life Offices who were required to pay fixed sums at future dates.[17]

The majority of home purchasers look to building societies for mortgage loans. Mortgage loans are generally restricted to ninety per cent of the valuation of the property and must not exceed a certain multiple of the household's gross annual income. The strength of building societies lies mainly in the safety and liquidity of their assets, which rose from £2000 million in 1955 to £9500 million in 1970. Thus their funds are greater than those in National Savings and close in size to insurance company assets and joint stock bank deposits. In the past, if mortgage funds fell short a relatively small adjustment of investment rates corrected the situation. Hutton[22] believes that the future is less certain, as mortgage demand may fall in the long-term and a high investment inflow cannot be guaranteed. Hence building societies are tending to enlarge their investment services by introducing life assurance linked schemes. The number of new mortgages issued by members of the Building Societies Association in recent years were: 1965: 382 000; 1966: 461 000; 1967; 504 000; and 1968: 498 000.[23]

The principal sources of funds for real estate investment are life assurance companies, pension funds, charities, property unit trusts and property bonds. In 1958 the amount invested by assurance companies in property was £494 million (8.9 per cent of total assets) and by 1968 this had increased to £1416 million (10.8 per cent of total assets). Occupational pension schemes had £230 million (five per cent of total assets) invested in property and within four years property unit trusts totalled £120 million and property bonds about £30 million.[24] Insurance companies and other institutional investors generally require gilt-edged investments or those involving a minimum of risk. They have accordingly recognised the desirability in an inflationary economy of investment in land and property as opposed to equities, which can be vulnerable to domestic, political and other short-term factors.[25]

Combination of Short-term and Long-term Finance

Where a development company is building for permanent investment, it is sometimes possible to combine short-term and long-term finance. Some institutional funds are prepared to provide money for development during the building period. They require an initial transfer of the developer's legal interest in the land, providing money during the building period against architect's certificates and also paying for other development costs, such as professional fees, legal expenses and other certified disbursements. On completion, rent is calculated at an agreed percentage on the total cost of development. During building operations, interest will be charged on the various capital sums advanced by the financier. These may either be paid by the developer or, in some cases, they may be capitalised. The institution may either grant to the developer an immediate lease of the site and enter into a separate building agreement, or enter into a building agreement under which the developer merely has a licence to enter on the site and erect the building, and the actual lease is not granted until the building has been completed. The method of combining short- and long-term finance is very convenient but is normally only available to companies with a well-established and successful record of property development.[17]

DEVELOPER'S BUDGET

General Philosophy and Approach

Prior to the purchase of a site for development, a developer must know what forms of development will be permitted on the site and have access to a financial appraisal or developer's budget (now sometimes referred to as a feasibility study). To decide whether the scheme is feasible, he will require advice on a fair price for the land; the probable building costs; and the probable rent or selling price of completed building(s). The developer's budget will include the following items.

Gross development value. This is based on the estimated total annual rent accruing from the completed building less the cost of outgoings (maintenance, repairs, management, etc.). The net annual income is capitalised by multiplying by an appropriate years' purchase and the total so obtained is referred to as *gross development value* (GDV). Rental values of properties vary widely as between the same class of building in different situations and between different classes of building in similar locations. Shops or offices in the centre of a city or large town will usually command much higher rents than similar premises in suburban locations, but even the suburban premises will vary in value according to the environment (industrial, local authority housing, good class private housing, etc.), as well as with the side of the street. For example, annual rental values of suburban shops in a large provincial city might average about £12/m² of zone A floor area (in front 6 m deep zone or strip of shops) and about £130/m² for central area shops. Annual rental values of bank premises on the south coast varied between £5 and £30/m² of usable floor area in 1971.

The valuer advises the developer on site values and future values of the development. Cost is generally associated with the supply element in the property market, while value relates to the demand side. The valuation surveyor gains his experience of values by observing and interpreting the evidence provided by the property market, in the form of sales and lettings. Thorncroft[26] suspects that the valuer may often be only faintly aware of the deeper causes which influence demand. The mathematics of valuation and the use of valuation tables are incidental to the valuer's calculations, for the basic figures are likely to be obtained quite intuitively, where comparable circumstances do not exist. Hence, the approach of the quantity surveyor and valuer can be quite different. In general, the quantity surveyor possesses far more cost data than the valuer has valuation data, and he can adapt it more readily from one situation to another.

The valuation surveyor is concerned with many value/cost relationships of a project. He is particularly concerned with securing a design which will achieve maximum value in proportion to cost, and is likely to be faced with a number of fundamental issues, such as 'If a cheaper finish is used in the building what will be its effect on rental value?' A valuer usually works from rental values and any discrepancy will be magnified about fifteen times when converted into capital terms. The property market is subject to many uncertainties; changes of government, overseas disturbances and new legislation can each have a significant effect on property values. Thorncroft[26] has described how valuing is too often a matter of opinion based on experience, and often valuations of eminent valuers diverge widely. For instance, opinions conflict on the relative values of different floors of a modern office block; some valuers believe the higher floors to be the more valuable, and others the lower. Eminent valuers have produced widely varying valuations of the site value of the headquarters building of the Royal Institution of Chartered Surveyors in Parliament Square, Westminster. A valuer would experience difficulty in assessing the reduction in value stemming from diminished daylighting at the centre of a large office block. These latter aspects serve to illustrate the very real difficulties which face valuers when assessing site and building values.

Cost of buildings. The cost of the proposed buildings will be assessed using one of the methods described in chapter 6, with the help of cost analyses, which were examined in chapter 8. The costs will normally be computed per square metre of floor area but the developer will generally be concerned only with the usable floor area on which a rent is payable.

Architect's and surveyor's fees. The normal allowance is ten per cent of the cost of erecting the building. This covers the preparation of all contract documents and supervision of and financial arrangements for the contract.

Legal and agency fees. An average allowance to cover legal costs on purchase of site, including stamp duty, and preparation and agreement of leases, agent's fees on letting the property and advertising costs, would be two to three per cent of gross development value (probably about four to six per cent of building costs). The sum involved will be influenced considerably by the number of lettings.

Developer's profit. An allowance of about ten to twenty per cent of the gross development value should be made to provide a return to the developer for devoting his skill and time to the project, and for the risks that he undertakes. The risks include rising costs, falling rents and inability to lease the property on completion. The actual allowance incorporated in a developer's budget will depend on the type of development and the degree of risk involved.

Cost of finance. To purchase a building site a developer will either have to borrow money, on which interest will be payable, or use his own capital and forego the return on it. The interest paid or revenue forfeited should be charged to the development, and will cover the period from date of purchase to the time when the completed building is let or sold. Financing of the building operations will proceed throughout the contract period as payments are made to the contractor based on periodic certificates. The building cost finance is usually calculated at an agreed rate of interest on half the building cost for the full contract period or the full building cost for half the contract period. A suitable rate of interest in 1971 was eight to nine per cent, using compound interest for lengthy contract periods.

Practical Application of Developer's Budget Approach

A number of worked examples follow to illustrate the application of the concept of the developer's budget to various practical development problems.

Example (a). Planning consent has been given for the erection of an office block of 10 000 m^2 on a vacant building site. It is estimated that the building will produce a net income of £90 000 pa and will cost £65/m^2 to build. Assuming it will take eighteen months to build, determine the present market value of the site.

Gross development value		
Net income from offices	£90 000 pa	
YP in perpetuity at seven per cent	14.3	£1 287 000
deduct costs		
Building – 10 000 m² at £65	650 000	
add architect's and surveyor's fees (ten per cent)	65 000	
	715 000	
One-and-a-half years building finance on $\frac{£715\ 000}{2}$ at eight per cent	42 900	
Legal and agency fees and advertising costs (two-and-a-half per cent of GDV)	32 175	
Developer's profit (fifteen per cent of GDV)	193 050	
		983 125
		303 875
less cost of site finance eight per cent on say £270 000 for one-and-a-half years		32 400
Site value		£271 475

(Rounded off to £270 000)

The anticipated return of the developer is usually assessed at about eight to nine per cent of development costs, and it would be useful to check the figures in this example by the anticipated return method.

Costs	
Building, including fees	£715 000
Building finance	42 900
Legal, agency and advertising costs	32 175
Site cost	270 000
Site finance	32 400
Total costs	£1 092 475

An eight per cent return on £1 092 475 would be £87 398 and a nine per cent return would be £98 323, so the annual net income of £90 000 falls within the range.

Example (b). A site is available at a purchase price of £50 000 and it is anticipated that planning permission could be obtained for a factory of £15 000 m² or an office block of 35 000 m². The estimated cost of the factory including site works is £600 000 and the comparable cost of the office block is £2 100 000. The contract periods are assessed at one year for the factory and two years for the office block. On completion the office block is likely to have an annual rental value of £260 000 with the landlord's annual outgoings of £60 000. The normal return on office blocks in this area is seven per cent. The factory is likely to sell for £900 000 on completion. Legal, agency and advertising costs are likely to be £20 000 for the factory and £80 000 for the office block. Architect's and surveyor's fees are to be taken as ten per cent of construction costs and finance is available at an interest rate of eight per cent. Determine which form of development is likely to be most profitable to the developer.

Factory

Cost of site		£ 50 000
Site finance for one year at eight per cent		4 000
Total cost of land		54 000
Building costs	£600 000	
add architect's and surveyor's fees (ten per cent)	60 000	
	660 000	
Building finance for one year on $\frac{£660\,000}{2}$ at eight per cent	26 400	
Legal, agency and advertising costs	20 000	
		706 400
Cost of development, excluding developer's profit		£760 400
Selling price		£900 000
less development costs		760 400
Development profit (received after one year)		£139 600

Office block

Cost of site		£ 50 000
Site finance for two years at eight per cent		8 000
		58 000
Building costs	£2 100 000	
add architect's and surveyor's fees (ten per cent)	210 000	
	2 310 000	
Building finance for two years on $\frac{£2310\,000}{2}$ at eight per cent	184 800	
Legal, agency and advertising costs	80 000	
		2 574 800
Cost of development, excluding developer's profit		£2 632 800
Annual rental		£ 260 000
less outgoings		60 000
Net annual income		200 000
YP in perpetuity at seven per cent		14.3
Selling price		2 860 000
less development costs		2 632 800
Development profit (received after two years)		£ 227 200

The factory would be the most profitable development as it shows a return in excess of eighteen per cent after one year, as compared with the return of 8.6 per cent after two years for the office block. The capital expenditure on the factory project is less than one-third that involved for the office block and, if the developer is using his own capital to float the project, this will release funds for investment in other schemes. Furthermore, as the return on the factory development is received one year earlier, the profit of £139 600 is available for investment throughout the second year. These calculations assume that there is ample demand for both forms of development.

Example (c). A developer is proposing to erect a block of six lock-up shops with two floors of offices above them. The shops are likely to let at an average annual rent of £12/m² and the offices at £10/m². The net floor areas will be 500 m² for the shops and 900 m² for the offices. Circulation areas will amount to about ten per cent of total floor area and will remain under the control of the landlord, whose annual outgoings are estimated at twenty-five per cent of income. The developer requires a twenty per cent profit on the development. The freehold site is available at a purchase price of £6000 and siteworks are estimated to cost £1500. The rate of return in this area for similar developments is around seven per cent. The contract period is likely to be one year and finance is available at eight per cent rate of interest. Legal, agency and advertising costs are likely to be about two-and-a-half per cent of gross development value. Determine the allowable building cost per m² of gross floor area and its feasibility.

Income from development		
Shops – £12 × 500		£ 6000
Offices – £10 × 900		9000
Gross income		15 000 pa
less annual outgoings (twenty-five per cent)		3750
Net income from shops and offices		11 250
YP in perpetuity at seven per cent		14.3
Gross development value		£160 875
deduct costs		
Cost of site	£ 6000	
Site finance – eight per cent on £6000 for one year	480	
Legal, agency and advertising costs – two-and-a-half per cent of £160 875 (GDV)	4022	
Developer's profit (twenty per cent of cost of development = $\frac{1}{6}$ of GDV)	26 812	
		37 314
Sum available to cover building costs, fees and finance		£123 561

Let x = building costs, including siteworks. Then architect's and surveyor's fees = $0.1x$ and finance for one year at eight per cent =

$$0.08\frac{(x + 0.1x)}{2}$$

$$x + 0.1x + 0.044x = £123\,561$$

$$1.144x = £123\,561$$

$$x = \frac{£123\,561}{1.44} = £108\,000$$

less siteworks	1 500
Building costs	£106 500

Building area	
Shops	500 m^2
Offices	900 m^2
Net usable area	1400 m^2
add circulation area ($\frac{1}{9}$)	156 m^2
Gross floor area	1556 m^2

$$\text{Allowable building cost/m}^2 = \frac{£106\ 500}{1556} = £68.5$$

An analysis of recently completed similar buildings in the district show this to be a suitable rate for the provision of shops and offices with good quality finishes and including allowance for increased costs.

Example (d). A developer requests advice on the value of a building site of 20 ha which he is considering purchasing. The land has outline planning permission for private residential development at a density of 120 persons per hectare. Similar building plots in the district are selling readily at £6/m^2.

It is assumed that the average size of dwelling will accommodate four persons, and the total number of building plots will be

$$20 \times \frac{120}{4} = 600$$

To arrive at the net area of the average plot, it is necessary to make allowance for roads and incidental open spaces. Assuming that roads occupy fifteen per cent of the site and incidental open spaces a further five per cent, then the effective area of the site for building purposes becomes 20 ha less twenty per cent, equalling 16 ha. The area of the average plot equals

$$\frac{16 \times 10\ 000}{600} = 267 \text{ m}^2$$

as there are 10 000 m^2 in a hectare.

The value of an average building plot = 267 × 6 = £1602 and the estimated income from the sale of all plots is 600 × £1602 = £961 200.

The programme and cost of preparing the site for development have now to be considered. It is likely to take two years to construct the roads and lay sewers and a further two years before all the plots are sold. Hence the income from the sale of plots will be deferred for $2 + (2 \times \frac{1}{2})$ = three years (average).

An outline layout of the development will now be prepared to determine the length of roads and sewers needed to develop the site. It is assumed that 3000 m of road and sewer at an all-in cost of £30/lin m will be needed, and as the constructional work is scheduled to take two years, the expenditure will be deferred an average of one year.

Income from sale of building plots 600 at £1602	£961 200	
PV of £1 in three years at six per cent	0.8396	
		£807 024
Development costs		
Roads and sewers – 3000 m at £30	90 000	
PV of £1 in one year (average) at six per cent	0.9434	
	84 906	
Legal, agency and advertising costs – four per cent of £807 024	32 280	
Developer's profit and risk – fifteen per cent on £807 024	121 053	
		238 239
Value of land (approximately £6650/hectare)		£568 785

CHOICE BETWEEN BUILDING LEASE OR PURCHASE

Sykes[27] has investigated the matters to be considered when making a decision as to whether to purchase a building or lease it for a twenty-one-year period. He shows how the alternatives must be reduced to a common time period, the simplest approach being to take the period of the lease and to evaluate the building in twenty-one-years time for the purchase calculations. The value should be the higher of replacement cost or resale value, and where there is doubt about the estimated value Sykes advocates adopting a range of values to determine how critical the figure is to the calculation. Tax allowances must be taken into account in the calculations, as purchase of the building will give rise to annual capital allowances and lease payments will be allowable against the company's taxable income. The annual capital allowance on an industrial building can be either one per cent per year on a straight-line basis or one-and-a-half per cent on a reducing-balance basis. With leases, the company is likely to secure a forty per cent reduction on lease payments due to tax relief, although the tax recovery will normally occur a year later.

REFERENCES

1. J. B. CANNELL. Development, the economy and the chartered surveyor. *The Chartered Surveyor* (November 1965)
2. T. ALLAN. The giants grow up. *The Guardian* (13 May 1971)
3. ROYAL INSTITUTION OF CHARTERED SURVEYORS WORKING PARTY. Co-operation between general practice surveyors and quantity surveyors. *The Chartered Surveyor* (July 1966)
4. INSTITUTION OF CIVIL ENGINEERS. *An Introduction to Engineering Economics* (1969)
5. T. ALLAN. Still a profitable game. *The Guardian* (23 April 1970)
6. ANONSUR. The role of the surveyor. *Architecture, East Midlands*, **34** (1971)
7. DEPARTMENT OF THE ENVIRONMENT. Fair deal for housing. Cmnd. 4728. HMSO (1971)
8. D. RELF. No bed of roses for developers. *The Financial Times* (8 June 1970)
9. E. ERDMAN. Shopping revolution makes it easier for the customer. *The Financial Times* (8 June 1970)
10. I. H. SEELEY. *Planned Expansion of Country Towns*. George Godwin (1968)
11. J. J. ATTENBURROW. Some impressions of the private sector of the housing market. *Building Research Station Current Paper*, **57/68** (1968)
12. P. NICHOLSON. House building in the dark. *Building Technology and Management*, **9.9** (1971)
13. H. P. MACNAUGHTON. Implications of housing and planning policies. *Building* (5 September 1969)
14. O. LUDER. Subside people not buildings. *Building* (17 October 1969)
15. CO-OPERATIVE PARTY. Council housing and co-operative ownership (1970)
16. W. LEAN and B. GOODALL. *Aspects of Land Economics*. Estates Gazette (1966)
17. E. R. YOUNG. Financing property development in the United Kingdom. International Federation of Surveyors (1968)
18. J. W. HUGHES. Building costs and valuations. Postgraduate cost planning course. Brixton School of Building (South Bank Polytechnic)/Royal Institution of Chartered Surveyors (1961)
19. P. HILLMORE. Increase in house prices accelerates. *The Guardian* (19 August 1971)
20. W. JAMES. The future of our profession and education and training for it. Address to RICS branch meeting, Cardiff (30 October 1967)

21. A. V. WARD. Economic aspects of building: finance for building. *Building* (25 June 1971)
22. J. HUTTON. The changing image of building societies. *The Financial Times* (8 June 1970)
23. BRICK DEVELOPMENT ASSOCIATION. The way we live now (1969)
24. N. BOWIE. Investment policy in the next decade and its effect on property development. *The Chartered Surveyor* (April 1970)
25. G. JAMES. The institutional investor's needs. *The Financial Times* (8 June 1970)
26. M. THORNCROFT. Cost and value: the valuation surveyor's approach. *The Chartered Surveyor* (February 1962)
27. A. SYKES. The economic evaluation of building lease or buy decisions. *The Building Economist* (*Australia*) (August 1970)
28. THE TREASURY, INFORMATION DIVISION. Economic progress report No. 22 (December 1971)
29. THE TREASURY, INFORMATION DIVISION. Economic progess report No. 27 (May 1972)

14 ENVIRONMENTAL ECONOMICS

IN THE FINAL chapter we examine the impact of both public and private investment and of government action generally on the construction industry, together with the structure of the industry and the relationship of its output to demand and available resources. The economic aspects of urban renewal and of new town and town development schemes are examined. Consideration is given to the philosophy and nature of cost benefit analysis and its application to a variety of environmental problems.

CONCEPT OF ENVIRONMENTAL ECONOMICS

Higgin[1] saw the need for quantity surveyors, in a future role as building economists, to be concerned with all aspects of the building process and the underlying forces behind the various activities. These would embrace the effects of public and private investment policies and aesthetic and planning factors, all of which play a part in determining the system of economic forces which underlie the building process. Coddington[2] has described how a study of economics in its broadest sense entails consideration of how people behave in their activities of producing, exchanging and consuming, and the motivating forces behind these activities.

Cooke[3] saw the 'environment' as embracing constructional works and their general surroundings and defined environmental economics as 'the study of the forces affecting the use of assets and resources in satisfying man's need for shelter and a properly managed environment'. He believes that the surveyor's skill should no longer be confined to the measurement of physical features or entities but should be extended to measure the forces which are at work in the deployment or changed use of the resources which shape our environment. Expressed in another way, we must deal with cause as well as effect. Cooke sees the need to move the base on which the surveying profession is founded from the land to the environment.

The quantity surveyor as a building cost consultant or building economist may be called upon to advise a building client on a number of separate but interrelated issues. The main issues are now listed, together with the broad field of knowledge and expertise that is needed to give effective advice in each case.

(1) *Why* the client should build. Requires a knowledge of economics and the ability to forecast trends in the economy, coupled with an appreciation of the client's financial and production problems and a knowledge of building costs.

(2) *Where* the client should build. Involves a knowledge of national and international markets, economics of transportation and communication, population trends, taxation benefits and building and planning legislation.

(3) *When* the client should build. Requires a knowledge of the alternative ways of financing buildings, legislation affecting capital and revenue expenditure and investment allowances.

(4) *How* the client should build. Requires expertise in contract planning and administration, selection of technical advisers and a construction team, and ability to maintain financial control of the contract.

(5) *What* the client should build. Involves co-ordination of the specialist advisers to ensure value for money and evaluation of capital and future costs of alternative design solutions.

PUBLIC AND PRIVATE INVESTMENT

It is important to realise the magnitude of the sums invested annually in building work in this country. Approximately forty per cent of the total value of building work is concerned with work of maintenance and repair, and

TABLE 14.1

GROSS DOMESTIC FIXED CAPITAL FORMATION IN THE UNITED KINGDOM AT CURRENT PRICES

	1960	1962	1964	1966	1968	1970
			£ million			
Public sector						
Dwellings	271	344	514	654	822	769
Other new buildings and works	646	831	1093	1276	1652	1884
(a) Total public building and civil engineering work	917	1175	1607	1930	2474	2653
(b) Total public gross fixed capital formation	1648	1962	2580	3131	3790	4004
(a) as percentage of (b)	56	60	62	62	65	66
Private sector						
Dwellings	479	547	697	680	765	696
Other new buildings and works	580	720	764	800	856	1050
(c) Total private building and civil engineering work	1059	1267	1461	1480	1621	1746
(d) Total private gross fixed capital formation	2472	2769	3283	3587	4094	4882
(c) as percentage of (d)	43	46	45	41	40	36
United Kingdom:						
(e) Total building and civil engineering work	1976	2442	3068	3410	4095	4399
(f) Total gross fixed capital formation	4120	4731	5863	6718	7884	8886
(e) as percentage of (f)	48	52	52	51	52	50

the remainder represents investment in new building work. Furthermore, the total invested in building and civil engineering work amounts to approximately one-half of the value of all capital goods produced annually in this country. Table 14.1, based on information extracted from the 'blue book' on National Income and Expenditure[4] provides some useful comparative figures for the period 1960–70. The proportion of total expenditure on capital formation spent on building and civil engineering work in the public sector in the last decade has increased significantly, while that in the private sector has tended to fall.

Table 14.2, also based on information extracted from the 'blue book', sets out to show the relative expenditure of the various categories of building client in both public and private sectors on gross capital formation and

TABLE 14.2

GROSS DOMESTIC FIXED CAPITAL FORMATION BY VARIOUS CATEGORIES OF DEVELOPER IN UNITED KINGDOM

Development initiated by	*Gross fixed capital formation £ million*			*Dwellings and other new buildings and works £ million*			*Dwellings, etc., as percentage of gross fixed capital formation*		
	1960	1965	1970	1960	1965	1970	1960	1965	1970
Central government	256	298	576	124	248	477	48	84	83
Local authorities	604	1185	1819	514	970	1501	84	82	82
Public corporations	788	1293	1609	239	351	458	30	27	29
Total: public sector	1648	2776	4004	877	1569	2436	53	57	61
Companies	1739	2624	3793	454	723	938	26	28	25
Persons	733	915	1089	590	914	949	80	100	87
Total: private sector	2472	3539	4882	1044	1637	1887	42	46	39
Total: all sectors	4120	6315	8886	1921	3206	4323	47	51	49

building and civil engineering work. In particular it highlights the large sums spent by local authorities on constructional work, such as houses, schools, roads and main drainage, and the high proportion of expenditure in the personal sector on new buildings and related works.

The total value of construction output in 1968 was estimated at £4524 million,[5] subdivided in the manner indicated below.

New housing	Public sector	£633 million
	Private sector	£623 million
Other new work	Public sector	£984 million
	Private industrial	£441 million
	Private nonindustrial	£409 million
Repairs and maintenance		£791 million
Work by direct labour in public sector		*£643 million
Total value		£4524 million

*NOTE: This represents the output of employees of government departments, local authorities and public utilities on construction work; there being no profit element, the value of output is not directly comparable with contractors' output.

There are some who believe that the volume of public development should be drastically reduced. Why not, they argue, leave development to private enterprise, resource allocation to the market, and the government to its traditional role of maintaining fair play? Clemens[6] points to the increasing size of projects in relation to the market, which renders the market test either impossible (because there is no established market for the project, for example, a space research programme), or inappropriate (because the project is not intended to earn a financial return, for example, a hospital) or potentially catastrophic (because the product cannot be 'brought to the market' in small amounts, but only as one enormous completed item, for example, a nuclear power station). Moreover, the increasing size of public projects, and the increasing total of public expenditure, means that public decisions have repercussive effects on the whole economy; on the one hand, resources that would have been available for private development are pre-empted, and on the other, new areas of opportunity are opened.

Another illustration of the enormity of public expenditure is provided by Blessley[7] when he described how the Greater London Council invested on average over £70 million per annum during the period 1965–8, in acquiring and redeveloping land, principally for housing, education and roads. In addition, the London boroughs and other public bodies have also made significant contributions, resulting in a very large scale of capital investment on renewal. Yet the return on this capital is seldom, if ever, satisfactory in terms of money alone, whether by income from rents or outright sale; but there are substantial benefits in environment, amenity and traffic flow which are difficult, if not impossible, to quantify.

STRUCTURE OF THE CONSTRUCTION INDUSTRY

The construction industry includes the constructors of both building and civil engineering works, and there are a considerable number of contractors who undertake both functions. Building work varies enormously in its scope and nature, from the erection of large multistorey blocks to the execution of works of minor repair and maintenance. It is also possible to distinguish between general contractors who undertake general building work and specialist contractors who concentrate on specific trades, such as plumbing, joinery, painting, plastering and electrical work.

In recent times there has also developed 'the lump' of labour-only subcontractors who undertake specific trades for the general contractor. The main contractor usually supplies most of the equipment and all of the materials, and the self-employed operatives quote him a 'lump sum' for the job, which they subsequently share between themselves. They have been heavily criticised by many building workers and the trade unions, as they are believed to manipulate national insurance and selective employment tax payments and evade income tax and training board levy. They sometimes produce poor quality work, adversely affect union structure in the industry, and their form of employment seems to offer no redress for industrial accidents. A government committee of inquiry,[12] whilst underlining the grave disadvantages inherent in this arrangement did, nevertheless, recognise that it combined the specialist skills of the subcontractor with a psychologically good form of wage incentive. Obviously some form of control is necessary.

It is possible to classify building firms accord to the type of work that they undertake, as below.

Civil engineering contractors. Those who undertake very large civil engineering projects such as motorways, bridges, dams, reservoirs and harbour works; these firms usually operate on a national or even international basis.

Building and civil engineering contractors. Those contractors who undertake large contracts in both fields, often at either national or regional level; these are medium to large-sized contractors.

General builders. These can be grouped into a number of separate categories.

(1) Large firms employing over 1000 employees operating on a national scale and undertaking the erection of a wide range of buildings from housing estates to offices, hospitals and factories.

(2) Medium-sized firms who operate at regional level and undertake all but the largest jobs.

(3) Speculative firms who specialise in building housing estates in advance of demand.

(4) Small firms who are generally restricted to the erection of single houses or small groups of dwellings, and improvements, conversions and maintenance work.

(5) Very small firms of jobbing builders who confine their activities to maintenance and repair work.

Specialist firms. Those who concentrate on a specific trade, such as roofing contractors and plasterers, or who offer a special service which may not be confined to the construction industry, such as heating and ventilating engineers and electrical contractors.

The construction industry employs about one worker in every fifteen in Great Britain and the output of the industry represents more than one-half of the country's capital investment. About 1.3 million construction workers have jobs in about 80 000 private firms. A further 300 000 are directly employed by public authorities and are mainly engaged on repairs and maintenance. About 200 000 more workers are self-employed but are not classified as one-man firms. Construction firms vary widely in size as well as in the type of work they perform. Table 14.3 shows how the total work force, including employers and office staff, is distributed in firms of different size.

TABLE 14.3

EMPLOYMENT IN THE CONSTRUCTION INDUSTRY BY SIZE OF FIRM IN 1968

Number of employees in firm	*Number of firms*	*Total number of employees*
0–13	65 699	243 083
14–34	8 870	186 151
35–114	4 034	238 381
115–299	953	168 334
300–599	282	119 860
600–1199	134	115 468
1200 and over	87	275 496
Totals	80 059	1 346 773

Source: DEA *progress report No. 52*[5]

Although there were about 66 000 firms with under fourteen employees each in 1968 (eighty-two per cent of all construction firms), these employed under one-fifth of all employees. The very large firms with over 1200 workers employed over one-fifth of the total labour force in the industry, and one-half of all employees were engaged by the larger firms with 115 or more employees. The trend in recent years has been for the really large firms (some have over 40 000 employees) to grow larger. The industry is also supported by manufacturers of building materials and components, many of which are stocked and distributed by builders and plumbers merchants. It is estimated that about 500 000 people are engaged in the production and distribution of these commodities.[5]

As Stone[8] has stated 'there is no general optimum size of firm or organisation, the best size depending on the nature of the work, the conditions under which it needs to be carried out, and the nature of the organisation and the ability of the management'. Although large firms give rise to co-ordination problems, nevertheless they secure many advantages such as increased mechanisation, fuller use of resources, economies from specialisation, improved techniques, and better financing and purchasing arrangements. On the other hand, the small firms are in an advantageous position for dealing with maintenance and repair works and small contracts for new work. The capital needs of small firms are extremely limited and they enjoy a comparatively high ratio of liquid to fixed capital. Furthermore plant requirements are small and much of this can be hired, and materials are mainly ordered to meet current needs and purchased on credit of sufficient length to permit recovery of the cost of the materials from the client prior to paying the builders merchant. The smaller construction firms are particularly vulnerable in periods of financial difficulty or reductions in building work, as was evidenced by the bankruptcies of 1027 building firms in 1970.

A government working party[13] investigating conditions on large industrial construction sites, found an 'industrial situation where at least one-half of all time spent on site is ineffective, where excess cost and lengthy commissioning delays are prevalent, and which ranks as among the most strike-prone sectors of the economy'. To put this in perspective, it involved 40 000 operatives working on power stations, chemical plants, refineries and similar major fixed capital assets valued at about £800 million per annum. The reasons for this pitiful state of affairs include poor management, weak labour relations, lack of stable employment and insistence on unrealistic lump contracts by clients who were unsure of their requirements and out of touch with the very large increase in size, complexity, cost and construction period which has occurred in the last decade.

VARIATIONS IN WORKLOAD ON THE CONSTRUCTION INDUSTRY

Lean and Goodall[9] have described how the construction industry covers a wide range of loosely integrated groups and organisations involved in the production, renewal, alteration, repair and maintenance of certain capital goods (building and civil engineering projects). Capital funds are required on a long-term basis to finance plant and equipment and short-term for the purchase of materials and components and hire of labour. Unlike most other production processes, production takes place essentially at the site where the product is to be used rather than in a factory. Furthermore, most constructional projects are unique with individual design, engineering and production characteristics, and are invariably of very large size and weight at fixed locations with relatively long lives.

Fluctuations in workload are essentially of a cyclical nature. The long-term fluctuations are often associated with general variations in business activity while short-term changes are generally on a seasonal basis resulting from variable weather conditions. The increased use of mechanical plant, heating apparatus, plastic-sheeted enclosures, prefabricated components and industrialised building are helping to reduce the loss in output during the winter months.

The rise and fall in building activity has in the past had a cyclical duration of about twenty years, considerably longer than the normal trade cycle, and has been described as a *building cycle*. In times of prosperity increases in income will, amongst other things, lead to increased demand for residential accommodation. This stems from a rise in marriage rates and the increase in demand for separate or improved accommodation. House prices and rents tend to rise and this encourages new contractors to undertake house building, whilst existing firms expand their output of new work. In addition, the higher incomes create greater demand for consumer goods which in its turn generates a need for more factory space and increased workload for the construction industry. The output of the construction industry is fairly inelastic, as it takes a considerable period of time to prepare development proposals and erect the buildings.

After about ten years the bulk of the essential demands will have been met and the downswing of building activity commences. This will be accelerated by increased building prices arising from shortages of factors of production (resources). The reducing demand will gather momentum until the major part of building activity is confined to replacement buildings and repairs and maintenance. During the downswing a surplus of buildings is likely to occur, resulting in a lowering of prices and rents, a number of construction firms will experience financial difficulties and many building operatives will leave the industry.

Up until 1940 building cycles had occurred continuously in this country over a period of 250 years and with an average duration of about twenty years. Factors influencing the length of cycle are the long production period, the sluggishness of the industry to respond to changes in demand and population changes. For instance, a high marriage rate now will result in another marriage hump in about twenty years time, with consequently higher demand for homes. There may also be an increased number of older buildings falling due for replacement. Lewis[10] has shown that each building cycle is unique and that other factors contributing to building cycles include the internal migration of persons, and external activities such as bad harvests and wars. As will be described later in the chapter, the government is able to exert a considerable influence on the output of construction work, and so can restrict the harmful effects of building cycles.

The Building and Civil Engineering Economic Development Committees' (little Neddies) forecast in 1971 that after an initial burst in the early nineteen-seventies, the growth rate will slacken during the second half of the decade back to two per cent. The forecast gives an average growth rate of 2.5 per cent for the period 1969–79 compared to 4.5 per cent for 1959–69. Public sector work in roads and hospitals is expected to increase rapidly while growth in housing will be relatively slow. House completions are expected to average about 380 000 dwellings per annum throughout the next decade but with a higher proportion of private houses. The private sector is also expected to secure a higher output of nonhousing work. This report assists in the provision of information that will permit the construction industry to plan ahead with greater certainty. Furthermore, the committees intend to update both medium(zero to five years) and longterm (five to ten years) forecasts every two years.

RELATIONSHIP OF OUTPUT OF CONSTRUCTION INDUSTRY TO AVAILABLE RESOURCES

Cooke[11] has emphasised the need to use our construction resources of land, labour and capital, to best advantage. This is vital, as construction is a key factor in an expanding economy and increasing demand will place a heavy strain on our resources. This country could conceivably receive an increase in population in the order of sixteen million by the end of the century and this is equivalent to about fifty times the population of Nottingham or one- and two-thirds the population of Nottingham each year. There are limitations to available resources and a need to match supply and demand. Cooke[11] believes that it is necessary to:

(1) continually predict demand for several years ahead, refine annual predictions and translate predictions into terms which allow resources to be assessed;
(2) continually assess resources needed and refine annual assessments;
(3) continually survey availability of resources; and
(4) compare demand with available resources and take action to match demand and resources.

It is now generally recognised that programmes need to be established for some years ahead and to be of a *rolling* nature (where the period of years ahead remains constant and the annual programmes are reviewed and revised periodically). Expenditure on national road building is generally planned for five years ahead; school building programmes are announced two-and-a-half to three years ahead; and hospitals for ten years ahead.

Many industrial firms also produce investment programmes for a number of years ahead. It frequently takes several years to increase the supply of a particular resource.

Cooke[11] has shown the need to express the demand for resources in terms which are readily identifiable, such as cubic metres of gravel, number of bricks, tonnes of steel, and number of operatives. A disadvantage of monetary values is that they do not stay constant for long periods of time. To exploit the resources economically requires accurate information on their quantity, availability and location. In like manner the possible demand must also be known – its nature, timing and geographical incidence. There is also a need to improve the prediction, assessment and survey techniques to secure effective matching of supply and demand.

EFFECT OF GOVERNMENT ACTION ON THE CONSTRUCTION INDUSTRY

The government has a crucial role in determining demand for the construction industry's output and its growth prospects, both because public authorities buy over half its output and because general economic measures have a powerful influence on the demand for private housing, and industrial and commercial building. The Department of Economic Affairs[5] has emphasised how a steady, rather than wildly, fluctuating growth of demand is particularly important for the industry if it is to plan its work ahead, make sure of supplies and deploy its resources effectively. Because the industry is undercapitalised it soon experiences difficulties when monetary policies are introduced to check the economy as a whole.

Over the years successive governments have found it necessary sometimes to restrain and at other times to stimulate the economy, and the construction industry is invariably caught up in this process. When policies of restraint operate, there is usually a reduction in the volume of public building work, particularly house building, and projects such as motorways and town centre redevelopment schemes are likely to be cut back or postponed. The adverse effects on contractors will not be immediate as contracts already in hand will normally be completed. In the longer term however the results can be serious, resulting in:

(1) unemployment of building operatives;
(2) smaller building firms being forced out of business;
(3) larger construction firms being reluctant to invest large sums in new plant and equipment or to experiment with new techniques;
(4) suppliers of materials and components being unlikely to extend their plant;
(5) recruitment of persons into the industry at all levels being made more difficult; and
(6) the lack of continuity of construction work which makes for increased building costs and reduced efficiency.

The various forms of government action and their likely repercussions on the construction industry are now considered in some detail.

The Government as a Client

The Government, either directly or through public agencies, such as local authorities, nationalised industries and other *ad hoc* bodies, purchases over one-half of the output of the construction industry each year, and the performance of the industry thus plays an important role in any government programme of economic growth and social advancement. Lean and Goodall[9] have described how, since the second world war, the construction industry has handled many large-scale reconstruction and development schemes designed to restore and improve the infrastructure of the country. Many of these are dependent on Government initiative, for instance the modernisation and electrification of railways, motorway programme, building of new and expanding towns and erection of local authority houses.

Relatively small changes in government policy may cause significant variations in the construction industry's workload, to which the industry will find difficulty in adjusting itself owing to the nature of its structure and product. Government spending on capital investment, such as schools, hospitals, roads and public housing can be curtailed with relative ease and these have the greatest impact on the work of the construction industry. A change of priorities by the government, such as a transfer of funds from housing to motorways, creates problems for the construction industry, as these two activities are largely undertaken by different types and sizes of firm using substantially different forms of plant. Hence, it is not only the level or volume of construction activity which needs considering but also its composition. For example, substantial increases in improvement grants will result in a significant rise in the volume of work available to small builders.

Other Forms of Public Control

In chapter 12 it was shown how the government seeks to influence the location of new industry through the issue of industrial development certificates and by giving grants, subsidies and tax incentives. New industry is therefore likely to be steered towards locations where the government wishes to see industrial development established, such as development areas, and new and expanding towns, and these may not necessarily be areas where there is a surplus of construction resources. The government also operates controls in the public interests, of which a typical example is the Building Regulations aimed at securing minimum standards of construction in buildings, which are more exacting than those on the Continent, and this must influence costs and hence demand for building work. Another example is the Factories Act 1961 which prescribes certain minimum requirements for temperature, ventilation and lighting in factories, and these requirements must affect constructional costs.

On occasions, social and political considerations can influence the form and pattern of a construction programme. For instance in 1971, the Department of Education and Science announced that all new school building in 1973–4 was to be devoted to primary schools, and this will affect the nature, scale and location of educational building in that particular financial year. Since the second world war the form and volume of the educational building programme has varied considerably from year to year and has been very much influenced by economic, social and political factors.

Monetary Policy

The government uses various controls, such as bank rate, open market operations and hire purchase restrictions, to alter the level of interest rates, and to control the amount of credit available and the terms on which it can be obtained. Credit restrictions strike at contractors both directly, through banks and lending institutions, and indirectly, as through builders merchants who are always a ready source of trade credit and could be the mainstay of many small builders.

In the short-term an increase in interest rates is unlikely to affect the demand for construction work. In the long-term it will result in higher prices for buildings and this may cause some developers and other building clients to refrain from investing capital in building projects, as they are no longer as profitable, and the clients themselves may experience difficulty in obtaining finance at the higher interest rates. Hence some projects are likely to be postponed or abandoned. Furthermore, contractors relying on bank overdrafts as a main source of working capital may find their loans curtailed and thus be compelled to reduce their output.

Taxation

Taxation is an important tool in a government's fiscal policy and has significant implications stemming from the alternative uses to which the money could have been put had it not been taken as tax, and the widespread distribution of the tax. Taxes are levied on both capital and income. An increase in capital tax on the value of property or on gains made from the sale of buildings or land may result in decreased demand, whilst an increase in tax levied on property income or use is likely to result in reduced demand for new buildings, unless it is offset by higher rents or profits. A practical example was the Land Commission Act 1967 (abolished in 1970), which introduced a betterment levy assessed at forty per cent of development value, and this reduced the supply of land for building.

Selective employment tax was introduced in the late nineteen-sixties as a levy on all non-manufacturing employment, to accelerate the transfer of surplus labour in service industry to manufacturing and, at the same time, to produce about £700 million per annum in additional revenue. Building was classed as an *assembly* industry and was accordingly liable for tax; this resulted in increased building costs, but subsequently the tax contribution was halved in 1971. It seems likely that a value added tax (VAT) will be linked with Britain's entry into the European Economic Community (the Common Market), where the tax already operates. This tax operates at each stage of manufacture with the tax being levied at a flat rate per cent on the increase in value of the article accruing from the manufacturing process. It seems likely that new building work will have a zero rating but that the surveying and architectural professions may be liable to tax.

Fiscal Policy

Apart from taxation, a government can influence the level of economic activity by regulating the amount of its spending. We have already seen that the level of public spending on construction work is very high. After the first world war it became a generally accepted facet of government policy that spending on public works needed adjusting to counteract changes in spending in the private sector of the economy. It was argued that in times of deflation and rising unemployment, an expansion of public works would increase employment and investment and revive the economy (pump priming). Furthermore, public works inevitably involve construction. It seems unlikely that the construction industry could absorb a sufficient proportion of the unemployed to correct the adverse level of economic activity, as many of the unemployed could not undertake construction work. In 1971 the number of unemployed in Great Britain rose to over one million partly due to inflation and general economic recession. The construction industry suffered severely through substantial cuts in both public and private sectors, although the government initiated limited programmes of additional construction work in the public sector.

Needs of the Industry

During the last two decades, successive governments have used both monetary and fiscal measures to alternatively restrain and stimulate the economy. This 'stop-go' policy has had serious repercussions for the construction industry, resulting in a continually fluctuating workload.[9] It has contributed to changes in the size-structure of construction firms, restricted additional capital expenditure on plant and equipment, reduced the effective labour force, contributed to a rise in labour-only subcontracting, retarded the use of new techniques and reduced efficiency. Ward[14] has rightly stressed the need to discard the concept of the industry as some form of economic regulator and to regard it as a medium for steady continuous development. The work on forecasting demand undertaken by the Economic Development Committees for Building and Civil Engineering with the help of the government and industry, which was commenced in 1969, should prove to be of great benefit in future planning.

Technologically the government has provided a lead through the Building Research Station, the National Building Agency, the Agrément Board and its own Department of the Environment. It must, however, pay more attention to the financial needs of the industry. For instance, it seems quite inequitable for the government to maintain its firm price policy against a background of continually increasing prices of labour and many building materials and components.

URBAN RENEWAL AND TOWN CENTRE REDEVELOPMENT

Basic Problems

One of the most urgent and complex problems facing many local authorities at the present time is that of urban renewal. Most towns and cities have evolved over a long period on a radial road pattern which is ill-suited for present-day traffic needs. Surrounding the pressurised inner core is often a girdle of mixed residential, commercial and industrial uses in varying stages of obsolescence, frequently termed twilight or 'blighted' areas. Comprehensive development of all these areas is essential if satisfactory layouts are to be achieved.[15]

Complete redevelopment of town centres will be necessary in some towns, particularly where extensive growth is anticipated. Redevelopment is needed to overcome present deficiencies, such as bad traffic congestion, lack of parking space, inadequate loading and unloading facilities, restrictive traffic regulations, excessive fumes and noise, dangers to pedestrians, and in some places there is a definite conflict arising from the use of particular roads by different kinds of traffic. In smaller towns, pedestrianised shopping streets and precincts with rear-loading facilities, combined with a suitable network of distribution roads and adequate car parking space will often provide the best long-term solution. In larger towns, however, two or three-level development may be desirable.[16]

Some of the conclusions contained in the Buchanan Report[17] are particularly relevant to town centre redevelopment and are included here.

> 'The freedom with which a person can walk about and look around is a very useful guide to the civilised quality of an urban area . . . . Nothing would be more dangerous at this critical stage in planning for the new mobility offered by the motor vehicle than to underestimate its potential. . . . Even as this report is being written the opportunities are slipping past, for in many places the old obsolete street patterns are being 'frozen' by piecemeal rebuilding and will remain frozen for another half-century or longer. . . . The choice facing society is between affording, for its own convenience, a road system of some elaboration, or staggering its hours of work all round the clock to its obvious inconvenience.'

The town centre, with its concentration of people and traffic, variety of land use, intensity of building, historic and civic interest, diversity of ownership and high land and property values, is the most vital part of any town. It is the social hub of the town, the centre of local business and civic life, the entertainment and shopping centre. It must not be permitted to disintegrate under the strain of intolerable traffic conditions or the impact of haphazard development.[18]

Blessley[7] has drawn attention to some important practical issues arising from redevelopment. They entail extensive and costly diversion of underground services, many of which were routed under existing roads, which themselves are to be diverted or closed. The twilight areas are characterised by multiple-occupation, overcrowding, lack of essential facilities, neglect, dilapidation, mixed industrial and commercial uses and general environmental decay. There is both a social problem and one of physical decay, but it must be questioned whether the community can afford to allow the wholesale demolition of the houses, most of which could still have many years of life. Blessley[7] also gives an indication of the dissatisfaction of property owners concerning the method of delineating unfit properties in a slum clearance area, whose compensation is restricted to site value. Urban renewal offers the opportunity to restore integrity and character to a depressed neighbourhood.[19]

Basic Appraisals and Objectives

Prior to formulating town centre redevelopment proposals it is essential to make an objective appraisal of the existing centre, having regard to its function, assets and deficiencies in terms of convenience and safety, usefulness by day and night as a shopping, commercial and social centre, and its civic character and architectural qualities.[18] The objectives are primarily concerned with function (future size and purpose of town centre), layout (distribution and extent of main uses), circulation (pedestrian and vehicular movement) and character (retention and enhancement of town's individuality).

Plans for the future should aim at satisfying the human need on a modern scale by retaining or introducing some of the more picturesque uses which give character and vitality to the urban scene, such as pedestrian ways, arcades, street markets, small places or squares, landscaped spaces for rest and relaxation and facilities for amusement. The massing of buildings, their height and silhouette, and in detail the outline, textures and colours of materials, are all matters of vital importance in composing a street scene which is both attractive and harmonious.[20]

Strathon[21] advocates carrying out an economic survey and appraisal to assess town centre demand. This would embrace a physical survey of the existing shopping centre, analysis of existing traders, examination of recent transactions in sale or renting of premises, determination of market region, analysis of existing and potential turnover, examination of car parking facilities, and study of traffic flows and bypassable traffic. From the results of the surveys it will be possible to make a realistic assessment of shopping and parking needs and of the required scale and rate of development.

Implementation by Partnership

Large-scale town centre redevelopment schemes are often undertaken on the basis of a partnership between the local authority and private enterprise. The local authority has powers of compulsory acquisition which are vital if the best pattern of development is to be achieved. Modern developments require large areas of land in order to make the best use of modern techniques of mixed development, multilevel circulation and traffic arrangements. It is essential, for this reason, to be able to combine together numbers of small awkwardly shaped freehold sites of the nineteenth-century town. It is also desirable for land ownership to be in as large units as possible to balance the more profitable uses of land against unremunerative ones such as open space.[20] The local authority also meets certain basic costs apart from land acquisition, such as site clearance and the provision of roads and services and execution of other public improvements.[22]

The private developer often has a vital role to play with regard to availability of capital, knowledge of the market and ability to exploit commercial opportunities. To be successful the development must be in the right location with adequate and readily accessible parking space, satisfactory service access, correct amount of retail space for purchasing power of area and reasonable rent levels. Split-level trading, as performed at Coventry, is no longer favoured except in department stores. A covered air-conditioned shopping centre needs a minimum supporting population of 50 000 to 70 000, but trading in such a centre is likely to exceed that in an open precinct by about twenty per cent.

A developer may be selected by one of the five methods listed in Planning Bulletin No. 3;[22] negotiation with selected developers; open auction; open tender; selective tender; or open invitation to submit outline proposals.

In all cases it is necessary to prepare a developers' brief, listing the essential feature of the scheme as a basis for competitive tendering or negotiation with developers. The brief would probably include such matters as the objectives of the scheme; population figures; town centre map; site plan; number and types of unit to be provided; special requirements such as pedestrian and vehicular access, parking provision, siting, height and architectural quality of buildings and landscaping; phasing; drawings required; disposal terms; and obligations with regard to displaced traders. It is essential that the local authority should retain the initiative in planning and guiding the redevelopment. The Royal Institution of Chartered Surveyors in its joint memorandum with the Royal Institute of British Architects on tendering methods has drawn attention to many of the problems encountered in disposing of sites acquired by local authorities.[23]

In disposing of land for development, the local authority usually has two principal aims in view. One is to secure the satisfactory development of the site in the interests of the public it represents, which in many cases will involve a major civic improvement and the solution of such problems as traffic congestion, car parking, lack of open space and other public amenities. The other is to obtain the maximum financial return from the developers, partly in fulfilment of the local authority's duty to the ratepayers, and partly in order to recover its own capital outlay in buying the land and constructing public services in connection with the development. These two aims tend to conflict and the solution will often involve a balance between them. It is essential to consider and reconcile the aesthetic and economic circumstances of each case.[23]

Where a local authority has done no more than prepare a master design, it may decide to arrange a competition which offers developers opportunities for submitting alternative architectural proposals. In that event, it may prefer to fix a sale price or ground rent, and so be free to judge the competition entirely on the architectural merits of the schemes submitted. The services of a valuer are needed in the assessment of the price or rent. In other cases the local authority may wish to combine a design competition with a formula for securing the best rent obtainable. A possible procedure, in these circumstances, would be for the local authority to fix a fair ground rent and then offer the site to selected developers, in a limited architectural competition, at the predetermined ground rent but with the proviso that the developer will be required to pay as additional ground rent, a specified percentage of any net rent derived from the development in excess of a specified return on the cost of development (after deducting the basic ground rent). The following simplified example will serve to illustrate the latter approach.

Example. A competition for a central area redevelopment scheme includes a fixed minimum ground rent of £60 000 per annum. The cost of the scheme is £3 million, the return to the developer is agreed at ten per cent and any surplus rent is to be shared equally between the landlord and the developer. Total annual rents from the buildings are estimated at £400 000.

Rent received (net)	£400 000
Return of ten per cent on £3 million	300 000
Surplus	100 000
less basic ground rent	60 000
Balance	£ 40 000
Additional ground rent = £40 000 × ½ =	£ 20 000

A scheme of this kind could have the initial ground rent adjusted within a short period of the commencement of the ground lease (say three years), and thereafter reviewed after possibly thirty-three and sixty-six years on a ninety-nine-year lease.

Another salient feature of the partnership arrangement is the collation into one or more ownerships of all those lands which are needed to carry out the redevelopment satisfactorily, and this process has been described as *land assembly* and *site assembly*. Longden[24] sees the importance of this procedure as stemming from the ability to develop the area in the best interests of the community including provision for some unprofitable uses, and yet at the same time ensuring an adequate return to the developer, coupled with the availability of the local authority's powers of compulsory purchase.

The procedure for land assembly is often undertaken on the following lines.

(1) The local authority defines a comprehensive development area and prepares proposals for development often in conjunction with the developer.

(2) The developer acquires as much land as possible by agreement and the local authority can compulsorily acquire the remaining land which is essential to the scheme.

(3) The developer transfers the freehold of his land to the local authority.

(4) The local authority grants a long lease of the land to the developer often with provision for periodic revision of the ground rent. The disposal terms should take account of the enhanced site values resulting from the local authority's share of the redevelopment work and of the long-term prospects. In practice, developers are anxious to secure a ten per cent minimum return on their capital to cover the risk element involved and the annual ground rent to the local authority may not exceed five to six per cent of the capital required for land acquisition. Furthermore, developers have recently been seeking lease periods of 125 to 150 years to enable their annual expenditure to be more favourably amortised. Obviously each case must be considered on its merits.

Financial Implications for the Local Authority

Blessley[7] has described how any public body, like a private developer, must have regard to the financial aspects of renewal, although this is only one of many significant factors. The private developer makes his appraisal, and is influenced in his decisions by the return which he will receive on capital investment. He will have regard to the cost of assembling the site, of obtaining possession, of clearance and development, and the ultimate rents to be obtained, making allowance for the period during which he will have to pay interest charges on his capital. The public authority does the same calculation but also considers its obligations, whether statutory or moral, in connection with rehousing or relocation of tenants or other people displaced by a scheme of renewal and with diversion of services. The authority also looks beyond the question of profit or loss and examines other issues such as planning, traffic, civic design and the general prosperity of an area. Many of these factors cannot be valued and in the final event the deciding issue may be the willingness of the public to meet the cost. A local authority will obtain its finance from various sources: rates, government grant, loans, subsidies and private capital. When interest rates are high, the burden to be carried may be heavy and frequently schemes of renewal carried out today will commit generations ahead to substantial payments.

Lichfield[25] has outlined the financial arrangements relating to a major reconstruction scheme at West Hartlepool. The estimated cost of land acquisition, site preparation and site works was about £2 million and it was estimated that it would take twenty years to complete. Loan periods were taken as sixty years for land, twenty years for roads and thirty years for sewers, with an interest rate of four per cent (the scheme was prepared in the nineteen-fifties). Ground rents on shops, offices and warehouses were estimated at £10 000 per annum. Lichfield calculated the total costs and receipts over a seventy-six-year period (sixty years after the raising of the last loan), made up of the total loan charges, less ground rents and government grants. The balance was a charge on the rates and was equivalent to an average annual charge of about £30 000.

Problems of Displaced Traders

The redevelopment work must be carefully phased and this becomes very difficult in practice, as it is most desirable to carry out the scheme as a continuous operation and this requires constant liaison between architect, surveyor, engineer and contractor. The phased programme is agreed with owners and occupiers to preserve the continuity of trading by those who are displaced and to minimise the compensation payable for disturbance. There remains the problem of the small displaced trader who frequently cannot afford to pay the rents commanded by modern shops, yet whose interests must be respected. Existing traders, both large and small, possess a wealth of experience and goodwill which can make a valuable contribution to the success of redevelopment. In particular, the small independent trader contributes variety and character to the shopping centre. Strathon has suggested the provision of small unit shops of up to 45 m^2 floor area letting at annual rentals of about £300 to £400 in covered shopping precincts, to satisfy this need.[21]

Town Centre Redevelopment Proposals

Proper segregation of town centre operations is essential for safety, maximum concentration of use and amenity and comfort. Furthermore, there are a variety of ways in which this can be undertaken – horizontal or vertical segregation of vehicles and pedestrians, segregation of commercial vehicles from other classes of vehicle, etc. For instance, the plan for the central area of Swindon provides for two main pedestrian ways each about 460 m in length and intersecting at right angles in the centre of the town. One of these proposed pedestrian ways is the existing main shopping street and the other is routed along the dry bed of an abandoned canal. Robertson[26] has emphasised the need for suitable 'magnets' located along and at the ends of each new pedestrian way, to ensure that sufficient numbers of persons are attracted into and drawn along the ways. With converted shopping streets the old-established shops will themselves perform this function. One of the biggest difficulties in converting existing streets in this way is the provision of rear access to 'land-locked' shops to provide adequate unloading space off the highway. Another difficulty stems from the varying depth of shops.

Another interesting approach to town centre redevelopment is at Basingstoke where, to ensure that shopping will be safe and attractive to the townspeople and visitors, the shopping areas in both the new and existing sections of the town will be free of vehicular traffic. In the new area the shopping platform is above service traffic and the main car parks are immediately alongside in multistorey decks. The original proposals showed layers of car parking under the shopping deck but the latest alternative is claimed to provide greater convenience for handling goods, easier pedestrian access from the sides and an architectural character of design more in harmony with the scale and character of the existing town.[27] In the existing shopping centre it is intended that the shopping streets shall be reserved for pedestrians only, this being achieved by traffic management measures or by the longer term possibility of providing rear service access and car parking at the rear of the premises.

Methods of assessing car parking needs in a town by means of street parking surveys are outlined in Planning Bulletin No. 7.[28] The Institution of Municipal Engineers estimated that there was an average deficiency of seven offstreet car spaces per 1000 population in 1961 and that a further fourteen spaces per 1000 population would be needed by 1971.[29] This gives an indication of the vast number of additional parking spaces which will have to be provided in the coming years mainly by local authorities, who have a direct responsibility of meeting the cost of public car parking. A useful description of the different methods of providing parking space is given in the author's *Municipal Engineering Practice*.[15]

NEW AND EXPANDING TOWNS

General Background

The post-war era has seen the emergence of new public development on a scale hitherto unrealised. The most outstanding and significant developments have been those of new and expanding towns. New towns are provided by government appointed development corporations under the provisions of the New Towns Act 1946, with the declared objective of establishing 'self-contained and balanced communities for work and living'. Their principal aim was to relieve congestion in the larger cities, particularly London, and eight of the first generation new towns formed a ring around London at a distance of about 32 to 48 km from it. A new town development corporation obtains its working capital from the Treasury and it is empowered to acquire, hold, manage and dispose of land and other property; carry out building and other operations; provide water, gas, electricity, sewerage and other services; and carry on any business or undertaking for the purposes of a new town. The first new towns to be designated had design populations of 50 000 to 60 000, but later new towns had larger populations culminating in Milton Keynes with a population of 250 000.

The dispersal policy which promoted the new town schemes was later extended to the expansion of widely dispersed smaller country towns by virtue of the Town Development Act 1952. Under this Act the large city authority concludes an agreement with a much smaller provincial authority for the transfer of population and industry. The large authority is termed the *exporting authority* and the smaller authority is called the *receiving authority*. The government can make contributions towards some of the development expenses of a receiving authority, in connection with housing, water supply and main drainage. In addition, the exporting authority, and sometimes the county council for the receiving area, also give financial assistance. In some of the later schemes the county council is a party to the agreement. There are two main types of town expansion scheme; and

(1) agency schemes, where the exporting authority carries out the bulk of the development work, and is subsequently reimbursed by the receiving authority when the work becomes revenue-producing; and
(2) nomination schemes, in which the receiving authority carries out all the development work but receives contributions from the exporting authority.

Apart from helping to relieve congestion in large cities, town development schemes infuse new life into small static or declining country towns which often have considerable history, character and charm. Expansion provides more and varied opportunitities for employment and improved public and other services. Industrialists can build, purchase or rent factories on new industrial estates and houses are available for employees within a reasonable distance of the factories. The biggest problems have been attracting sufficient suitable industries and the synchronisation of the housing and industrial construction programmes, whilst the provision of social services and amenities barely keeps pace with development in the early stages.[15] Economic and human factors dominate many town development activities. First and foremost, it is vital that the selected site possesses a real potential for development and that a sound and diversified industrial structure is secured.[16] It is doubtful whether small-scale expansion schemes can ever be sound economic ventures, and the latest proposals aim at doubling the size of much larger towns like Peterborough, Northampton and Ipswich. The practical difficulties of implementing town development schemes have been described by the author in *Planned Expansion of Country Towns*.[16]

Development Costs

Lichfield[25] has described how a new town development corporation is a unique kind of developer, being dependent on the government for finance and with no other powers of borrowing. Indeed, it starts operations with no assets and in the early years must borrow to meet revenue expenditure. Unlike a private developer, a development corporation undertakes many different kinds of development, many of which are non-remunerative in themselves; and in relating cost and yield it is concerned with the new town as a single financial unit and not with individual sections of it. Industrial and commercial development will be sold or rented and the cost of trading undertakings will be met out of charges. A development corporation also provides non-remunerative services like sewage works, roads and open spaces which benefit all townspeople and would normally be provided by local authorities. Apart from any local authority contributions the costs of these services will need to be met from the yield accruing from other services with a financial return.

In the case of an expanding town, a local authority is faced with a large annual programme of high-cost houses. These costs can be recovered in one or more of the methods listed below, none of which is really satisfactory.

(1) To spread the increased costs over all local authority houses whereby all the tenants of older dwellings are continually being subjected to rent increases.

(2) To apply a rent surcharge to all new houses to cover additional costs resulting in the payment of differential rents for the same type of dwelling in different parts of the town.

(3) To cover the increased housing costs by increased rates, so spreading the additional costs over all ratepayers and avoiding large rent increases on local authority dwellings.

Blessley[7] has described the way in which one expanding town authority is embarking on town centre renewal. It is proposed to remove existing industry and residential uses from the scheduled area of 5 ha and to redevelop with shops, offices, car parking and a road pattern which facilitates pedestrian segregation. The estimated cost of acquisition was £700 000 and full development costs will be around £1½ million. Annual rental income from shops and offices is estimated at £200 000, showing a net return of about nine per cent on capital. This illustrates how finely balanced such a project may be and how any significant alteration in the basic elements of the calculation can easily result in a loss.

TABLE 14.4

CAPITAL EXPENDITURE OVER FIFTEEN YEARS AT BASINGSTOKE AND PETERBOROUGH

Expenditure	*Town and population increase*			
	Basingstoke 50 000		*Peterborough 76 000*	
Total costs are authentic estimates adjusted to 1965 prices. The apportionments are approximate only and may not be consistent	*Total cost over fifteen years*	*Cost per additional head of population*	*Total cost over fifteen years*	*Cost per additional head of population*
1. Residential (including estate roads, services and open spaces)	£ 53 800 000	£1076	£ 77 572 000	£1021
2. Industry and commerce	23 800 000	476	35 632 000	469
3. Town centre (including roads and residential development)	19 600 000	392	48 357 000	636
4. County services (education, health, welfare, children, fire and police)	5 608 000	112	9 328 000	123
5. Hospitals and other public buildings	5 800 000	116	8 462 000	111
6. Public utility services	9 000 000	180	15 379 000	202
7. Main roads (principal town roads, and main distributors or equivalents)	8 800 000	176	24 510 000	322
Totals	£126 408 000	£2528	£219 240 000	£2884

Table 14.4 shows the total investment for the Basingstoke scheme compared with the corresponding estimates for the expansion study for Peterborough, both adjusted to 1965 prices. The figures for both schemes are broadly similar, except for the town centre and main roads which are more expensive at Peterborough because of more difficult site conditions.[27]

It is interesting to note that expenditure on main roads at Basingstoke amounts to £176 per head of additional population (£116 per head of the total population), and this represents nearly seven per cent of the total cost of the scheme compared with a little over eleven per cent at Peterborough. Yet the normal increase in car ownership would have required improvements to the road system in Basingstoke even without an increased population, and it is likely that these improvements would have represented a substantial portion of the expenditure contemplated. It is a reasonable assumption that town development will accelerate as well as expand the programme of improvements and so Basingstoke will possess an efficient road system much earlier than might otherwise have been possible, with all the resultant advantages. A similar argument could be advanced in respect of the town centre redevelopment and extension.

Steel[27] calculated the cost of roads, car parks and garages per car at Basingstoke as an interesting yardstick of expenditure, although this ignores the fact that the roads are also used by vehicles from outside the town, service traffic and public transport. The total cost of new roads, car parks and garages of about £20 million divided by the estimated number of additional cars (24 750) gives a cost of £912 per car, and this possibly justifies a higher contribution by car owners towards the provision and upkeep of town highways.

It is apparent that questions of town size may be particularly relevant when considering the costs of expanding existing communities. Studies have been made to assess the relative costs of projected expansion schemes at Ipswich,[30] Peterborough[31] and Worcester,[32] and comparable data has been published for the expansion of Basingstoke, which is in hand.[33] These studies point to certain characteristics that give rise to differences in total cost, such as the higher cost of town centre development at Peterborough, because of higher existing values of property to be redeveloped, and higher road costs owing to problems of developing a more difficult site.

Furthermore, theses studies indicate that within certain brackets of expansion the cost per head of additional population decreases as the scale of expansion increases, so that larger expansions are generally more economical than smaller ones. It is also relevant that the expenditure involved in carrying out an expansion would normally improve local conditions of traffic congestion and improve amenities for the town as a whole, all of which would enhance the effective return on the capital investment involved.

The estimated gross costs per head of additional population (adjusted to 1965 levels) are shown in table 14.5.

TABLE 14.5

COMPARATIVE TOWN DEVELOPMENT COSTS

Town	*Estimated cost per head of additional population*	*Increase in population*
Ipswich	£2140	120 000
Peterborough	2884	76 000
Worcester	2647	66 000
Basingstoke	2528	50 000

Source: *Steel and Slayton, Urban renewal proposals*[33]

Town expansion was also discussed at the 1965 Annual Conference of the Royal Institution of Chartered Surveyors, in the wider context of whether there was any method of relating efficiency to the total size of a town and at what size the closest economic relationship between population and cost might be achieved. It was generally agreed that this might be a fruitless investigation because of the impossibility of ascertaining reliable economic data and of evaluating the many other factors, such as amenity and social and cultural advantages, that are relevant to the size of a community. This Conference served to emphasise how little is known about many of the fundamental influences on town design.[34]

Comparison of New Town and Expanding Town Costs

Stone undertook investigations into the comparable costs and came to the conclusion that town expansion offers a potentially cheaper solution than new town development.[35] This conclusion is based on general town amenities for town expansion projects costing £1 to £2 million for each 10 000 of new population as against £2½ million for new town schemes. On overall costs, Stone arrives at a figure of £1100 per person for town expansion schemes and £1180 per person for new town schemes.[36] Stone's findings are largely supported by Lichfield.[25]

The Minister of Housing and Local Government commissioned private consultants to make expansion studies of Ipswich, Peterborough and Worcester.[30,31,32] Table 14.6 indicates the estimated costs of expansion based on 1963 prices, with adjusted costs for the suggested new town at Hook by way of comparison.

TABLE 14.6

ESTIMATED TOWN DEVELOPMENT COSTS

	Town							
	Ipswich		*Worcester*		*Peterborough*			*Hook New Town*
	Extent of expansion (per cent)							*adjusted prices*
	50	100	50	100	50	100	150	
Transport and services	£ 668	£ 627	£1621	£1106	£2703	£1685	£1360	£1325
Central area	1118	840	268	281	3730	2133	1400	550
Industry	561	588	1041	988	1558	1460	1430	680
Housing	2839	2846	3670	3518	2919	2925	2929	2430
Other	659	844	482	375	601	593	585	515
Total cost per dwelling	£6873	6397	7081	6281	11 511	8796	7700	5500
Total cost per person	£1980	1830	2000	1780	2875	2375	2075	1780

On the basis of these figures, expanding town schemes are considerably more expensive than new town projects. They also indicate reducing costs per head of additional population with a larger scale of expansion. Unfortunately, different sources are producing differing costs and arriving at conflicting conclusions. It is very difficult to obtain accurate information on all the costs of expanding a town – the services are provided by a variety of organisations both public and private, and all the constituent costs are continually changing but not at a uniform rate, and this gives rise to difficulties in making comparisons.[16]

COST BENEFIT ANALYSIS

Nature of Technique

Cost benefit analysis has its origins in a paper presented by a French economist, Dupuit, in 1884 on the utility of public works. The technique has been further developed in the United States, where its sphere of operation has been extended into many aspects of society including river and harbour projects and flood control schemes.[37] Cost benefit analysis aims at setting out the factors which need to be taken into account in making economic choices. Most of the choices to which it has been applied involve investment projects and decisions – whether or not a particular project is worthwhile financially, which is the best of several alternative projects, or even when to undertake a particular project. The aim is generally to maximise the present values of all benefits less that of all costs, subject to specified restraints. Hender[28] has defined cost benefit analysis as 'a technique of use in either investment appraisal or the review of the operation of a service for analysing and measuring the costs and benefits to the community of adopting specified courses of action and for examining the incidence of these costs and benefits between different sections of the community'.

It has the basic objective of identifying and measuring the costs and benefits which stem from either the investment of monies or the operation of a service, but in particular it is concerned with examining not only those costs and benefits which have a direct impact on the providing authority but also those which are of an external nature and accrue to other persons. Furthermore, the costs and benefits to be measured are those which accrue throughout the life of the project.

Prest and Turvey[29] considered that the principal criteria to be determined were:

(1) Which costs and benefits are to be included?
(2) How are they to be valued?
(3) At what interest rate are they to be discounted?
(4) What are the relevant constraints?

Wilson[40] has described the methods to be used in a cost benefit study, and the methodology can be usefully summarised as:

(1) define the problems to be studied;
(2) identify the alternative courses of action;
(3) identify the costs and benefits, both to the providing authority and to external parties;
(4) evaluate the costs and benefits; and
(5) draw conclusions as to the alternative to be adopted.

Enumeration of Costs and Benefits

Some of the more commonly used expressions are defined and described.

Social costs may be defined as the sum total of costs involved as the result of an economic action. *Private costs* are those which affect the decisions of the performers; hence production costs include those of labour, materials, land and capital. There may also be *external costs*, for instance damage to buildings or decline of property values through smoke emanating from a factory; these costs are not met by the industrialist.

Similar effects may occur on the other side of the equation – benefits are reflected in the amount paid by consumers for goods produced; but in addition, favourable externalities might also accrue to society. An example could be a dam which in addition to generating electricity for sale in the market, gives flood protection to others for which they may not pay.[41]

Where there are strong relationships on either the supply or the demand side, allowance must be made for these in cost benefit calculations. Thus where an authority responsible for a long stretch of river constructs a dam at a point upstream, this will affect the water level and hence the operation of existing or potential dams downstream. The construction of a fast motorway which in itself speeds up traffic and reduces accidents, may lead to more congestion or more accidents on feeder roads if they are left unimproved.

Externalities

These are the costs and benefits which accrue to bodies other than the one sponsoring a project. The promoters of public investment projects should take into account the external effects of their actions in so far as they alter the physical production possibilities of other producers or the satisfactions that consumers can obtain from given resources; they should not take *side-effects* into account if the sole effect is through the prices of products or factors.[42]

An example of an external effect to be taken into account would be the construction of a reservoir by the upstream authority of a river basin which results in more dredging by a downstream authority. An example of a side-effect is where the improvement of a road leads to a greater profitability of the garages and restaurants on that road and the employment of more labour by them, higher rent payments, etc. Any net difference in profitability and any net rise in rents and land values is simply a reflection of the benefits of more journeys being undertaken, and it would be double counting if these were included too.[39]

Constraints

Projects are frequently subject to a variety of constraints or restricting factors and these have been classified by Eckstein.[43]

Physical constraints. The most common is the production function which relates the physical inputs and outputs of a project. Where a choice is involved between different projects or concerning the size or timing of a particular project, external physical restraints may also be relevant.

Legal constraints. Restrictions such as rights of access and time needed for public inquiries may be encountered.

Administrative constraints. These might possibly limit the size of the project.

Uncertainties. These may result from possible unreliability of estimates on future trends, etc.

Distributional and budgetary constraints. For instance, tolls on a motorway will affect the volume of traffic and may influence the width of the carriageways.

Applications of Cost Benefit Analysis

Cost benefit analysis techniques are being applied to a wide range of projects.

In this country one of the first cost benefit studies was that conducted by Coburn, Beesley and Reynolds[44] on the economic assessment of the London–Birmingham motorway (M1). This attempted to provide the economic justification for the expenditure of large sums of public money on motorway construction by showing the benefits which would flow from their development. It was restricted inasmuch as it was concerned with the benefits directly attributable to the construction of motorways and did not take account of the effect these new roads would have on neighbouring communities.

A further important study was undertaken by Beesley and Foster[45] on the construction of the Victoria Underground line in London. This study took place after the decision to construct the line had been taken and attempted to show the benefits which would accrue to different sectors of the population when it became operational. It is a particularly useful study for the way it illustrates the difficulty of placing measurements on certain intangible items, such as time savings during leisure hours, and also for the way it emphasises the importance of exercising extreme care in deciding the cut-off points in a practical situation.

The action of public authorities often has a ripple effect – the costs and benefits spread out from the centre and become more diffuse and difficult to measure as they become more remote from the direct action taken by the authority. Hender[38] gives the example of a local authority providing a housing estate which affects the tenants, shopkeepers and others who provide them with services. It could also affect servicing arrangements in neighbouring communities and a decision has to be taken as to the extent to which attempts are made to measure these indirect effects. Determining cut-off points is often one of the most difficult problems because if it is too tightly circumscribed major effects will be omitted from the study.

Lichfield has evolved a methodology known as the *planning balance sheet* by which he has applied cost benefit analysis to a wide range of town and regional planning problems.[46,47,48] A brief account of the methodology of analysis will explain its distinctive features.

An initial step is to enumerate the sectors of the community which are affected by the alternative proposals, treating them on the one hand as producers and operators of the investment to be made in the new project and on the other hand as consumers of the goods and services arising from that project. Then for each sector the question is asked: 'What would be the difference in costs and in benefits which would accrue under the respective schemes under examination?'

The costs and benefits comprise all those which are of relevance to the planning decision. Thus they include those which are direct as between the parties to the transaction and those which are indirect and come within the conventional definition of social costs and benefits; it includes those which relate to real resources and those which are transfers; it includes those which cannot be measured as well as those which can. Thus it is possible to evolve and summarise a set of social accounts for each sector of the community showing clearly the differences in costs and benefits which will accrue to them under the alternative plans. This final summary of social accounts does not produce the decision itself any more than any other economic calculus, but is the basis for the judgement leading to the decision. In some studies the judgement that ought to be made is apparent; in others the issues are more finely balanced.

Cost benefit analysis carried out by or on behalf of government departments include the Cambrian coast railway (1968), congestion costs in London (1968), Thamesmead river crossing (1969), dispersal of government offices (1969), rural water supplies (1968), recreation at Graffham Water (1969), improvement areas (1969–70), development in the Clyde Estuary (1968) and the third London airport (1969). A series of cost benefit studies prepared under the auspices of the Institute of Municipal Treasurers and Accountants[49] included alternative forms of housing (Coventry), industrial site development (Durham) and city centre redevelopment (Norwich).

Application to Development Decisions

Holmans[54] believes that cost benefit studies can make a useful contribution in arriving at decisions on a wide range of development problems, and he has given the following examples.

(1) Should a new town programme be carried out as a small number of large developments or a larger number of more modest size?

(2) Should a new town be built on a 'green-field' site or be grafted on to an existing town of substantial size?

(3) What funds should be allocated to redeveloping a central area?

(4) How should the choice be made between improving old houses and clearing the site and building anew?

(5) What economic criteria are there for choosing which areas of old housing to improve when improvement is the preferred course rather than replacement?

The general principles of choice between replacement and improvement of old housing have been defined by Needleman.[55] But Holmans[54] draws attention to the major practical problem of evaluating the difference in the standard of accommodation provided by new houses or flats and old houses improved. Scoring the accommodation according to the presence or absence of specific features would be one way, albeit not very precise. Another approach would be to use free-market rents (fair rents) as a basis for evaluating the difference in standards, but there are rarely sufficient to do this. Furthermore, as incomes rise occupiers are likely to become more exacting in their requirements and attach increased importance to features which do not exist in the improved older dwelling.

Stone[53] has described how town development is a complex process with far-reaching physical, social and economic consequences. Its impact varies as between public and private developers and the community at large, and so a method is needed for determining the consequences of decisions in this field from the various points of view. Thus the need arises for evaluating alternative solutions to the same problem and alternative ways of investing the same resources, with due regard to appearance, amenity and costs. The term amenity as used in this context includes the factors of safety and convenience. It is much easier to reach rational decisions when the consequences of alternative solutions have been quantified and due consideration must be given to all social and economic benefits. Both costs and benefits are spread over long periods of time involving difficult future predictions and discounting, some costs and benefits are indirect and are not easily costed and some intangible benefits become a matter of opinion.[53]

A cost benefit study of the provision of a tunnel or bridge to cross the Thames at Thamesmead indicated a close balance between the quantified factors (traffic and housing benefits being roughly equal to the higher cost of a tunnel). The unquantified environmental and delay factors favoured a tunnel, so that the final decision would have to be based on subjective matters, always provided that the extra money for the tunnel was available.

Application to Building Proposals

There is little evidence of cost benefit analysis being applied to building projects to any great extent. One very useful example was a study carried out by the Department of the Environment and published in 1971,[58] to compare the costs and benefits of planned open offices with air-conditioning, and traditional cellular offices without air-conditioning. In the study it was assumed that the open offices would be planned to certain essential standards of space, lighting, acoustic treatment, layout and furnishing for functional reasons. Differences were calculated on a cost per capita basis, and discounted to a net present value (NPV) in the year of building at eight per cent, and also at three per cent and sixteen per cent. It was assumed that the life of the buildings would be sixty years and 1966 prices were used throughout. An allowance was made for a relative increase in the cost of labour-intensive industries in calculating maintenance and cleaning costs. The following costs and benefits were quantified.

Costs. Capital costs of building, mechanical and electrical depreciation and running costs, and initial provision, maintenance and renewal of furniture.

Staff benefits. Increased productivity due to teamwork and better communications (one per cent staff salaries and overheads); increased productivity from air-conditioning (1.5 per cent staff salaries, etc.), less sick leave because of better working conditions (two per cent of sick leave); saving on messenger services (two messengers per 100 staff); increased productivity from better lighting (one per cent staff salaries, etc.); and better recruiting and staff satisfaction (0.1 per cent of noncareer grade staff salaries).

Building benefits. Greater flexibility – avoid remodelling internally after forty years (at cost of ten per cent of original cost of building), more economical use of space (valued at 0.1 m^2/head rising to 0.5 m^2/head during forty years before remodelling), no removal of partitions (valued at one per cent of partitions moved each year); low maintenance materials – reduction in replacement costs and decoration and in staff disturbance; and reduction in cleaning costs (owing to easier and fewer surfaces and building kept cleaner through air-conditioning).

The study concluded that the higher costs of open plan arrangements with air-conditioning were more than offset by the additional benefits at interest rates of eight per cent and three per cent; at sixteen per cent the cellular offices were slightly better. At eight per cent interest air-conditioned planned open offices were calculated to give a capitalised saving of about £260 per head compared with traditional nonair-conditioned cellular offices and £200 per head compared with air-conditioned cellular offices, at 1966 prices.

Conclusions

Wide divergences of view have been expressed about the role and usefulness of cost benefit analysis. Peston[50] and Stocks[51] both stress the perplexities stemming from uncertainties about the consequences of various courses of action and the difficulties of measuring the costs and benefits. Diverse types of benefit, avoidance of double counting, dealing with externalities and choice of discount rates pose a formidable range of problems.

On the other hand, is there a better alternative? The situation was well summed up in the Resources for Tomorrow Conference at Montreal in 1961:[52] 'There was general agreement that benefit cost analysis is a basically useful tool in project evaluation. While it has certain limitations and is sometimes difficult to apply, it is, nevertheless, an objective approach to the selection of projects.' It was emphasised that benefit cost analysis should be regarded only as a tool to be used in the decision-making process but not as a substitute for that process.

An important advantage of cost benefit study is that it compels those responsible to quantify costs and benefits as far as possible, rather than resting content with vague qualitative judgements or personal hunches. Furthermore, quantification and evaluation of benefits, however rough, does give some indication of the charges which consumers are willing to pay.[39] Its limitations are clearly shown in the Roskill Commission's report on the third London airport, where after costing all the intangibles relating to the four sites and arriving at total figures in excess of £4000 million, the cheapest site at Cublington is only five per cent less than the most expensive at Foulness. This is hardly a large enough margin to be conclusive in making a choice.

Even where 'shadow prices' cannot be computed for all the main costs and benefits, cost benefit analysis may still prove useful to the decision maker. In some cases, the valued and unvalued factors may both point in the same direction. If not, and if there is only one unvalued factor, such as noise, it is possible to calculate the value which the decision maker would have to put on this to justify a particular decision. Where there are several unvalued costs or benefits, the problem becomes more complex, but it may still be possible to show that such projects are worse than others on all counts; for instance, one transport strategy might be more expensive, save less travelling time, and have worse distributional effects on the environment than another. In some cases it might be more meaningful to assign limiting values (maxima and minima) to factors that cannot be valued more precisely.[56]

EFFECT OF ENTRY INTO THE COMMON MARKET

At the time of writing, the effect on the quantity surveying profession of entry into the Common Market was still rather uncertain. Nisbet[57] described how on the one hand there appeared to be an enormous potential for the expansion of the profession and tremendous opportunities for quantity surveyors to practise throughout the EEC countries, while on the other hand there appeared to be many factors which might operate to the profession's disadvantage and which might even inhibit its progress in Britain. Because of these apparent contradictions, the Royal Institution of Chartered Surveyors decided in 1971 to carry out several surveys to obtain information about the situation in the EEC countries.

Expenditure in 1969 on construction in Britain was about £4700 million, whilst that in Common Market countries amounted to £19 500 million. There appear to be serious deficiencies in the cost control methods employed in the EEC countries. In recent years missions from a number of European countries have visited Britain to examine quantity surveying practice, as there is no comparable profession on the Continent. Some doubts have been expressed as to whether the quantity surveying profession will be recognised in the EEC countries, and engineers, in particular, might offer strong opposition. The EEC countries at present rely mainly upon tenders based on specifications and drawings and the Treaty of Rome might prohibit the use of negotiated contracts and similar contractual processes. Nisbet[57] sees the main difficulties in registration, approximation of laws and general lack of knowledge of the profession. It is evident that the EEC clients must be satisfied that the economies the quantity surveyor can achieve through cost advice in briefing and design stages, improved tender procedures with better safeguards against collusion and better financial control, will more than offset the cost of his fees.

REFERENCES

1. G. HIGGIN. The future of quantity surveying. *The Chartered Surveyor* (October 1964)
2. A. CODDINGTON. Economics of the environment. *The Guardian* (9 August 1971)
3. J. E. COOKE. Environmental economics and the chartered surveyor. *The Chartered Surveyor* (May 1966)
4. CENTRAL STATISTICAL OFFICE. *National Income and Expenditure.* HMSO (1971)
5. DEPARTMENT OF ECONOMIC AFFAIRS. DEA progress report No. 52 (1969)
6. K. CLEMENS. Economic surveys for development: Paper 908. International Federation of Surveyors (1968)
7. K. BLESSLEY. Appraisal for urban renewal in the United Kingdom: Paper 916. International Federation of Surveyors (1968)
8. P. A. STONE. *Building Economy: Design, Production and Organisation – A Synoptic View.* Pergamon Press (1966)
9. W. LEAN and B. GOODALL. *Aspects of Land Economics.* Estates Gazette (1966)
10. J. P. LEWIS. *Building Cycles and Britain's Growth.* Macmillan (1965)
11. J. E. COOKE. The economic use of resources in the construction process. *Proceedings of Annual Conference of Royal Institution of Chartered Surveyors, University of Nottingham* (1966)
12. GOVERNMENT COMMITTEE OF INQUIRY (Phelps Brown). Into certain matters concerning labour in building and civil engineering. Cmnd. 3714. HMSO (1968)
13. NATIONAL ECONOMIC DEVELOPMENT COUNCIL, Report of working party on large industrial construction sites. HMSO (1970)
14. A. V. WARD. Economic aspects of building: Inflation, the economy and the building industry. *Building* (9 July 1971)
15. I. H. SEELEY. *Municipal Engineering Practice.* Macmillan (1967)
16. I. H. SEELEY. *Planned Expansion of Country Towns.* Godwin (1968)
17. MINISTRY OF TRANSPORT. Traffic in towns. HMSO (1963)
18. MINISTRIES OF HOUSING AND LOCAL GOVERNMENT AND TRANSPORT. Planning bulletin no. 1, Town centres: approach to renewal. HMSO (1962)
19. NATIONWIDE BUILDING SOCIETY. The prospect for housing (1971)
20. H. J. MANZONI. The evolution of city centres. Convention on town centre redevelopment. Institution of Municipal Engineers (1962)
21. E. C. STRATHON. Economic and estate management aspects of redevelopment. Convention on town centre redevelopment. Institution of Municipal Engineers (1962)
22. MINISTRIES OF HOUSING AND LOCAL GOVERNMENT AND TRANSPORT. Planning bulletin no. 3, Town centres: cost and control of redevelopment. HMSO (1963)

23. ROYAL INSTITUTE OF BRITISH ARCHITECTS AND ROYAL INSTITUTION OF CHARTERED SURVEYORS. Methods of tendering for the redevelopment of central areas (1962)
24. F. LONGDEN. Site assembly and the local authority. *Urban Renewal.* University of Salford (1967)
25. N. LICHFIELD. *Economics of Planned Development.* Estates Gazette (1966)
26. L. ROBERTSON. Pedestrian shopping centres – some factors in the design. *Journal of Institution of Municipal Engineers,* **89.3** (1962)
27. R. STEEL. Town development – roads and economics. *Journal of the Institution of Highway Engineers* (September 1965)
28. MINISTRIES OF HOUSING AND LOCAL GOVERNMENT AND TRANSPORT, Planning bulletin no. 7, Parking in town centres. HMSO (1965)
29. INSTITUTION OF MUNICIPAL ENGINEERS. Provision of car parks in shopping and commercial centres (1963)
30. VINCENT AND GORBING. Ipswich – a study in town development. HMSO (1963)
31. H. M. WELLS. Peterborough – an expansion study. HMSO (1963)
32. J. MADIN and PARTNERS. Worcestershire expansion study. HMSO (1963)
33. R. STEEL and W. L. SLAYTON. Urban renewal – proposals for Britain and experience in America. Royal Institution of Chartered Surveyors (1965)
34. ROYAL INSTITUTION OF CHARTERED SURVEYORS. Urban renewal. *Proceedings of Annual Conference* (1965)
35. P. A. STONE. The economics of housing and urban development. *Journal of the Royal Statistical Society – Series A (General),* **122A** (1959)
36. P. A. STONE. *Housing, Town Development, Land and Costs.* Estates Gazette (1963)
37. E. J. MISHAN. A survey of welfare economics, 1939–1959. *Economic Journal,* **70** (June 1960)
38. J. D. HENDER. Introduction to cost benefit analysis. *Cost Benefit Analysis.* Institute of Municipal Treasurers and Accountants (1968)
39. A. R. PREST and R. TURVEY. Cost benefit analysis – a survey. *Economic Journal,* **75** (December 1965)
40. A. G. WILSON. The technique of cost benefit analysis. *Cost Benefit Analysis.* Institute of Municipal Treasurers and Accountants (1968)
41. G. H. PETERS. Cost benefit analysis and public expenditure: Eaton paper 8. Institute of Economic Affairs (1966)
42. R. N. MCKEAN. *Efficiency in Government Through Systems Analysis.* John Wiley, New York (1958)
43. O. ECKSTEIN. A survey of the theory of public expenditure criteria. *Public Finances: Needs, Sources and Utilization.* Princeton University Press (1961)
44. T. M. COBURN, M. E. BEESLEY and D. J. REYNOLDS. The London–Birmingham motorway: traffic and economics. Road Research Laboratory technical paper 46. HMSO (1960)
45. M. E. BEESLEY and C. D. FOSTER. The Victoria line: social benefits and finances. *Journal of Royal Statistical Society,* **128.1** (1965)
46. N. LICHFIELD. Cost benefit analysis in town planning: a case study of Cambridge. Cambridgeshire County Council (1966)
47. N. LICHFIELD. Cost benefit analysis in town planning – a case study: Swanley. *Urban Studies,* **3.3** (1966)
48. N. LICHFIELD. Cost benefit analysis in urban expansion – a case study: Peterborough. *Regional Studies,* **3.2** (1969)
49. INSTITUTE OF MUNICIPAL TREASURERS AND ACCOUNTANTS. *Cost Benefit Analysis* (1968)
50. M. PESTON. Cost benefit values. *Town and Country Planning* (December 1969)
51. N. R. STOCKS. Cost benefit analysis. *The Chartered Surveyor* (April 1966)
52. RESOURCES FOR TOMORROW CONFERENCE. Water workshop B, Benefit cost analysis. *Proceedings of Resources for Tomorrow Conference, Montreal* (1961)
53. P. A. STONE. Decision techniques for town development. *Operational Research Quarterly,* **15** (1964)
54. A. E. HOLMANS. Applications of cost benefit analysis to new towns and urban redevelopment policy. *Proceedings of Seminar on Treatment of Land, Infrastructure and Amenity in Cost Benefit Studies, Sunningdale.* Management Accounting Unit, HM Treasury (1969)
55. L. NEEDLEMAN. The comparative economics of improvement and new building. *Urban Studies,* **6.2** (1969)
56. TREASURY, INFORMATION DIVISION. Costs and benefits, Economic progress report 4 (1970)
57. J. NISBET. Chartered quantity surveyors and the Common Market. *Building* (11 June 1971)
58. DEPARTMENT OF THE ENVIRONMENT, WHITEHALL DEVELOPMENT GROUP. Planned open offices: cost benefit analysis (1971)

APPENDIXES

APPENDIX

AMOUNT OF

Rate per cent

Years	2	$2\frac{1}{2}$	3	$3\frac{1}{2}$	4	$4\frac{1}{2}$	5	$5\frac{1}{2}$
1	1.0200	1.0250	1.0300	1.0350	1.0400	1.0450	1.0500	1.0550
2	1.0403	1.0506	1.0608	1.0712	1.0815	1.0920	1.1024	1.1130
3	1.0612	1.0768	1.0927	1.1087	1.1248	1.1411	1.1576	1.1742
4	1.0824	1.1038	1.1255	1.1475	1.1698	1.1925	1.2155	1.2388
5	1.1040	1.1314	1.1592	1.1876	1.2166	1.2461	1.2762	1.3069
6	1.1261	1.1596	1.1940	1.2292	1.2653	1.3022	1.3400	1.3788
7	1.1436	1.1886	1.2298	1.2722	1.3159	1.3608	1.4071	1.4546
8	1.1716	1.2184	1.2667	1.3168	1.3685	1.4221	1.4774	1.5346
9	1.1950	1.2488	1.3047	1.3628	1.4233	1.4860	1.5513	1.6190
10	1.2189	1.2800	1.3439	1.4105	1.4802	1.5529	1.6288	1.7081
11	1.2433	1.3120	1.3842	1.4599	1.5394	1.6228	1.7103	1.8020
12	1.2682	1.3448	1.4257	1.5110	1.6010	1.6958	1.7958	1.9012
13	1.2936	1.3785	1.4685	1.5639	1.6650	1.7721	1.8856	2.0057
14	1.3194	1.4129	1.5125	1.6186	1.7316	1.8519	1.9799	2.1160
15	1.3458	1.4482	1.5579	1.6753	1.8009	1.9352	2.0789	2.2324
16	1.3727	1.4845	1.6047	1.7339	1.8729	2.0223	2.1828	2.3552
17	1.4002	1.5216	1.6528	1.7946	1.9479	2.1133	2.2920	2.4848
18	1.4282	1.5596	1.7024	1.8574	2.0258	2.2084	2.4066	2.6214
19	1.4568	1.5986	1.7535	1.9225	2.1068	2.3078	2.5269	2.7656
20	1.4859	1.6386	1.8061	1.9897	2.1911	2.4117	2.6532	2.9177
21	1.5156	1.6795	1.8602	2.0594	2.2787	2.5202	2.7859	3.0782
22	1.5459	1.7215	1.9161	2.1315	2.3699	2.6336	2.9252	3.2475
23	1.5768	1.7646	1.9735	2.2061	2.4647	2.7521	3.0715	3.4261
24	1.6084	1.8087	2.0327	2.2833	2.5633	2.8760	3.2250	3.6145
25	1.6406	1.8539	2.0937	2.3632	2.6658	3.0054	3.3863	3.8133
26	1.6734	1.9002	2.1565	2.4459	2.7724	3.1406	3.5556	4.0231
27	1.7068	1.9477	2.2212	2.5315	2.8833	3.2820	3.7834	4.2444
28	1.7410	1.9964	2.2879	2.6201	2.9987	3.4296	3.9201	4.4778
29	1.7758	2.0464	2.3565	2.7118	3.1180	3.5840	4.1161	4.7241
30	1.8113	2.0975	2.4272	2.8067	3.2433	3.7453	4.3219	4.9839
35	1.9998	2.3732	2.8138	3.3335	3.9460	4.6673	5.5160	6.5138
40	2.2080	2.6850	3.2620	3.9592	4.8010	5.8163	7.0399	8.5133
45	2.4378	3.0379	3.7815	4.7023	5.8411	7.2482	8.9850	11.1265
50	2.6915	3.4371	4.3839	5.5849	7.1066	9.0326	11.4673	14.5419
55	2.9717	3.8887	5.0821	6.6331	8.6463	11.2563	14.6356	19.0057
60	3.2810	4.3997	5.8916	7.8780	10.5196	14.0274	18.6791	24.8397
65	3.6225	4.9779	6.8299	9.3566	12.7987	17.4807	23.8398	32.4645
70	3.9995	5.6321	7.9178	11.1128	15.5716	21.7841	30.4264	42.4299
75	4.4158	6.3722	9.1789	13.1985	18.9452	27.1469	38.8326	55.4542
80	4.8754	7.2095	10.6408	15.6757	23.0497	33.8300	49.5614	72.4764
85	5.3828	8.1569	12.3357	18.6178	28.0436	42.1584	63.2543	94.7237
90	5.9431	9.2288	14.3004	22.1121	34.1193	52.5371	80.7303	123.8002
95	6.5616	10.4416	16.5781	26.2623	41.5113	65.4707	103.6346	161.8019
100	7.2446	11.8137	19.2186	31.1914	50.5049	81.5885	131.5012	211.4686

1

£1 TABLE

compound interest

6	6½	7	7½	8	9	10
1.0600	1.0650	1.0700	1.0750	1.0800	1.0900	1.1000
1.1235	1.1342	1.1448	1.1556	1.1663	1.1880	1.2099
1.1910	1.2079	1.2250	1.2422	1.2597	1.2950	1.3309
1.2624	1.2864	1.3107	1.3354	1.3604	1.4115	1.4640
1.3382	1.3700	1.4025	1.4356	1.4693	1.5386	1.6105
1.4185	1.4591	1.5007	1.5433	1.5868	1.6771	1.7715
1.5036	1.5589	1.6057	1.6590	1.7138	1.8280	1.9487
1.5938	1.6549	1.7181	1.7834	1.8509	1.9925	2.1435
1.6894	1.7625	1.8384	1.9172	1.9990	2.1718	2.3579
1.7908	1.8771	1.9671	2.0610	2.1589	2.3673	2.5937
1.8982	1.9991	2.1048	2.2156	2.3316	2.5804	2.8531
2.0121	2.1290	2.2521	2.3817	2.5181	2.8126	3.1384
2.1329	2.2674	2.4098	2.5604	2.7196	3.0658	3.4522
2.2609	2.4148	2.5785	2.7524	2.9371	3.3417	3.7974
2.3965	2.5718	2.7590	2.9588	3.1721	3.6424	4.1772
2.5403	2.7390	2.9521	3.1807	3.4259	3.9703	4.5949
2.6927	2.9170	3.1588	3.4193	3.7000	4.3276	5.0544
2.8543	3.1066	3.3799	3.6758	3.9960	4.7171	5.5599
3.0255	3.3085	3.6165	3.9514	4.3157	5.1416	6.1159
3.2071	3.5236	3.8696	4.2478	4.6609	5.6044	6.7274
3.3995	3.7526	4.1405	4.5664	5.0338	6.1088	7.4002
3.6035	3.9966	4.4304	4.9089	5.4365	6.6586	8.1402
3.8197	4.2563	4.7405	5.2770	5.8714	7.2578	8.9543
4.0489	4.5330	5.0723	5.6728	6.3411	7.9110	9.8497
4.2918	4.8276	5.4274	6.0983	6.8484	8.6230	10.8347
4.5493	5.1414	5.8073	6.5557	7.3963	9.3991	11.9181
4.8223	5.4756	6.2138	7.0473	7.9880	10.2450	13.1099
5.1116	5.8316	6.6488	7.5759	8.6271	11.1671	14.4209
5.4183	6.2106	7.1142	8.1441	9.3172	12.1721	15.8630
5.7434	6.6143	7.6122	8.7549	10.0626	13.2676	17.4494
7.6860	9.0622	10.6765	12.5688	14.7853	20.4139	28.1024
10.2857	12.4160	14.9744	18.0442	21.7245	31.4094	45.2592
13.7646	17.0110	21.0024	25.9048	31.9204	48.3272	72.8904
18.4201	23.3066	29.4570	37.1897	46.9016	74.3575	117.3908
24.6503	31.9321	41.3149	53.3906	68.9138	114.4082	189.0591
32.9876	43.7498	57.9464	76.6492	101.2570	176.0312	304.4816
44.1449	59.9410	81.2728	110.0398	148.7798	270.8459	490.3706
59.0759	82.1244	113.9893	157.9764	218.6063	416.7300	789.7468
79.0569	112.5176	159.8760	226.7956	321.2045	641.1908	1271.8952
105.7959	154.1589	224.2343	325.5945	471.9547	986.5515	2048.4000
141.5788	211.2110	314.5002	467.4330	693.4564	1517.9319	3298.9687
189.4645	289.3774	441.1029	671.0605	1018.9149	2335.5264	5313.0221
253.5462	396.4721	618.6696	963.3942	1497.1203	3593.4969	8556.6753
339.3020	543.2012	867.7162	1383.0771	2199.7612	5529.0406	13780.6110

APPENDIX

PRESENT VALUE OF

	Rate per cent							
Years	2	2½	3	3½	4	4½	5	5½
1	0.98039	0.97560	0.97087	0.96618	0.96153	0.95693	0.95238	0.94786
2	0.96116	0.95181	0.94259	0.93351	0.92455	0.91573	0.90702	0.89845
3	0.94232	0.92859	0.91514	0.90194	0.88899	0.87629	0.86383	0.85161
4	0.92384	0.90595	0.88848	0.87144	0.85480	0.83856	0.82270	0.80721
5	0.90573	0.88385	0.86260	0.84197	0.82192	0.80245	0.78352	0.76513
6	0.88797	0.86229	0.83748	0.81350	0.79031	0.76789	0.74621	0.72524
7	0.87056	0.84126	0.81309	0.78599	0.75991	0.73482	0.71068	0.68743
8	0.85349	0.82074	0.78940	0.75941	0.73069	0.70318	0.67683	0.65159
9	0.83675	0.80072	0.76641	0.73373	0.70258	0.67290	0.64460	0.61762
10	0.82034	0.78119	0.74409	0.70891	0.67556	0.64392	0.61391	0.58543
11	0.80426	0.76214	0.72242	0.68494	0.64958	0.61619	0.58467	0.55491
12	0.78849	0.74355	0.70137	0.66178	0.62459	0.58966	0.55683	0.52598
13	0.77303	0.72542	0.68095	0.63940	0.60057	0.56427	0.53032	0.49856
14	0.75787	0.70772	0.66111	0.61778	0.57747	0.53997	0.50506	0.47256
15	0.74301	0.69046	0.64186	0.59689	0.55526	0.51672	0.48101	0.44793
16	0.72844	0.67362	0.62316	0.57670	0.53390	0.49446	0.45811	0.42458
17	0.71416	0.65719	0.60501	0.55720	0.51337	0.47317	0.43629	0.40244
18	0.70015	0.64116	0.58739	0.53836	0.49362	0.45280	0.41552	0.38146
19	0.68643	0.62552	0.57028	0.52015	0.47464	0.43330	0.39573	0.36157
20	0.67297	0.61027	0.55367	0.50256	0.45638	0.41464	0.37688	0.34272
21	0.65977	0.59538	0.53754	0.48557	0.43383	0.39678	0.35894	0.32486
22	0.64683	0.58086	0.52189	0.46915	0.42195	0.37970	0.34184	0.30792
23	0.63415	0.56669	0.50669	0.45328	0.40572	0.36335	0.32557	0.29187
24	0.62172	0.55287	0.49193	0.43795	0.39012	0.34770	0.31006	0.27665
25	0.60953	0.53939	0.47760	0.42314	0.37511	0.33273	0.29530	0.26223
26	0.59757	0.52623	0.46369	0.40883	0.36068	0.31840	0.28124	0.24856
27	0.58586	0.51339	0.45018	0.39501	0.34681	0.30469	0.26784	0.23560
28	0.57437	0.50087	0.43707	0.38165	0.33347	0.29157	0.25509	0.22332
29	0.56311	0.48866	0.42434	0.36874	0.32065	0.27901	0.24294	0.21167
30	0.55207	0.47674	0.41198	0.35627	0.30831	0.26700	0.23137	0.22064
35	0.50002	0.42137	0.35538	0.29997	0.25341	0.21425	0.18129	0.15351
40	0.45289	0.37243	0.30655	0.25257	0.20828	0.17192	0.14204	0.11746
45	0.41019	0.32917	0.26443	0.21265	0.17119	0.13796	0.11129	0.08987
50	0.37152	0.29094	0.22810	0.17905	0.14071	0.11070	0.08720	0.06876
55	0.33650	0.25715	0.19676	0.15075	0.11565	0.08883	0.06832	0.05261
60	0.30478	0.22728	0.16973	0.12693	0.09506	0.07128	0.05353	0.04025
65	0.27605	0.20088	0.14641	0.10687	0.07813	0.05720	0.04194	0.03080
70	0.25002	0.17755	0.12629	0.08998	0.06421	0.04590	0.03286	0.02356
75	0.22645	0.15693	0.10894	0.07576	0.05278	0.03683	0.02575	0.01803
80	0.20510	0.13870	0.09397	0.06379	0.04338	0.02955	0.02017	0.01379
85	0.18577	0.12259	0.08106	0.05371	0.03565	0.02372	0.01580	0.01055
90	0.16826	0.10835	0.06992	0.04522	0.02930	0.01903	0.01238	0.00807
95	0.15239	0.09577	0.06032	0.03807	0.02408	0.01527	0.00970	0.00618
100	0.13803	0.08464	0.05203	0.03206	0.01980	0.01225	0.00760	0.00472

2

£1 TABLE

compound interest

6	6½	7	7½	8	9	10
0.94339	0.93896	0.93457	0.93023	0.92592	0.91743	0.90909
0.88999	0.88165	0.87343	0.86533	0.85733	0.84168	0.82644
0.83961	0.82784	0.81629	0.80496	0.79383	0.77218	0.75131
0.79209	0.77732	0.76289	0.74880	0.73502	0.70842	0.68301
0.74725	0.72988	0.71208	0.69655	0.68058	0.64993	0.62092
0.70496	0.68933	0.66634	0.64796	0.63016	0.59626	0.56447
0.66505	0.64350	0.62274	0.60275	0.58349	0.54703	0.51315
0.62741	0.60423	0.58200	0.55070	0.54026	0.50186	0.46650
0.59189	0.56735	0.54393	0.52158	0.50024	0.46042	0.42409
0.55839	0.53272	0.50834	0.48519	0.46319	0.42241	0.38554
0.52678	0.50021	0.47509	0.45134	0.42888	0.38753	0.35049
0.49696	0.46968	0.44401	0.41985	0.39711	0.35553	0.31863
0.46883	0.44101	0.41496	0.39056	0.36769	0.32617	0.28966
0.44230	0.41410	0.38781	0.36331	0.34046	0.29924	0.26333
0.41726	0.38882	0.36244	0.33796	0.31524	0.27453	0.23939
0.39364	0.36509	0.33873	0.31438	0.29189	0.25186	0.21762
0.37136	0.34281	0.31657	0.29245	0.27026	0.23107	0.19784
0.35034	0.32188	0.29586	0.27204	0.25024	0.21199	0.17985
0.33051	0.30224	0.27650	0.25306	0.23171	0.19448	0.16350
0.31180	0.28379	0.25841	0.23541	0.21454	0.17843	0.14864
0.29415	0.26647	0.24151	0.21898	0.19865	0.16369	0.13513
0.27750	0.25021	0.22571	0.20371	0.18394	0.15018	0.12284
0.26179	0.23494	0.21094	0.18949	0.17031	0.13778	0.11167
0.24697	0.22060	0.19714	0.17627	0.15769	0.12640	0.10152
0.23299	0.20713	0.18424	0.16397	0.14601	0.11596	0.09229
0.21981	0.19449	0.17219	0.15253	0.13520	0.10639	0.08390
0.20736	0.18262	0.16093	0.14189	0.12518	0.09760	0.07627
0.19563	0.17147	0.15040	0.13199	0.11591	0.08954	0.06934
0.18455	0.16101	0.14056	0.12278	0.10732	0.08215	0.06303
0.17411	0.15118	0.13136	0.11422	0.09937	0.07537	0.05730
0.13010	0.11034	0.09366	0.07956	0.06763	0.04898	0.03558
0.09722	0.08054	0.06678	0.05541	0.04603	0.03183	0.02209
0.07265	0.05878	0.04761	0.03860	0.03132	0.02069	0.01371
0.05428	0.04290	0.03394	0.02688	0.02132	0.01344	0.00851
0.04056	0.03131	0.02420	0.01872	0.01451	0.00874	0.00528
0.03031	0.02285	0.01725	0.01304	0.00987	0.00568	0.00328
0.02265	0.01668	0.01230	0.00908	0.00672	0.00369	0.00203
0.01692	0.01217	0.00877	0.00633	0.00457	0.00239	0.00126
0.01264	0.00888	0.00525	0.00440	0.00311	0.00155	0.00078
0.00945	0.00648	0.00445	0.00307	0.00211	0.00101	0.00048
0.00706	0.00473	0.00317	0.00213	0.00144	0.00065	0.00030
0.00527	0.00345	0.00226	0.00149	0.00098	0.00042	0.00018
0.00394	0.00252	0.00161	0.00103	0.00066	0.00027	0.00011
0.00294	0.00184	0.00115	0.00072	0.00045	0.00018	0.00007

APPENDIX

AMOUNT OF £1

Rate per cent

Years	2	2½	3	3½	4	4½	5	5½
1	1.000	1.000	1.000	1.000	1.000	1.000	1.000	1.000
2	2.019	2.024	2.029	2.034	2.039	2.044	2.049	2.054
3	3.060	3.075	3.090	3.106	3.121	3.137	3.152	3.168
4	4.121	4.152	4.183	4.214	4.246	4.278	4.310	4.342
5	5.204	5.256	5.309	5.362	5.416	5.470	5.525	5.581
6	6.308	6.387	6.468	6.550	6.632	6.716	6.801	6.888
7	7.434	7.547	7.662	7.779	7.898	8.019	8.142	8.266
8	8.582	8.736	8.892	9.051	9.214	9.380	9.549	9.721
9	9.754	9.954	10.159	10.368	10.582	10.802	11.026	11.256
10	10.949	11.203	11.463	11.731	12.006	12.288	12.577	12.875
11	12.168	12.483	12.807	13.141	13.486	13.841	14.206	14.583
12	13.412	13.795	14.192	14.601	15.025	15.464	15.917	16.385
13	14.680	15.140	15.617	16.113	16.626	17.159	17.712	18.286
14	15.973	16.518	17.086	17.676	18.291	18.932	19.598	20.292
15	17.293	17.931	18.598	19.295	20.023	20.784	21.578	22.408
16	18.639	19.380	20.156	20.971	21.824	22.719	23.657	24.641
17	20.012	20.864	21.761	22.705	23.697	24.741	25.840	26.996
18	21.412	22.386	23.414	24.499	25.645	26.855	28.132	29.481
19	22.840	23.946	25.116	26.357	27.671	29.063	30.539	32.102
20	24.297	25.544	26.870	28.279	29.778	31.371	33.065	34.863
21	25.783	27.183	28.676	30.269	31.969	33.783	35.719	37.786
22	27.298	28.862	30.536	32.328	34.247	36.303	38.505	40.864
23	28.844	30.584	32.452	34.460	36.617	38.937	41.430	44.111
24	30.421	32.349	34.426	36.666	39.082	41.689	44.501	47.537
25	32.030	34.157	36.459	38.949	41.645	44.565	47.727	51.152
26	33.670	36.011	38.553	41.313	44.311	47.570	51.113	54.965
27	35.344	37.911	40.709	43.759	47.084	50.711	54.669	58.989
28	37.051	39.859	42.930	46.290	49.967	53.993	58.402	63.233
29	38.792	41.856	45.218	48.910	52.966	57.423	62.322	67.711
30	40.568	43.902	47.575	51.622	56.084	61.007	66.438	72.435
35	49.994	54.928	60.462	66.674	73.652	81.496	90.320	100.251
40	60.401	67.402	75.401	84.550	95.025	107.030	120.799	136.605
45	71.892	81.516	92.719	105.781	121.029	138.849	159.700	184.119
50	84.579	97.484	112.796	130.997	152.667	178.503	209.347	246.217
55	98.586	115.550	136.071	160.946	191.159	227.917	272.712	327.377
60	114.051	135.991	163.053	196.516	237.990	289.497	353.583	433.450
65	131.126	159.118	194.332	238.762	294.968	366.237	456.797	572.083
70	149.977	185.284	230.594	288.937	364.290	461.869	588.528	753.271
75	170.791	214.888	272.630	348.529	448.631	581.044	756.653	990.076
80	193.771	248.382	321.362	419.306	551.244	729.557	971.228	1299.571
85	219.143	286.278	377.856	503.367	676.090	914.632	1245.086	1704.068
90	247.156	329.154	443.348	603.204	827.983	1145.268	1594.607	2232.730
95	278.084	377.664	519.271	721.780	1012.784	1432.684	2040.693	2923.670
100	312.232	432.548	607.287	862.611	1237.623	1790.855	2610.025	3826.702

3

PER ANNUM TABLE

compound interest

6	6½	7	7½	8	9	10
1.000	1.000	1.000	1.000	1.000	1.000	1.000
2.059	2.064	2.069	2.074	2.079	2.089	2.099
3.183	3.199	3.214	3.230	3.246	3.278	3.309
4.374	4.407	4.439	4.472	4.506	4.573	4.640
5.637	5.693	5.750	5.808	5.866	5.984	6.105
6.975	7.063	7.153	7.244	7.335	7.523	7.715
8.393	8.522	8.654	8.787	8.922	9.200	9.487
9.897	10.076	10.259	10.446	10.636	11.028	11.435
11.491	11.731	11.977	12.229	12.487	13.021	13.579
13.180	13.494	13.816	14.147	14.486	15.192	15.937
14.971	15.371	15.783	16.208	16.645	17.560	18.531
16.869	17.370	17.888	18.423	18.977	20.140	21.384
18.882	19.499	20.140	20.805	21.495	22.953	24.522
21.015	21.767	22.550	23.365	24.214	26.019	27.974
23.275	24.182	25.129	26.118	27.152	29.360	31.772
25.672	26.754	27.888	29.077	30.324	33.003	35.949
28.212	29.493	30.840	32.258	33.750	36.973	40.544
30.905	32.410	33.999	35.677	37.450	41.301	45.599
33.759	35.516	37.378	39.358	41.446	46.018	51.159
36.785	38.825	40.995	43.304	45.761	51.160	57.274
39.992	42.348	44.865	47.552	50.422	56.764	64.002
43.392	46.101	49.005	52.118	55.456	62.873	71.402
46.995	50.098	53.436	57.027	60.893	69.531	79.543
50.815	54.354	58.176	62.304	66.764	76.789	88.497
54.864	58.887	63.249	67.977	73.105	84.700	98.347
59.156	63.715	68.676	74.070	79.954	93.323	109.181
63.705	68.856	74.483	80.631	87.350	102.723	121.099
68.528	74.332	80.697	87.679	95.338	112.968	134.209
73.639	80.164	87.346	95.255	103.965	124.135	148.630
79.058	86.374	94.460	103.399	113.283	136.307	164.494
111.434	124.034	138.236	154.251	172.316	215.710	271.024
154.761	175.631	199.635	227.256	259.056	337.882	442.592
212.743	246.324	285.749	332.064	386.505	525.858	718.904
290.335	343.179	406.528	482.529	573.770	815.083	1163.908
394.172	475.879	575.928	698.542	848.923	1260.091	1880.591
533.128	657.689	813.520	1008.656	1253.213	1944.791	3034.816
719.082	906.785	1146.755	1453.865	1847.247	2998.288	4893.706
967.932	1248.068	1614.134	2093.019	2720.079	4619.223	7887.468
1300.948	1715.655	2269.657	3010.608	4002.556	7113.232	12708.952
1746.599	2356.290	3189.062	4327.926	5886.934	10950.572	20474.000
2342.981	3234.016	4478.575	6219.107	8655.705	16854.798	32979.687
3141.075	4436.576	6287.185	8934.140	12723.936	25939.182	53120.221
4209.103	6084.187	8823.852	12831.922	18701.503	39916.632	85556.753
5638.367	8341.557	12381.661	18427.694	27484.515	61422.673	137796.110

APPENDIX

ANNUAL SINKING

Rate per cent

Years	2	2½	3	3½	4	4½	5	5½
1	1.00000	1.00000	1.00000	1.00000	1.00000	1.00000	1.00000	1.00000
2	0.49505	0.49382	0.49261	0.49140	0.49019	0.48899	0.48780	0.48661
3	0.32675	0.32513	0.32353	0.32195	0.32034	0.31877	0.31720	0.31565
4	0.24262	0.24081	0.23902	0.23725	0.23549	0.23374	0.23201	0.23029
5	0.19215	0.19024	0.18835	0.18648	0.18462	0.18279	0.18097	0.17917
6	0.15852	0.15654	0.15459	0.15266	0.15076	0.14867	0.14701	0.14517
7	0.13451	0.13249	0.13050	0.12894	0.12660	0.12470	0.12281	0.12096
8	0.11650	0.11446	0.11245	0.11047	0.10852	0.10660	0.10472	0.10286
9	0.10251	0.10045	0.09843	0.09644	0.09449	0.09257	0.09069	0.08883
10	0.09132	0.08925	0.08723	0.08524	0.08329	0.08137	0.07950	0.07766
11	0.08217	0.08010	0.07807	0.07609	0.07414	0.07224	0.07038	0.06857
12	0.07455	0.07248	0.07046	0.06848	0.06655	0.06466	0.06282	0.06102
13	0.06811	0.06604	0.06402	0.06206	0.06014	0.05827	0.05645	0.05468
14	0.06260	0.06053	0.05852	0.05657	0.05466	0.05282	0.05102	0.04927
15	0.05782	0.05576	0.05376	0.05182	0.04994	0.04811	0.04634	0.04462
16	0.05565	0.05159	0.04961	0.04768	0.04582	0.04401	0.04226	0.04058
17	0.04996	0.04792	0.04595	0.04404	0.04219	0.04041	0.03869	0.03704
18	0.04670	0.04467	0.04270	0.04081	0.03899	0.03723	0.03554	0.03391
19	0.04378	0.04176	0.03981	0.03794	0.03613	0.03440	0.03274	0.03115
20	0.04115	0.03914	0.03721	0.03536	0.03358	0.03187	0.03024	0.02867
21	0.03878	0.03678	0.03487	0.03303	0.03128	0.02960	0.02799	0.02646
22	0.03663	0.03464	0.03274	0.03093	0.02919	0.02754	0.02597	0.02447
23	0.03460	0.03269	0.03081	0.02901	0.02730	0.02568	0.02413	0.02266
24	0.03287	0.03091	0.02904	0.02727	0.02558	0.02398	0.02247	0.02103
25	0.03122	0.02927	0.02742	0.02567	0.02401	0.02243	0.02095	0.01954
26	0.02969	0.02776	0.02593	0.02420	0.02256	0.02102	0.01956	0.01819
27	0.02829	0.02637	0.02456	0.02285	0.02123	0.01971	0.01829	0.01695
28	0.02698	0.02508	0.02329	0.02160	0.02001	0.01852	0.01712	0.01581
29	0.02577	0.02389	0.02211	0.02044	0.01887	0.01741	0.01604	0.01476
30	0.02464	0.02277	0.02101	0.01937	0.01783	0.01639	0.01505	0.01380
35	0.02000	0.01820	0.01653	0.01499	0.01357	0.01227	0.01107	0.00997
40	0.01655	0.01483	0.01326	0.01182	0.01052	0.00934	0.00827	0.00732
45	0.01390	0.01226	0.01078	0.00945	0.00826	0.00720	0.00626	0.00543
50	0.01182	0.01025	0.00886	0.00763	0.00655	0.00560	0.00477	0.00406
55	0.01014	0.00865	0.00734	0.00621	0.00523	0.00438	0.00366	0.00305
60	0.00876	0.00735	0.00613	0.00508	0.00420	0.00345	0.00282	0.00230
65	0.00762	0.00628	0.00514	0.00418	0.00339	0.00273	0.00218	0.00174
70	0.00666	0.00539	0.00433	0.00346	0.00274	0.00216	0.00169	0.00132
75	0.00585	0.00465	0.00366	0.00286	0.00222	0.00172	0.00132	0.00101
80	0.00516	0.00402	0.00311	0.00238	0.00181	0.00137	0.00102	0.00076
85	0.00456	0.00349	0.00264	0.00198	0.00147	0.00109	0.00080	0.00058
90	0.00404	0.00303	0.00225	0.00165	0.00120	0.00087	0.00062	0.00044
95	0.00359	0.00264	0.00192	0.00138	0.00098	0.00069	0.00049	0.00034
100	0.00320	0.00231	0.00164	0.00115	0.00080	0.00055	0.00038	0.00026

4

FUND TABLE

compound interest

6	6½	7	7½	8	9	10
1.00000	1.00000	1.00000	1.00000	1.00000	1.00000	1.00000
0.48543	0.48426	0.48309	0.48192	0.48076	0.47846	0.47619
0.31410	0.31257	0.31105	0.30953	0.30803	0.30505	0.30211
0.22859	0.22690	0.22522	0.22356	0.22192	0.21866	0.21547
0.17739	0.17563	0.17389	0.17216	0.17045	0.16709	0.16379
0.14336	0.14156	0.13979	0.13804	0.13631	0.13291	0.12960
0.11913	0.11733	0.11555	0.11380	0.11207	0.10869	0.10540
0.10103	0.09923	0.09746	0.09572	0.09401	0.09067	0.08744
0.08702	0.08523	0.08348	0.08176	0.08007	0.07679	0.07364
0.07586	0.07410	0.07237	0.07068	0.06902	0.06582	0.06274
0.06679	0.06505	0.06335	0.06169	0.06007	0.05694	0.05396
0.05927	0.05756	0.05590	0.05427	0.05269	0.04965	0.04676
0.05296	0.05128	0.04965	0.04806	0.04652	0.04356	0.04077
0.04758	0.04594	0.04434	0.04279	0.04129	0.03843	0.03574
0.04296	0.04135	0.03979	0.03828	0.03682	0.03405	0.03147
0.03895	0.03737	0.03585	0.03439	0.03297	0.03029	0.02781
0.03544	0.03390	0.03242	0.03100	0.02962	0.02704	0.02466
0.03235	0.03085	0.02941	0.02802	0.02670	0.02421	0.02193
0.02962	0.02815	0.02675	0.02541	0.02412	0.02173	0.01954
0.02718	0.02575	0.02439	0.02309	0.02185	0.01954	0.01745
0.02500	0.02361	0.02228	0.02102	0.01983	0.01761	0.01562
0.02304	0.02169	0.02040	0.01918	0.01803	0.01590	0.01400
0.02127	0.01996	0.01871	0.01753	0.01642	0.01438	0.01257
0.01967	0.01839	0.01718	0.01605	0.01497	0.01302	0.01129
0.01822	0.01698	0.01581	0.01471	0.01367	0.01180	0.01016
0.01690	0.01569	0.01456	0.01349	0.01250	0.01071	0.00915
0.01569	0.01452	0.01342	0.01240	0.01144	0.00973	0.00825
0.01459	0.01345	0.01239	0.01140	0.01048	0.00885	0.00745
0.01357	0.01247	0.01144	0.01049	0.00961	0.00805	0.00672
0.01264	0.01157	0.01058	0.00967	0.00882	0.00733	0.00607
0.00897	0.00806	0.00723	0.00648	0.00580	0.00463	0.00368
0.00646	0.00569	0.00500	0.00440	0.00386	0.00295	0.00225
0.00470	0.00405	0.00349	0.00301	0.00258	0.00190	0.00139
0.00344	0.00291	0.00245	0.00207	0.00174	0.00122	0.00085
0.00253	0.00210	0.00173	0.00143	0.00117	0.00079	0.00053
0.00187	0.00152	0.00122	0.00099	0.00079	0.00051	0.00032
0.00139	0.00110	0.00087	0.00068	0.00054	0.00033	0.00020
0.00103	0.00080	0.00061	0.00047	0.00036	0.00021	0.00012
0.00076	0.00058	0.00044	0.00033	0.00024	0.00014	0.00007
0.00057	0.00042	0.00031	0.00023	0.00016	0.00009	0.00004
0.00042	0.00030	0.00022	0.00016	0.00011	0.00005	0.00003
0.00031	0.00022	0.00015	0.00011	0.00007	0.00003	0.00001
0.00023	0.00016	0.00011	0.00007	0.00005	0.00002	0.00001
0.00017	0.00011	0.00008	0.00005	0.00003	0.00001	0.00000

APPENDIX

PRESENT VALUE OF £1 PER ANNUM

Rate per cent

Years	2	2½	3	3½	4	4½	5	5½
1	0.9803	0.9756	0.9708	0.9661	0.9615	0.9569	0.9523	0.9478
2	1.9415	1.9274	1.9134	1.8996	1.8860	1.8726	1.8594	1.8463
3	2.8838	2.8560	2.8286	2.8016	2.7750	2.7489	2.7232	2.6979
4	3.8077	3.7619	3.7170	3.6730	3.6296	3.5875	3.5459	3.5051
5	4.7134	4.6458	4.5797	4.5150	4.4518	4.3899	4.3294	4.2702
6	5.6014	5.5081	5.4171	5.3285	5.2421	5.1578	5.0756	4.9955
7	6.4719	6.3493	6.2302	6.1145	6.0020	5.8927	5.7863	5.6829
8	7.3254	7.1701	7.0196	6.8739	6.7827	6.5958	6.4632	6.3345
9	8.1622	7.9708	7.7861	7.6076	7.4353	7.2687	7.1078	6.9521
10	8.9825	8.7520	8.5302	8.3166	8.1108	7.9127	7.7217	7.5376
11	9.7868	9.5142	9.2326	9.0015	8.7604	8.5289	8.3064	8.0925
12	10.5753	10.2577	9.9540	9.6633	9.3850	9.1185	8.8632	8.6185
13	11.3483	10.9831	10.6349	10.3027	9.9856	9.6828	9.3935	9.1176
14	12.1062	11.6909	11.2960	10.9205	10.5631	10.2228	9.8996	9.5896
15	12.8492	12.3813	11.9379	11.5174	11.1183	10.7395	10.3796	10.0375
16	13.5777	13.0550	12.5611	12.0941	11.6522	11.2340	10.8377	10.4621
17	14.2918	13.7121	13.1661	12.6513	12.1656	11.7071	11.2740	10.8646
18	14.9920	14.3533	13.7535	13.1896	12.6592	12.1599	11.6895	11.2460
19	15.6784	14.9788	14.3237	13.7098	13.1339	12.5932	12.0853	11.6076
20	16.3514	15.5891	14.8774	14.2124	13.5903	13.0079	12.4622	11.9503
21	17.0112	16.1845	15.4150	14.6979	14.0291	13.4047	12.8211	12.2752
22	17.6580	16.7654	15.9369	15.1671	14.4511	13.7844	13.1630	12.5831
23	18.2922	17.3321	16.4436	15.6204	14.8568	14.1477	13.4885	12.8750
24	18.9139	17.8849	16.9355	16.0583	15.2469	14.4954	13.7986	13.1516
25	19.5234	18.4243	17.4131	16.4815	15.6220	14.8282	14.0939	13.4139
26	20.1210	18.9506	17.8768	16.8903	15.9827	15.1466	14.3751	13.6824
27	20.7068	19.4640	18.3270	17.2853	16.3295	15.4513	14.6430	13.8980
28	21.2812	19.9648	18.7641	17.6670	16.6630	15.7428	14.8981	14.1214
29	21.8443	20.4535	19.1884	18.0357	16.9837	16.0218	15.1410	14.3331
30	22.3964	20.9302	19.6004	18.3920	17.2920	16.2888	15.3724	14.5837
35	24.9986	23.1451	21.4872	20.0006	18.6646	17.4610	16.3741	15.3905
40	27.3554	25.1027	23.1147	21.3550	19.7927	18.4015	17.1590	16.0461
45	29.4901	26.8330	24.5187	22.4954	20.7200	19.1563	17.7740	16.5477
50	31.4236	28.3623	25.7297	23.4556	21.4821	19.7620	18.2559	16.9315
55	33.1747	29.7139	26.7744	24.2640	22.1086	20.2480	18.6334	17.2251
60	34.7608	30.9086	27.6755	24.9447	22.6234	20.6380	18.9292	17.4498
65	36.1974	31.9645	28.4528	25.5178	23.0466	20.9509	19.1610	17.6217
70	37.4986	32.8978	29.1234	26.0003	23.3945	21.2021	19.3426	17.7533
75	38.6771	33.7227	29.7018	26.4066	23.6804	21.4036	19.4849	17.8539
80	39.7445	34.4518	30.2007	26.7487	23.9153	21.5653	19.5964	17.9309
85	40.7112	35.0962	30.6311	27.0368	24.1085	21.6951	19.6838	17.9898
90	41.5869	35.6657	31.0024	27.2793	24.2672	21.7992	19.7522	18.0348
95	42.3800	36.1691	31.3226	27.4835	24.3977	21.8827	19.8058	18.0694
100	43.0983	36.6141	31.5989	27.6554	24.5049	21.9498	19.8479	18.0958

5

OR YEARS' PURCHASE TABLE

of compound interest

6	6½	7	7½	8	9	10
0.9433	0.9389	0.9345	0.9302	0.9259	0.9174	0.9090
1.8333	1.8206	1.8080	1.7956	1.7832	1.7591	1.7355
2.6730	2.6484	2.6243	2.6003	2.5770	2.5312	2.4868
3.4651	3.4257	3.3872	3.3493	3.3121	3.2397	3.1698
4.2123	4.1556	4.1001	4.0458	3.9927	3.8896	3.7907
4.9178	4.8410	4.7665	4.6938	4.6228	4.4859	4.3552
5.5823	5.4845	5.3892	5.2966	5.2063	5.0329	4.8684
6.2097	6.0887	5.9712	5.8573	5.7466	5.5348	5.3349
6.8016	6.6561	6.5152	6.3788	6.2468	5.9952	5.7590
7.3600	7.1888	7.0235	6.8640	6.7100	6.4176	6.1445
7.8868	7.6890	7.4986	7.3154	7.1389	6.8051	6.4950
8.3838	8.1587	7.9426	7.7352	7.5360	7.1607	6.8136
8.8526	8.5997	8.3576	8.1258	7.9037	7.4869	7.1033
9.2949	9.0138	8.7454	8.4891	8.2442	7.7861	7.3666
9.7122	9.4026	9.1079	8.8271	8.5594	8.0606	7.6060
10.1058	9.7677	9.4465	9.1415	8.8513	8.3125	7.8237
10.4772	10.1105	9.7632	9.4339	9.1216	8.5436	8.0215
10.8276	10.4324	10.0590	9.7060	9.3718	8.7556	8.2014
11.1581	10.7347	10.3355	9.9590	9.6035	8.9501	8.3649
11.4690	11.0185	10.5940	10.1944	9.8181	9.1285	8.5135
11.7640	11.2849	10.8355	10.4134	10.0168	9.2922	8.6486
12.0415	11.5351	11.0612	10.6171	10.2007	9.4424	8.7715
12.3033	11.7701	11.2721	10.8066	10.3710	9.5802	8.8832
12.5503	11.9907	11.4693	10.9829	10.5287	9.7066	8.9847
12.7833	12.1978	11.6535	11.1459	10.6747	9.8225	9.0770
13.0031	12.3923	11.8257	11.2994	10.8099	9.9289	9.1609
13.2105	12.5749	11.9867	11.4413	10.9351	10.0265	9.2372
13.4061	12.7464	12.1371	11.5733	11.0510	10.1161	9.3065
13.5907	12.9074	12.2776	11.6961	11.1584	10.1982	9.3696
13.7648	13.0586	12.4090	11.8103	11.2577	10.2736	9.4269
14.4982	13.6869	12.9476	12.2725	11.6545	10.5668	9.6441
15.0462	14.1455	13.3317	12.5944	11.9246	10.7573	9.7790
15.4558	14.4802	13.6055	12.8186	12.1084	10.8811	9.8628
15.7618	14.7246	13.8007	12.9748	12.2334	10.9616	9.9148
15.9905	14.9028	13.9399	13.0836	12.3186	11.0139	9.9471
16.1614	15.0329	14.0391	13.1593	12.3765	11.0479	9.9671
16.2891	15.1279	14.1099	13.2121	12.4159	11.0700	9.9796
16.3845	15.1972	14.1603	13.2489	12.4428	11.0844	9.9873
16.4558	15.2478	14.1963	13.2745	12.4610	11.0937	9.9921
16.5091	15.2848	14.2220	13.2923	12.4735	11.0998	9.9951
16.5489	15.3117	14.2402	13.3048	12.4819	11.1037	9.9969
16.5786	15.3314	14.2533	13.3134	12.4877	11.1063	9.9981
16.6009	15.3458	14.2626	13.3194	12.4916	11.1080	9.9988
16.6175	15.3562	14.2692	13.3236	12.4943	11.1091	9.9992

APPENDIX 6: METRIC CONVERSION TABLE

Length	1 in = 25.44 mm (approximately 25 mm) $\left(\text{then } \frac{\text{mm}}{100} \times 4 = \text{inches}\right)$
	1 ft = 304.8 mm (approximately 300 mm)
	1 yd = 0.914 m (approximately 910 mm)
	1 mile = 1.609 km (approximately $1\frac{3}{5}$ km)
	1 m = 3.281 ft = 1.094 yd (approximately 1.1 yd)
	(10 m = 11 yd approximately)
	1 km = 0.621 mile ($\frac{5}{8}$ mile approximately)
Area	1 ft^2 = 0.093 m^2
	1 yd^2 = 0.836 m^2
	1 acre = 0.405 ha (1 ha or hectare = 10 000 m^2)
	1 mile^2 = 2.590 km^2
	1 m^2 = 10.764 ft^2 = 1.196 yd^2 (approximately 1.2 yd^2)
	1 ha = 2.471 acres (approximately $2\frac{1}{2}$ acres)
	1 km^2 = 0.386 mile^2
Volume	1 ft^3 = 0.028 m^3
	1 yd^3 = 0.765 m^3
	1 m^3 = 35.315 ft^3 = 1.308 yd^3 (approximately 1.3 yd^3)
	1 ft^3 = 28.32 litres (1000 litres = 1 m^3)
	1 gal = 4.546 litres
	1 litre = 0.220 gal (approximately $4\frac{1}{2}$ litres to the gallon)
Mass	1 lb = 0.454 kg (kilogramme)
	1 cwt = 50.80 kg (approximately 50 kg)
	1 ton = 1.016 tonnes (1 tonne = 1000 kg = 0.984 ton)
	1 kg = 2.205 lb (approximately $2\frac{1}{5}$ lb)
Density	1 lb/ft^3 = 16.019 kg/m^3
	1 kg/m^3 = 0.062 lb/ft^3
Velocity	1 ft/s = 0.305 m/s
	1 mile/h = 1.609 km/h
Energy	1 therm = 105.506 MJ (megajoules)
	1 Btu = 1.055 kJ (kilojoules)
Thermal conductivity	1 Btu/ft^2h°F = 5.678 W/m^2 °C
	(where W = watt)
Temperature	x°F = $\frac{5}{9}(x - 32)$°C
	x°C = $\frac{9}{5}x + 32$°F
	0°C = 32°F (freezing)
	5°C = 41°F
	10°C = 50°F (rather cold)
	15°C = 59°F
	20°C = 68°F (quite warm)
	25°C = 77°F
	30°C = 86°F (very hot)

Pressure

$1\ \text{lbf/in}^2 = 0.0069\ \text{N/mm}^2 = 6894.8\ \text{N/m}^2$
$(1\ \text{MN/m}^2 = 1\ \text{N/mm}^2)$
$1\ \text{lbf/ft}^2 = 47.88\ \text{N/m}^2$ (newtons/square metre)
$1\ \text{tonf/in}^2 = 15.44\ \text{MN/m}^2$ (meganewtons/square metre)
$1\ \text{tonf/ft}^2 = 107.3\ \text{kN/m}^2$ (kilonewtons/square metre)

For speedy but approximate conversions:

$$1\ \text{bf/ft}^2 = \frac{\text{kN/m}^2}{20}, \text{ hence } 40\ \text{lbf/ft}^2 = 2\ \text{kN/m}^2$$

$$\text{and tonf/ft}^2 = \text{kN/m}^2 \times 10, \text{ hence } 2\ \text{tonf/ft}^2 = 20\ \text{kN/m}^2$$

Floor loadings

office floors – general usage: $50\ \text{lbf/ft}^2 = 2.50\ \text{kN/m}^2$
office floors – data/processing equipment: $70\ \text{lbf/ft}^2 = 3.50\ \text{kN/m}^2$
factory floors: $100\ \text{lbf/ft}^2 = 5.00\ \text{kN/m}^2$

Safe bearing capacity of soil

$1\ \text{tonf/ft}^2 = 107.25\ \text{kN/m}^2$
$2\ \text{tonf/ft}^2 = 214.50\ \text{kN/m}^2$
$4\ \text{tonf/ft}^2 = 429.00\ \text{kN/m}^2$

Stresses in concrete

$100\ \text{lbf/in}^2 = 0.70\ \text{MN/m}^2$
$1000\ \text{lbf/in}^2 = 7.00\ \text{MN/m}^2$
$3000\ \text{lbf/in}^2 = 21.00\ \text{MN/m}^2$
$6000\ \text{lbf/in}^2 = 41.00\ \text{MN/m}^2$

Costs

£1/m^2 = £0.092/ft^2
1 shilling/ft^2 = £0.538/m^2
£1/ft^2 = £10.764/m^2 (approximately £11/m^2)
£2.50/ft^2 = £27/m^2
£5/ft^2 = £54/m^2
£7.50/ft^2 = £81/m^2
£10/ft^2 = £108/m^2

INDEX